SOLAR DECATHLON

SOLAR DECATHLON
Building a Renewable Future

Melissa DiGennaro King
Richard James King

JENNY STANFORD
PUBLISHING

Solar Decathlon: Building a Renewable Future

First published in paperback in 2024

Published by

Jenny Stanford Publishing Pte. Ltd.
101 Thomson Road
#06-01, United Square
Singapore 307591

Email: editorial@jennystanford.com
Web: www.jennystanford.com

Copyright © 2024 by Jenny Stanford Publishing Pte. Ltd.

All rights reserved. No part of this book may be reprinted or reproduced or utilized in any form or by any electronic, mechanical, or other means, now known or hereafter invented, including photocopying and recording, or in any information storage or retrieval system, without permission in writing from the publishers.

Trademark notice: Product or corporate names may be trademarks or registered trademarks, and are used only for identification and explanation without intent to infringe.

For photocopying of material in this volume, please pay a copying fee through the Copyright Clearance Center, Inc., 222 Rosewood Drive, Danvers, MA 01923, USA. In this case permission to photocopy is not required from the publisher.

Unless otherwise noted, all photos in this book were taken by Melissa and Richard King. Photos may not be used without permission in advance.

British Library Cataloguing-in-Publication Data

A catalogue record for this book is available from the British Library.

Publisher's note: The publisher has gone to great lengths to ensure the quality of this reprint but specifies that some imperfections may be visible in it with respect to the original print edition.

ISBN 978-981-5129-47-2 (Paperback)
ISBN 978-981-5129-13-7 (Hardcover)
ISBN 978-1-003-47759-4 (eBook)

DOI: 10.1201/9781003477594

We dedicate this book to our loving sons,
Bryan and Nathan,

their wonderful wives,
Kiwon and Keeghan, and

our amazing grandchildren,
Nate, Lily, Logan, and Tyler,

who enrich our lives beyond measure.

Contents

Contributing Writers		ix
Foreword		xi
Acknowledgments		xv
1.	**Call to Action**	1
2.	**Competitions as Motivators and Educational Tools**	7
3.	**Moving Mountains to Carry the Torch**	35
4.	**The Dream Unfolds**	**79**
	Solar Decathlon 2002	79
	Solar Decathlon 2005	110
5.	**Expectations Surpassed**	**143**
	Solar Decathlon 2007	144
	Solar Decathlon 2009	188
	Who Took Top Honors in 2009?	208
	Special Award: Civic Innovators	215
	Solar Decathlon 2011	221
	An Abrupt Change	222
	Newcomer to the 10 Contests	225
	Top Three Teams for SD 2011	246
6.	**Europeans Take Up the Charge**	**261**
	Solar Decathlon Europe 2010	261
	Solar Decathlon Europe 2012	289
	Solar Decathlon Europe 2014	315
	Solar Decathlon Europe 2019	336
	Solar Decathlon Europe 2021/22	344
7.	**Newfound Momentum**	**371**
	Solar Decathlon 2013	371
	Winning Teams	388

Solar Decathlon 2015	395
Winning Teams	409

8. International Expansion — **417**

The Beginning: SD China 2013	419
SD China 2018	435
SD China 2022	454
SD Latin America and Caribbean 2015	462
SD Latin America and Caribbean 2019	482
Special Tribute to Carlos Rodriguez-Marin	494
SD Middle East 2018	495
SD Middle East 2020/21	512
SD Africa 2019	515

9. The Multiplier Effect — **543**

U.S. DOE Richard King Award	549

Appendices — 553

Index — 565

Contributing Writers

Marianna Angelini	p. 544
Betsy Black	p. 136
Michael J. Brandemuehl	p. 70
Wendy Butler Burt	p. 88
Dan Eberle	p. 10
Amy Gardner	p. 252
Barbara Gehrung	p. 175
Carolynne Harris	p. 121, 413
Louise Holloway	p. 332
Marcela Huertas Figueroa	p. 464
Samir Idrissi Kaitouni	p. 540
Badr Ikken	p. 519
Lawrence L. Kazmerski	p. 31, 35, 84
Linus Knappe	p. 354
Allison Kopf	p. 208
Souad Lalami	p. 516
Joyce Mason	p. 149
Edward Mazria	p. 102
Thomas Meyers	p. 184
Ruby Theresa Nahan	p. 38
Michel Orlhac	p. 328
Derek Ouyang	p. 385
Susan Piedmont-Palladino	p. 61, 218
Carlos Rodriguez-Marin	p. 480
Edwin Rodriguez-Ubinas	p. 509, 548
Pascal Rollet	p. 302
Peter Russell	p. 367
David Schieren	p. 132

Robert Schubert	p. 253
Javier Serra	p. 262
Yuan Tian	p. 461
Bobby Vance	p. 546
Sergio Vega	p. 283
Karsten Voss	p. 364
Mike Wassmer	p. 66
Brittany L. Williams	p. 172, 434
Anita Worden and James Worden	p. 16
Hongxi Yin	p. 436
Pingrong Yu	p. 422

Foreword

Cultural visionary and acclaimed architect Sarah Susanka is the bestselling author of nine books, including The Not So Big House, The Not So Big Life, *and* Not So Big Remodeling. *Through her "build better, not bigger" approach to residential design, she has demonstrated that the sense of "home" we seek has to do with quality, not quantity. As a leading advocate for the re-popularization of residential architecture, Susanka has improved the quality of home design while countering the elitist image of architects so commonly held by the public. Her books have sold well over one million copies. Join her online at www.susanka.com*

Every once in a while, there are events that take place in the world, amongst all the other events of our everyday lives, that change us completely. These events introduce new visions for the future that change both those involved in the making of the event, as well as those who have the privilege of participating in what the event presents to the world. This is unequivocally the case for the Solar Decathlon. The book you have in your hands right now gives access to what I can only refer to as the phenomenal happening called the Solar Decathlon—a gathering every two years of collegiate teams from around the world in competition with one another to create the most innovative and beautifully designed self-sustaining house of the show. Open to the public and to the jurors of the homes for just ten days, the impact of these events continues to ripple through the lives of all those of us who have directly participated. And now, with this book, the Solar Decathlon will continue to influence many more who are looking for inspiration for a more sustainable and renewable future.

When Richard and Melissa King asked if I'd be interested in writing the foreword for the book, I jumped at the opportunity. Having served twice as an architectural juror for the Solar Decathlon, once in 2005, and once in 2009, I can declare without hesitation that these events helped color my own vision of the future, but also inspired me in ways that are hard to put into words, but which have fueled my enthusiasm ever since for multi-generational learning.

What I wasn't expecting when I agreed to be a juror the first time around, was that we jurors ourselves were in for an extraordinary education by the coming generation of architects, engineers, and inventors, all of whom were at that time still furthering their education at institutions of higher learning. They carried within them a new and vibrant vision for a renewable future, and were deeply involved in developing the possibilities for shaping a better world for themselves and their decedents. The innovation, inventiveness, and overall exuberance put forth by each of the teams of young university students from across the globe was beyond infectious. It was as though each team were saying to all of us who visited their works of art, invention, and artistry—the astonishing and decidedly *Not So Big* houses they had each assembled for the competition—"Look what we see as possible! Look what we can do if we work together! Look at how a sustainable and renewable world is completely within reach!" We looked, and were awed by what we saw. And I suspect you will be, too.

The success of the Solar Decathlon over the past two decades has, I suspect, astonished no one more than its founder Richard King—a quiet and somewhat self-effacing man with a great big vision that came along at exactly the right time to help shift public perceptions. Where he had hoped for at least a small amount of press for the ideas he'd asked the student teams to focus upon, what came instead was a veritable tsunami of interest. Where he had hoped to bring some attention to his belief that it was possible to imagine a future in which every house and its accompanying vehicle or vehicles are powered entirely by solar energy, what resulted instead was a tidal wave of enthusiastic embrace of the concepts, not only of solar-powered energy independence, but of many other renewable energy and sustainable design techniques as well. And best of all, because of the amount of publicity the event generated every two years, they have helped to make the feasibility of renewable energy adoption a popular and attainable objective all around the world.

When a good idea takes hold, it spreads quickly. Today we use the phrase "going viral" to describe the phenomenon, but the first two

Solar Decathlons were held before the advent of the iPhone, or any of the social media tools for spreading ideas, that we take so much for granted today. But go viral it did nonetheless, via the media of the day. And in this book, you'll hear from the wide variety of people whose lives have been dramatically impacted by their participation in the Solar Decathlon.

If we were able to track down the youngest members of the general public who walked through those early houses from the first two Solar Decathlons, I suspect we'd find yet another generation of inspired architects, engineers, and renewable energy scientists and advocates who caught the sustainability bug that day of their visit to the National Mall, where those first few Solar Decathlons took place.

The most important part of the success of this phenomenal happening is that everyone involved is seen as an important contributor, whether they be 18, 48, or 80, and whether they be from the United States, from Portugal, or from Singapore. It is the ability to work together as a team, across the decades and the continents, that brings about the magic of collaboration revealing itself at these events. It's an experience of full-immersion learning on steroids, and no one escapes without catching the excitement and thrill of imagining a better future for all of us who live here on Planet Earth.

The Solar Decathlon offers a rare opportunity for the visionary youth of today to provide a portal into their imaginations for those of us of earlier generations who are presently running the world. Our educations, decades ago, put us into positions that helped shape the world we currently inhabit. But from this point of orientation, we tend to build on what we already know, attempting to solve today's problems based on the concepts that were innovative several decades ago, but are no longer so. By tapping into and giving a platform to those who see the possibilities of tomorrow with new eyes, we open up so much that would remain invisible but for this sharing across the generations by those who are part of the next wave of humanity's presence on Earth. They see things that don't occur to those whose vision was directed at an earlier era's challenges. And when their turn comes, I suspect that with this event as precedent, they'll pass the torch forward, and create the forum for a similar sharing across the generations for those who come after them.

So, my hat's off to all those teams of Solar Decathletes of the past 20+ years who've poured themselves into their creations to help us share with them their visions for a healthier, more collaborative, and more renewable tomorrow. I hope you'll enjoy this journey through pictures, words, and stories, and that you'll consider visiting a future Solar Decathlon yourself, if and when the opportunity arises. I can guarantee that the experience will be at the very least mind expanding, and very possibly life changing.

May this phenomenal happening continue for many decades to come!

Sarah Susanka, FAIA
Autumn 2023

Acknowledgments

First and foremost, we express sincere gratitude for our parents, Emily and Adam DiGennaro and Barbara and Max King, for their abiding love and guidance from our earliest moments. We are also grateful for the many family members and relatives who have supported and encouraged us throughout life.

Along our pathway, some outstanding mentors have enhanced our professional capacity by believing in our potential to go above and beyond: Dr. Morton Prince, Dr. Lloyd Herwig, and Jim Rannels (U.S. Department of Energy); Herbert W. Ware (Arlington Public Schools); Tom DiGiovanni (K12 Inc.); and above all, Joan and Walter F. Mondale, whose unwavering generosity, kindness, and encouragement over many years helped shape our future.

Inspiration is a key driver for novel ideas. We want to acknowledge a handful of individuals who fueled our desire to charge ahead with confidence in crafting a brand-new vision for moving the needle toward clean energy and sustainable solutions: Anita and James Worden, Hans Tholstrup, Paul MacCready, Art Boyt, Dan Eberle, and Cecile Warner. Their influence cannot be overstated.

To the contributing writers who have freely given time and energy to this effort, we thank you for your reflections and extraordinary insight, which enrich this historical narrative.

To the thousands of Decathletes, faculty advisors, organizers, volunteers, jury members, sponsors, journalists, and friends who participated in Solar Decathlon events covered in this book, we thank you for being part of this journey. You have made a difference on the road to a brighter future.

Finally, we thank Jenny Rompas and Stanford Chong for reaching out to propose this book, which has been a true labor of love for us. It's been a pleasure to work with the team at Jenny Stanford Publishing to make this book a reality.

With God's grace, we have been blessed with this opportunity.

Melissa and Richard King

Chapter 1

Call to Action

Global climate change has now become a familiar phrase that evokes an acute sense of urgency. According to NASA, human actions have led to irreversible effects on planet earth. Unless we find ways to reduce the continual addition of atmospheric gases that trap heat, deleterious conditions that result from rising temperatures on earth are on a steady course to get worse. The Intergovernmental Panel on Climate Change (IPCC), which analyzes scientific data about climate change, contends that the current rate of relatively rapid global warming is unprecedented. Furthermore, scientists predict that this trend of steadily rising temperatures will continue. This is an unacceptable risk.

A review of early concerns with carbon dioxide levels in the atmosphere reveals that some researchers investigated the effects of human activities on atmospheric content way back in 1938. Guy Stewart Callender, a British steam engineer with an inquiring mind, collected records from 147 weather stations around the world for more than 5 years. Relying on manual calculations, he made an astonishing discovery: Temperatures on earth had risen 0.3°C since the beginning of the 20th century. He also collected data on carbon dioxide levels and reported that the amount of CO_2 in the atmosphere had risen by 10% during that same period. Putting this information together, he theorized that by burning fossil fuels, humans had

---

Solar Decathlon: Building a Renewable Future
Melissa DiGennaro King and Richard James King
Copyright © 2024 Jenny Stanford Publishing Pte. Ltd.
ISBN 978-981-5129-47-2 (Paperback), 978-981-5129-13-7 (Hardcover), 978-1-003-47759-4 (eBook)
www.jennystanford.com

increased greenhouse gases in the atmosphere, thereby trapping heat and leading to warming temperatures on our planet. However, to a large extent, scientists at the time were skeptical of Callender's findings.

A stable set of climatic conditions helps to protect our natural resources and preserve life on earth. Photo credit: Pixabay.

Two decades later, the Scripps Institution of Oceanography in La Jolla, California, hired Dr. Charles David Keeling to collect atmospheric data at a remote location far from any industrial activity. Keeling went to Mauna Loa in Hawaii and gathered measurements of carbon dioxide four times per day for 18 months. Keeling's subsequent analyses showed increasingly higher levels of CO_2 in the atmosphere, with a definitive upward trend during the period of his study. That outcome, which confirmed Callender's earlier results, was a surprise to many researchers.

For more than 60 years after that ground-breaking realization, scientists have continued to collect daily readings from Mauna Loa, revealing the "Keeling Curve," which shows a verifiable pattern of increasing CO_2 concentration. Keeling's work also spawned additional studies at various locations. Scientists have continued to measure the amount of carbon dioxide, as well as other greenhouse gases such as methane, ozone, and chlorofluorocarbons, in earth's atmosphere. By the mid-1970s, it was clear that human activity had a measurable impact on global warming.

Landscapes are changing in dramatic ways due to the effects of rising temperatures and changing climatic conditions. Photo credit: Pixabay.

With global warming, shrinking and retreating glaciers are a common sight on earth. Photo credit: Pixabay.

Then came the Organization of the Petroleum Exporting Countries (OPEC) oil embargo of 1973 and the ensuing energy crisis, which rocked the world. With limited supplies, gas rationing, and long lines at the pump, it seemed like the perfect storm to encourage consumers to move away from dependence on fossil fuels. At the time, many people did not believe or understand the dynamics of global warming and its relationship with carbon emissions. The notion that simply driving a car down the street could have a noticeable impact on temperatures worldwide seemed absurd. American leaders asked the public to conserve energy, but staggering inflation and a sluggish economy had led to growing anxiety in our nation. Transitioning from oil and coal to alternative sources of comparatively higher-priced "green" energy seemed unrealistic when people were struggling to pay their bills and get to and from work.

As we know so well, "Timing is everything."

It just so happens that in the 1970s, Richard King, a college student at the American University in Washington, DC, was paying close attention to these converging issues. Initially interested in pursuing political science, he changed his major to physics. He wanted to learn more about the science of energy and photovoltaics (solar electricity), to better comprehend the complexities of the energy crisis. An optimist and dreamer by nature, Richard began a long-term quest for new ways to address the lingering problem of fossil-fuel dependence.

Subsequently, Richard's professional career leading R&D efforts in the Photovoltaics Program at the U.S. Department of Energy (1986–2016) reinforced his steadfast belief that solar applications were a viable alternative to fossil fuels. In the 1980s and 1990s, productive research and new developments in the solar industry had moved forward by leaps and bounds, but the public had not yet embraced clean energy technology. For various reasons, people were reluctant to reduce their consumption of nonrenewable energy resources, such as coal, natural gas, and oil.

With the 1992 publication of *Earth in the Balance* (Houghton-Mifflin), climate change became a hot topic. But there were two distinct sides in the conversation: those who accepted the science and felt a strong need to address climate change, and naysayers who rejected the concept and scientific data that indicated global warming. By this time, Richard King was thoroughly convinced that public education and broad-based marketing campaigns were essential keys for a future that relied on green energy. Our nation had the necessary solar technology but could not seem to move the needle toward consumer buy-in. This puzzling juxtaposition motivated Richard to seek a novel way of transforming that misalignment.

Fast forward to the 21st century. Burgeoning datapoints attest to the existential threat of climate change. It is real, and it is not going away anytime soon. This pervasive, vexing problem is now front-and-center on the international stage, but a critical question remains: How do you get the public on board to attack such a troubling predicament for planet earth?

These thoughts came to light in Richard's mind:

- First of all, view the scenario as an opportunity, rather than a misfortune. Accept reality, no matter how harsh or frightening, with a positive attitude, in order to imagine the possibilities up ahead. Take aim at the problem and explore novel pathways to hit the target.
- Next, enlist energetic, young minds. Motivate the next generation to envision and create fresh, innovative ideas. Pass the torch willingly to embolden unconventional approaches. Give students a chance to demonstrate their understanding through experimentation.
- Finally, communicate with passion and invite everyone to discover first-hand "what can be." Shared experiences in stimulating settings boost our capacity to be open minded. Expect the unexpected. Grow the future by shining a light on those who will lead us toward brighter tomorrows.

Building on those premises, Richard King had an idea of his own. Discover the story that follows and enjoy a fascinating journey!

Chapter 2

Competitions as Motivators and Educational Tools

Most of us understand the desire to excel, to be the best, and ultimately, to win. Think back to playing tag with neighbors, running races with peers, playing chess with friends, and giving presentations in front of classmates. What pushed you to give 110% in those situations? How did those outcomes impact your preparation for the next game of tag, foot race, chess match, or class presentation? If you were successful, how did that affect your performance the next time? How much did the behavior and performance of those around you influence your own actions and responses?

Perhaps it is human nature to be competitive. Perhaps the push to be the "top-dog" motivates people to work harder and put forth more effort. Perhaps in the process of preparing to win a competition, people gain new skills and understanding. When all is said and done, most participants in any competition come out ahead, one way or another, even if they do not achieve first place. While talking about the genesis of his unique idea for "Solar Decathlon," Richard talks about the value of competitions in general.

"Competitions appeal to most of us. If you're a competitor, you give it your all to be the best. You want to win, and you believe you can win. When you compete, you strive to run faster and jump higher than all your competitors, in order to prevail. If you're a fan, or if you have some connection to competitors, you follow their progress

---

Solar Decathlon: Building a Renewable Future
Melissa DiGennaro King and Richard James King
Copyright © 2024 Jenny Stanford Publishing Pte. Ltd.
ISBN 978-981-5129-47-2 (Paperback), 978-981-5129-13-7 (Hardcover), 978-1-003-47759-4 (eBook)
www.jennystanford.com

closely and become an enthusiastic cheerleader. Of course, you want your chosen favorite to win. Everyone is invested in the process, as well as the outcome. In other words, everyone gets fired up."

Richard points out that racing of any sort is a distinct type of competition. "A race is one type of competition. In a racing event, there's a starting point and an ending point. Each competitor wants to be the first to cross the finish line. Racers push themselves to the max to make it to the winner's circle. Often, they'll do whatever it takes to become better than everyone else—to rise above. Along the way, competitors make important new discoveries. They learn about their own strengths, their level of confidence, ways to improve, actions or behaviors that lead to greater achievement, tools that might support success, and the importance of an appropriate mindset. Racing participants lean in and strive to get better and better as they prepare and then compete."

"For racing fans, the anticipation, followed by being part of a racing event, is great entertainment. Fans are interested in learning about the competitors, and they make outcome predictions based on what they know about racers. Some place bets on who is most likely to win, and everyone gets excited about what might happen. Fans scream and holler during a race, cheering for their favorites. The unforeseen results of any competition can cause plenty of stress and anxiety, but the actual involvement is quite compelling. Of course, it's also lots of fun!"

Solar car racing is a unique sport with historical roots on several continents, but plenty of credit goes to Hans Tholstrup, a bold, visionary entrepreneur with a fervent desire to challenge the status quo. No stranger to adventure, Tholstrup had circumnavigated Australia in an open boat, driven a motorcycle around the world, and completed a solo flight around the globe before he turned 40 years old. However, the energy crisis of the 1970s was a sudden wake-up call that motivated him to take bold action. Tholstrup decided to organize "economy races" for trucks and commercial vehicles in Australia to prove the value of fuel efficiency. In his mind, determining *how far* a vehicle could go, rather than *how fast* it could go, was the crucial issue for energy conservation.

Intrigued by the process of using solar cells to convert sunlight to electricity, a light bulb went on inside Tholstrup's head: Why not use solar power for transportation? In pursuit of that goal, he built *The Quiet Achiever* (aka BP Solar Trek) in 1982, one of the first solar cars ever to be assembled. He then drove that one-of-a-kind vehicle from

Perth to Sydney, Australia, a distance of 2519 miles (4052 km) across rugged terrain in just 20 days, to show it could be done. In his own words, Tholstrup said, "We hope that by making this BP Solar Trek we will motivate people to solve whatever problems are before us. If it will motivate just one more idea and thought in the development of solar power, then the venture will be well worthwhile."

In 1987, Tholstrup expanded on this effort to promote sun-powered vehicles by creating an international solar car race called the World Solar Challenge (WSC). This daunting race went from Darwin to Adelaide on the Stuart Highway, right across the unforgiving Australian Outback. General Motors (GM) had partnered with AeroVironment and Hughes Corporation to beat the odds and take first place. To accomplish that goal, they built the Sunraycer, a stunning solar-powered race car that smashed expectations with a huge lead over all competitors. Referred to as the "pace car of the future," Sunraycer's impressive win handed GM a new tool to "showcase innovative car design and stimulate interest in scientific and technical education."

Adventurer Hans Tholstrup with scale model of *The Quiet Achiever*, the solar car he drove across Australia in 1982.

Richard King caught the buzz of solar car racing in 1988. While attending a conference in Las Vegas, Nevada, he got a first-hand look at that oversized beetle on wheels covered with solar cells, called the Sunraycer. In his words, "I was hooked immediately. I was working at the U.S. Department of Energy at the time and knew the solar field needed something wild and crazy to get attention. It dawned on me that weird-looking cars like this might be just the right ticket. After talking with a representative at the Sunraycer display, I found out that GM was organizing a solar car race for university teams in the United States. I wanted to jump right in, so I offered to help. As a result of that encounter, I became part of the steering committee for GM SUNRAYCE USA, joining organizers to plan and execute a cross-country solar car race in the United States."

In July 1990, 32 North American collegiate teams competed in that 1650-mile (2656 km) journey from Orlando, Florida to Detroit, Michigan passing through eight states in 11 days, powered only by sunlight. During that exhilarating competition, students demonstrated refreshing creativity and astonishing innovation, going far beyond anyone's expectations. In Richard's words, "What started out as a race of students ended up as a race of scientists and engineers who were better prepared for the future. Convinced that this type of student-based competition for university students was sure to invigorate greater interest in solar energy, I became actively involved with solar car racing for the next 15 years."

Reflections from Dan Eberle, educator, clean energy enthusiast, and director of the American Solar Challenge for many years.

"I am not sufficiently schooled in anthropology to know the degree to which competition is inherent in human societies in general. Still, from my observations, competition for its own sake seems rampant everywhere I've been. Without wading into the weeds of pressure created by parental and cultural expectations, many children exhibit, almost innately, a propensity to challenge each other to run faster or further, climb higher, lift more, or to be the quietest (per mom's suggestion). I think it's in our genes.

Absent a human competitor; we will test ourselves against nearly anything in nature; animals, mountains, oceans, glaciers, microbes, or even the universe. Both science and art are manifestations of our apparent biological need to push toward the edges to see and master what's there. We are a species that thrives on challenge.

I've never been a fan of dog-eat-dog, winner-take-all, watch-your-back types of competitions. In the winter of 1982–83, Hans Tholstrup and the Quiet Achiever, his solar-powered vehicle, crossed the entire continent of Australia from west to east. Wow! His idea was to challenge the real world of weather, roadways, traffic, and terrain. The point here is that a challenge which includes components of the unknown or exotic tends to be more appealing. Crossing a continent fit the bill. A year later, I was lucky to be part of the Crowder College *TSAR project* that built a solar car which culminated with the first successful cross-continental trip of a sun-powered vehicle from San Diego, California to Jacksonville, Florida. We could hardly believe that we made it the entire distance! There again, ideas were thrown up against the wall of reality to see what would stick.

Two years later, Tholstrup invited the world to come to Australia and compete against the Great Australian Outback by traveling north to south from Darwin in the Northern Territory, to Adelaide, the jewel of South Australia. The field that showed up for the first World Solar Challenge (WSC) in October 1987 was a complete mixed bag of university teams, tinkerers, scientists, automotive and electronics experts, solar enthusiasts, and engineers. Among them was one small community college from Neosho, Missouri called Crowder College (that was my team) and one major automobile company, General Motors (GM). We all came to sprint 1,118 miles (1,800 kilometers) across the desert.

The World Solar Challenge raised the stakes. That's what a competition of ideas is all about. It thrives on the possibility of pitching *our own ideas* and *my team's creations and innovations* against the challenge itself and also against *your* ideas. The GM Sunraycer wiped the floor with its solar racing competitors, finishing in two-thirds the time of the nearest competitor. The societal and environmental effects of all those minds trying to solve an energy efficiency problem should not be overlooked. Before they arrived at the starting line in Darwin, every team member had gained volumes of knowledge and experience that couldn't help but affect solutions to issues they would later face in their private and professional lives. The follow-on effects of their efforts are the real profit we gain from such endeavors. The actual head-to-head or us-against-them nature of a project is essential to the creative process as a stimulus and a goal to keep in sight, but the process itself is where the magic is. That's where the rest of us reap the benefits. Richard King, of the U.S. Department of Energy (DOE), and Howard Wilson, the seminal GM executive for the GM Sunraycer Project, shared a vision of the potential benefits from this kind of competition.

GM and the U.S. DOE developed the SUNRAYCE program in 1990 that challenged university teams from across the United States to create a solar-powered vehicle that would travel from Walt Disney World in Orlando, Florida to the GM Tech center in Warren, Michigan. Since its inception, the SUNRAYCE program (rebranded in 2000 as the American Solar Challenge) has engaged thousands of young engineers directly in the process of designing and building competitive solar vehicles. In addition, millions of spectators have been mesmerized by these futuristic vehicles as they traveled across the U.S. Over the past three decades, those students and solar car team members have influenced nearly every significant development in energy efficiency, solar technology, aerospace innovations, battery technology, hybrid and electric transportation, and power electronics. That's a noteworthy outcome!"

Reflecting on his many years of engagement with solar car racing, Richard says, "I saw the tremendous educational power of those competitions for students who were involved with designing, creating, and participating in such stimulating events. They learned by leaps and bounds, and their output influenced automotive R&D in valuable ways. However, I noticed that solar car racing was not overly effective in convincing policymakers and the public that solar power was a credible alternative for fossil fuels. Sure, spectators who witnessed those races with unusual space-age looking vehicles were captivated and intrigued, but they were not sufficiently convinced to join a renewable energy movement. In other words, we needed something more."

Richard had a new goal: A distinctly different type of competition that would move the needle past solar car racing as entertainment and technology demonstrations. Up ahead, he imagined a more expansive competition that would excite, educate, and inspire thousands of people. He envisioned an audacious, adventurous competition to showcase the innovative prowess of collegiate teams that would also motivate consumers and government leaders to take bold action steps in support of renewable energy. In his words: "It struck me that a competition related to residential housing might grab people's attention. After all, everyone needs shelter, and just about everyone wants to economize how much it costs to operate a personal residence. Unfortunately, that seemed somewhat of a pipe-dream that might be too cumbersome to actualize, so I put the idea in my back pocket for a little while."

"For 12 years, I led R&D efforts in the Photovoltaics Program at the U.S. Department of Energy in Washington, DC. Yes, there was progress in the advancement of solar cell technology, but there were persistent, major barriers to widespread adoption: the manufacture of PV panels was slow; PV panels were expensive; efficiency was low (8–12 percent for commercial or residential applications); PV relied on unpredictable sunlight (intermittent in daytime and unavailable at night); battery storage was problematic (lead acid batteries were bulky and had a short lifetime). Large-scale applications of PV were not economically feasible, and solar power just didn't seem like the best option for our nation's energy needs at the time. Solar technology worked well in pocket calculators, but the general public couldn't seem to make the quantum leap from those handheld devices with tiny solar cells to more expansive rooftop PV systems that could power a whole house. A recurring question was, "What do you do when the sun isn't shining?"

Photovoltaics proved to be reliable and highly effective for the space program in the mid-to-late 20th century. Thanks to Dr. Hans Ziegler and his unrelenting efforts to demonstrate the capacity of solar cells to power communication satellites that orbit the earth, NASA put aside its doubts and accepted the unrivaled value of PV for the space race. The U.S. Department of Defense recognized Ziegler's "exceptional contributions as a world pioneer in communications satellites and solar energy systems to power them" with the Meritorious Civilian Service Award in 1963. People watched with excitement as NASA launched more and more remote satellites and then went on to create the space shuttle, which featured PV modules for power. Without realizing that this would be a clairvoyant statement about the future, Ziegler spoke these words in 1955: "In the long run, mankind has no choice but to turn to the sun if it wants to survive."

Richard's narrative continues: "Despite those developments, the predominant belief that persisted was this: solar power is not an effective solution for most terrestrial applications, such as on homes and buildings. To move the needle and prove that PV had serious potential for everyday applications, a novel approach was sorely needed. Somehow, decision-makers and the public had to be convinced that solar could work, and work well."

"Thinking back to cross-country solar car racing, I remember how excited people were along the race route. They'd line up on both sides of the road just to catch a glimpse of those sleek, aerodynamic vehicles gliding along silently on sun-power. Curious onlookers were captivated, but I kept hearing this same question over and over again: 'Will I be driving a car like that in the future?'"

Unfortunately, the answer was, "Probably not." Lightweight solar race cars are designed primarily for efficiency and speed, in order to cover a predetermined distance in the least amount of time. Most of them are one-person vehicles with a cramped driver's seat, minimal legroom, and no extra storage space. Any additional weight will compromise a car's ultra-efficient design and require greater energy expenditure. PV panels are mounted on or built into the body of the vehicle, so they can capture as much sunlight as possible when they race. To put this into perspective, solar cars run on approximately one horsepower (1 hp = 745.7 watts), which is less than the typical 1,000 watts of a personal hair dryer, to traverse distances greater than 1,600 miles (2576 km). And that's an impressive accomplishment!"

Some might say, "If solar cars are not for everyday driving, why make them?" The critical fact to remember is that solar cars are always on the leading edge. Starting with the Tour de Sol organized in Switzerland (1985), followed by the World Solar Challenge in Australia (1987) and GM SUNRAYCE USA (1990), and continuing today, teams of engineers all over the world have been inspired to dream big in designing cars powered solely by the sun's light. Outside-the-box thinking and clever ingenuity have pushed automotive technology forward, driving remarkable technical advancements and improved safety measures. A few examples of innovations that emerged from the development of solar cars include: lightweight carbon composite materials, sleek aerodynamic shapes to decrease drag, highly efficient solar cells for optimal power, and less bulky lithium-ion batteries. Furthermore, most solar car races have few restrictions on design elements, such as body style, which has spawned tremendous variety in their structure and appearance.

When comparing standard gasoline-powered race cars with solar race cars, Richard points out that the design goals go in completely opposite directions. "The main objective for a gasoline powered race car is to *produce more power*, which is essentially unlimited, in order to make the car go faster. That often requires more energy and larger motors. In stark contrast, the main objective for a solar race car is to *optimize efficiency and reliability*, so cars can go farther with less energy." Solar cars are optimized to capture as much sunlight as possible, which is limited on any given day. That said, minimum equipment combined with maximum efficiency will require strategic planning and careful execution. No wonder Tholstrup called solar car racing the 'brain sport'. To succeed, scientists and engineers must use concepts and technologies slightly ahead of their time. Or, in the words of Dr. Paul MacCready, an energy minimalist with a background in physics and aeronautics who dedicated his professional life to taking on competitive challenges that showed how to conserve energy, "**Doing more with less.**"

Paul MacCready masterminded the 163-mile flight (262 km) of the PV-powered Solar Challenger across the English Channel at 11,000 feet with 16,128 solar cells mounted on its wings in 1981. His objective? To demonstrate that what seemed practically impossible—the successful flight of a 217-pound aircraft for five hours and 23 minutes on just 3000 watts of power—was actually achievable. MacCready believed photovoltaics held great promise as a renewable energy resource with tremendous potential for many practical applications. He hoped the Solar Challenger's flight would generate more interest in pushing PV technology forward. Its predecessor, the ultra-lightweight Gossamer Albatross, won the coveted Kremer Prize for human-powered flight in 1979 when it crossed the English Channel on the raw muscle power of a single person. MacCready won accolades and international recognition for that impressive design, and he was awarded the highly acclaimed Collier Trophy for aeronautics achievement with the success of the Gossamer Albatross.

Paul MacCready was the founder of AeroVironment, the company that partnered with GM and Hughes to build the GM Sunraycer, the sleek, shiny, silent race car that handily won the inaugural World Solar Challenge in 1987. In MacCready's words, "Sunraycer was a breakthrough in aerodynamic engineering." He believed the overarching value of that success was to stimulate the development of "more efficient transportation while making fewer demands on the earth's resources."

Richard King met Paul MacCready in 1989 at a steering committee meeting for GM SUNRAYCE USA in Flagstaff, Arizona. As two like-minded professionals who envisioned a bright future for sun power, they connected right away. Richard had tremendous respect for what MacCready had accomplished, especially his unabashed pursuit of winning strategies to pursue incredibly difficult challenges. Richard recognized MacCready's deep understanding of how to strategically balance trade-offs in the design process. He also admired MacCready's bold steps to tackle initiatives that pushed beyond "**what is**" to arrive at "**what could be**."

Richard King (third from left) with Paul MacCready (third from right) and others at a planning meeting in 1990 for GM SUNRAYCE USA.

Reflections from James Worden and his partner at Lightspeed Energy, Anita Worden, who study electric vehicles, building science, renewable energy, and energy storage systems. James and Anita are early pioneers of solar car racing; they are currently involved in non-profit work in their community, where they lend their time, talent, and treasures.

How Solar Car Racing Set Our Life's Work in Motion

(James) From an early age, I dreamed of building solar electric cars for people to commute in, and I built two tiny one- and two-seater cars in high school. However, when I arrived as a freshman at MIT in the fall of 1985, I thought I was putting those dreams on hold for a few years. That soon changed course when my friend Megan Smith grabbed me for a hallway conversation that was to transform my life.

Megan had heard about this solar car race called the Tour de Sol. It would be held in June, and she thought I should consider building a car to compete in the event. I was instantly smitten with the idea: a race for cars powered only by the sun around Switzerland and through the Alps. But it was already March, and the race was in June—just three months away.

I spared no time getting to work. I talked to as many undergrads, graduate students, professors, and MIT lab and shop staff members as I could and started designing a car with all the systems it needed while juggling a full load of classes. They have a metaphor at MIT that studying there is like taking a drink of water from a fire hose. I learned what that meant because I was living it.

Rohan Rajan (Anita's nephew) drinking from fire hose at MIT on October 19, 2021; photo credit: James Worden.

This couldn't be a solo effort. Within a few weeks, I was able to assemble a group of incredibly bright, highly-motivated people at the university and even a high school student who helped me design and build the race car. They became lifelong mentors and friends who went on to work with me in the years to come.

My mechanical engineering design professors and a couple of doctoral students were incredibly gracious when I peppered them with questions and gave me much-needed encouragement. They helped me create an incredible, adrenalin-filled, hands-on learning experience for a few dozen students over several years. People outside MIT pitched in as well. Companies and shops donated time, funds, and materials, in some cases manufacturing specialty parts that simply weren't available on the market. On one occasion, I was at a local bike shop in Arlington, MA asking whether they could make the custom race wheels I needed. A local high school senior, Cathy Anderson, overheard me and said she could build them for me. Sure enough, she delivered on her word. One year later she joined us at MIT and became a star designer and builder on the team.

All these people came together and worked around the clock to make the impossible happen. We were scrambling right down to the wire. While studying for finals, I was dashing from lab to lab to shops in different areas of campus getting a frame built, buying aircraft motors, motorcycle sprockets, bicycle and moped parts, structural foam, and aircraft aluminum tubing.

In the last few days, the vehicle still didn't have a fuselage nose-cone. With no time left, my brother John, a mathematics student at Harvard, came up with the idea of cutting a lightweight rubber soccer ball in half, turning it inside out, and attaching it to the front of the car.

The solar car, Solectria III, was packed up in a wood crate, and John and I rushed to Swissair Cargo in the nick of time. I followed in a passenger plane one day later, soldering solar cell tabs onto solar cells while on the flight. Luckily, smoking was allowed in airplanes back then, so a butane pen soldering iron and a little wafting smoke wasn't seen as a problem to flight attendants.

The race was a crazy, surreal experience, from landing in Zurich not realizing how I was going to get to the starting line hundreds of miles away in Southern Germany, to being surrounded by nearly 100 teams of like-minded, passionate solar car designers and builders. Observing the diversity of interesting and awesome machines with the same purpose blew my mind. I had never been more excited, albeit exhausted from being sleep-deprived over the previous days and weeks. How did my classes and grades go? Well, thank goodness MIT had a pass-fail system for freshmen at the time! I'll leave it at that.

MIT Solectria IV with driver James Worden in 1987; photo credit: James Worden.

The solar race car movement grew quickly over the next few years. After Urs and Sigrid Muntwyler started the Tour de Sol in Europe, Hans Tholstrup launched a cross-continental race in Australia in 1987, and Rob Cotter organized the American Solar Cup in 1988. Nancy Hazard and Rob Willis began the annual American Tour de Sol on the East Coast in 1989, and in 1990 Richard King organized GM Sunrayce USA under the auspices of the Department of Energy. Thousands of students, faculty, and staff from around the world became involved in these races. The Sunrayce, now known as the American Solar Challenge, and the World Solar Challenge still exist today.

I credit solar car racing with introducing me to my future wife, Anita Rajan Worden. After seeing one of the MIT solar cars on the streets of MIT and meeting Cathy Anderson, Anita quickly became an integral part of the team—building cars, organizing the team, working with sponsors, and managing classes that we held for new students.

As with every team, we had to learn and do so many different things in order to pull everything together. Many of those experiences became important life lessons that taught us valuable skills. Many were critical learning episodes that helped us gear up to start our own business. We had to figure out how to "sell the concept," to raise funds, attract talent, and secure free or discounted parts for car builds. We also needed to understand rules and how to apply innovative design concepts to the rules. On another dimension, we had to negotiate with town officials and police to test-drive unregistered car prototypes on main streets like Mass Avenue in Arlington, MA in the wee hours of the night!

Why Solar Car Racing is Important

Our motivation for this work was to advance the technology towards practical uses for the world, promote clean vehicles, and develop alternative concepts for transportation. We both strongly believed in the direct application of technologies for real-world uses. Together, we went on to create two successful companies in succession, one in the electric vehicle space, *Solectria Corporation* (before EVs became cool), and another in the grid-tied solar inverter and electronics industry, *Solectria Renewables, LLC* (which happened to be timed pretty well).

James Worden (center), Arvind Rajan, Rohan Rajan, Kim Vandiver (Edgerton Center) with 2021 MIT Solar Electric Vehicle Team members, Jenny Zhang and Sydney Paige Kim, October 2021; photo credit: Edgerton Center.

Like a new team recruit in the Netflix show, Ted Lasso said, "Football is life." I absolutely believed that "Solar car racing is life!" Solar racing was a formative part of our lives, and the aspirational hard work of race visionaries, founders, staff, and volunteers helped transform the lives of many thousands of bright, forward-thinking young people across the globe. There is no way to express in words the excitement of working with a tight-knit team driven toward the goal of winning and helping change the world to get us all to a better place. This is why, even today, we enjoy supporting the MIT Solar Electric Vehicle Team, which just won the 2021 American Solar Challenge.

Solar Car Racing: "Doing More with Less"

Participating in solar car racing taught me to do more with less in several different ways. Energy and architectural visionary Paul MacCready, the founder of AeroVironment, is famous for human-powered flight and the GM Sunraycer (among other things). He used a term that resonated with me from an early age: "Doing more with less." Paul's obsession with efficiency and going the extra mile to save ounces and watts was something we took to heart. Because Anita and I diligently applied this thinking to every design challenge, we typically ended up with the lightest and most efficient race cars in most competitions.

"Doing more with less" also informed how we used resources. We always operated on a shoestring budget, unlike some of our competitors who had large automakers supporting them. We were proudly sponsored by Toscanini's, our beloved local ice cream shop. I strongly believe that scarcity drives discipline and innovation—we learned how to be scrappy and come up with creative solutions because we simply had no other choice. Because of that, while spending around $30,000 a year, we were able to compete toe-to-toe with teams with six- and seven-figure budgets. As Anita and I went on to become entrepreneurs, this mindset of financial and resource efficiency helped us immensely. Our businesses became profitable quickly, and we were able to have some control over our destiny.[1]

Looking back on the legacy and imprint of solar car racing, it is clear that the design-and-build model for these competitive events is highly effective as an educational tool. Enthusiastic collegiate teams are immersed in the process for up to two years as they prepare, design, and create. Academic coursework for college credit is implemented to support content related to solar car design and

---

[1] With great sadness, we note that Anita Rajan Worden lost her courageous battle with cancer while this book was in development. Her impressive 35-year career included being named one of the Top 100 Female Business Leaders in Massachusetts. She is fondly remembered as a loving wife and mother of three children who was also a champion of charitable causes.

construction. Students form teams, eager to push the envelope with novel ideas. They tinker together and debate with each other about various approaches to creating a reliable, efficient vehicle. The group effort is an essential aspect of that dynamic, along with the recognition of the value of trade-offs in team decision-making. The design process combined with hands-on engineering in this collaborative endeavor has had resounding success. Solar car events can boast a proven track record on several continents for more than three decades.

During solar car races, such as the 2005 American Solar Challenge, teams would stop mid-day for a 15-minute break to change drivers. Crowds of curious onlookers were eager to get a close-up look at these aerodynamic vehicles during the daily pit stops.

In addition, solar car racing encompasses the complexities of organizing and participating in long-distance competitions that take place in public settings on our nation's highways. To win, teams must collaborate to envision, construct, and drive a reliable, road-worthy vehicle that can withstand multiple, unpredictable challenges while in active transit. On top of that, teams have to figure out how to deal with less-than-ideal weather conditions along the route, such as cloudy or rainy days without sunshine. They must be both practical and ingenious, in order to conserve energy and keep their cars going when abundant sunlight (needed to power their vehicles) is not present. Teams must be adept, resourceful planners who can quickly mobilize to execute responsive strategies that keep their solar cars charged up and ready to cross the finish line.

Richard hoped solar car races would help people realize that the PV cells powering those odd-looking vehicles could also provide power for a fully-equipped house. Unfortunately, that turned out to be a difficult translation. At the time, concerns about climate change were not yet on the front burner, and most consumers were content to "stay the course" with conventional fossil fuels to meet everyday energy needs. For many, the concept of a solar house conjured up images of geodesic domes, earth-sheltered dwellings, or earth-ship models that emerged during the counter-culture movement of the 1960s and the 1973 energy crisis.

Those non-standard designs typically emphasized passive solar strategies, but to some, those homes represented styles too far beyond the mainstream. Given various options for residences, most people tend to stick with more traditional, recognized standards, rather than embracing less familiar, avant-garde models. For example, living below ground in a bunker-like structure is not very appealing, even though it may be more energy efficient, due to the insulation of surrounding earth, which keeps the temperature inside that structure more consistent.

Jump forward to the 1980s, and the general population still had little exposure to contemporary models of homes equipped with photovoltaics. Beyond architectural and engineering circles, most people had almost no opportunity to view homes with active PV systems, so they simply did not understand the design features, such as why more windows might be positioned on the south side or why no trees should obstruct sunlight from reaching rooftop PV panels. It did not help that some homeowner associations frowned upon or denied resident requests for permission to install PV on their homes. Furthermore, many people lacked fundamental knowledge about how PV technology worked or how it could be incorporated into residential architecture.

In the 1990s, the United States was moving toward gradual acceptance of solar, but progress was at a snail's pace. The Florida

Solar Energy Center, commissioned by the Florida State Legislature in 1974, became the first state-supported energy organization. The U.S. Department of Energy (DOE) was established in 1977, which then authorized the creation of the Solar Energy Research Institute (SERI), which later became known as the National Renewable Energy Laboratory (NREL). This was a big deal because it meant that public funds could be dedicated to solar R&D. Solar Energy Tax Credits (set at 30% of investment) began in 1978 to make renewable energy more cost effective and encourage businesses and consumers to implement solar technology. At the highest level of government, President Carter had solar panels installed on the White House to show that photovoltaics could replace fossil fuels as a source of electrical power. Carter also established an ambitious goal: to provide for 20% of the nation's energy needs with solar power by the year 2000.

At the time, the scientific community was well aware of the significant benefits of PV in providing electricity to remote, rural areas, especially in the developing world. Engineers realized that rooftop solar systems made it possible for individual homes and buildings to become their own electrical power plants, bypassing the need for cumbersome, centralized systems with extensive transmission lines. This micro-approach was economically advantageous and much easier to execute, but a thorny issue remained: "How to connect the dots to motivate homebuilders and consumers in highly-developed countries to seriously consider more widespread applications of photovoltaics."

The search for an innovative approach tested Richard's patience. He says, "Honestly, I was somewhat discouraged. It seemed like PV might continue as a non-starter without some truly novel, attention-grabbing method to stimulate greater interest. To succeed with educating the public about the value of solar technologies and convince the construction industry that PV systems were *the right stuff* for homes and buildings, a new approach was definitely needed."

Fortunately, R&D in the PV field made remarkable progress in the last decade of the 20th century, with great strides in solar cell reliability and efficiency. A major breakthrough occurred in 1992 at the University of South Florida. The manufacture of a thin-film solar cell made of cadmium telluride proved to be 15.9% efficient.

The next year, Pacific Gas & Electric developed and installed a 500-kW system at a substation in Kerman, California. This was a major milestone because it was the first demonstration of "distributed" PV power on a large scale.

In 1994, the construction of a new facility for the National Renewable Energy Laboratory (NREL) featuring PV systems and passive solar design was completed. This structure was recognized as the most energy-efficient U.S. government building at that time. That same year, NREL demonstration projects with solar cells made of gallium indium phosphide and gallium arsenide verified conversion efficiency that exceeded 30%, which was about 15% beyond typical solar cell efficiency at the time. This was testimony to significant progress for PV, but policymakers, industry, and the public were not yet on board with scientific researchers.

In the meantime, successful solar car racing continued in the United States with two more events, Sunrayce '93 and Sunrayce '95, funded by the U.S. DOE as the lead sponsor. The 1993 race included 36 participating universities and covered 1000 miles (1609 km) from Dallas, Texas to Minneapolis, Minnesota. The 1995 race included 28 universities and covered 1150 miles (1852 km) from Indianapolis, Indiana to Denver, Colorado. Richard was encouraged with growing media attention for these trailblazer events and energized with the high caliber of output from student teams. This was a robust model for developing leading-edge technology and real-world promotion of technical careers. It proved advantageous for all who were directly involved. This smash-hit showcase for solar energy was a notable step forward, but it was not a huge game-changer for widespread adoption of renewable energy strategies. Richard persisted in his search for how to make bolder headlines for practical applications of solar power.

With President Clinton in office in the 1990s, the photovoltaic R&D budget increased substantially, a step in the right direction. In addition, nations were beginning to come together in order to plan for global efforts that addressed climate change. The 1992 Rio Earth Summit led to the adoption of some broad environmental agreements to help curb greenhouse gas emissions. At the first Conference of Parties (COP 1) in 1995, some nations wanted to establish specific targets and time frames for reducing emissions from developed countries, while not requiring targeted commitments from developing countries. The United States refused to sign this climate agreement. Richard was exasperated, expressing his feelings this way: "How could Al Gore, the sitting Vice President and author of *"Earth in the Balance"* (c. 1992), turn away from such a crucial international agreement? How could the U.S. government refuse to acknowledge that we already had the technology to mitigate negative effects of climate change? Why didn't our government want to admit that we needed to begin on a path toward decarbonization? I suddenly realized the crucial need to tackle this issue was from the bottom up, rather than striving for a top-down approach, and that meant public education."

In the summer of 1997, Richard attended the European Photovoltaic Solar Energy Conference and Exhibition in Barcelona, Spain. He took a walk around the impressive site where the 1992 Olympics XXV were held. The spectacular setting overlooked the Mediterranean Sea, and Richard remembers a sudden burst of inspiration during that evening stroll. "While on that site, I was struck by a whole new idea. Why not create an *Energy Olympics®*? I was totally excited about launching an epic, worldwide competition focused on clean energy and energy efficiency. When I returned home, I got right to work developing a framework for this concept. I focused on this idea, but realized it was overly ambitious. The few people I talked to about this concept humored me with polite but noncommittal responses such as, 'hmmm, that sounds interesting…….' I could tell they thought I was crazy."

"As Team Leader of the Photovoltaic R&D Program at the U.S. DOE, one of my responsibilities was to brief the Budget Committee in the U.S. Senate and House of Representatives. I remember one trip to Capitol Hill where I showcased notable progress with PV: increased performance, greater reliability, and lower costs. The Budget Committee was not impressed, and when I presented a higher budget request, they declined to agree on the funding increase. I was upset. We were on the cusp of some breakthrough developments in solar energy, but decision-makers were reluctant to support this critical research that promised a cleaner, brighter future. I was frustrated and spent a lot of time wondering how I might change the mindset in and around Washington, DC."

"Later that year I participated in a PV reliability workshop at the Florida Solar Energy Center in Cocoa Beach, Florida. I took a walk on the beach one evening, listened to the waves rolling in, and observed the beach homes lined up behind sand dunes. I reflected on my overly-ambitious *Energy Olympics* idea and toned it down a bit and pulled the earlier residential housing idea out of my back pocket. How about a competition for solar-powered homes? I was intrigued by this possibility for a set of contests, but wasn't sure it would be exciting enough. Unlike solar cars, houses don't move. Perhaps this wouldn't be stimulating enough to attract attention. I put the idea in my front pocket and kept revisiting the idea because I believed it had strong potential for success."

In 1997, another milestone occurred in the PV world: the U.S. Department of Energy introduced the "Million Solar Roofs Program" to grow investment in residential PV systems across the nation. In the following year, Subhendu Guha, a brilliant pioneer who had been experimenting with amorphous silicon in the manufacture of thin-film solar cells, succeeded in creating a new type of material: flexible solar shingles. This innovative development made it possible for PV to be incorporated directly into roofing material as a replacement for commonly used residential asphalt shingles. Richard recalls how that invention had a strong influence on the design of buildings with photovoltaic systems. "Building integrated photovoltaics panels (BIPV) refers to the process of incorporating solar panels directly into various parts of buildings, such as façades, roof overhangs, and exterior walls. This ushered in novel and strategic approaches to the design of solar-powered structures and led to greater architectural creativity in homes and buildings that featured PV. That helped architects go beyond rooftop placement and consider putting PV on other areas of a building's envelope. For example, an unusual new skyscraper in midtown New York City made a big splash in 1999. Called '4 Times Square,' this skyscraper had BIPV panels built into the south and west-facing exterior walls of the six upper-level floors. Called 'photovoltaic skin' (in place of glass cladding), these clearly visible panels generated some of the electricity to power that large structure. This new arrival garnered plenty of public attention in Manhattan, but most people viewed the PV application simply as an eye-catching display, rather than an option they might consider for their own dwellings. Many in the industry thought that BIPV might be a game-changer for photovoltaics, but it took years for this idea to be widely adopted."

An additional and important piece of the PV puzzle was the inverter, a specialized equipment that converts direct current (DC) produced by solar systems to alternating current (AC) for standard household use of electricity. In 2000 there was a historic breakthrough when Sandia National Laboratories developed a device called the "non-islanding inverter." The pivotal change was that the inverter had automatic shut-off capability whenever an electrical distribution line went down. Prior to this, utility companies were wary of PV because solar systems were unable to detect when the power went out. They would continue delivering electricity, which endangered line workers or others who might be exposed to "live" wires and get electrocuted. The new Sandia-developed inverter was a huge step forward for safety and protection of equipment. This groundbreaking achievement was a boost to residential solar systems and sparked Richard's idea for a housing competition to come alive.

Richard wrestled with how to attract attention and mobilize bold steps to get solar and renewable technologies at the forefront as viable energy solutions for the present and future. He continued to play around with the idea of creating a collegiate competition focused on houses instead of cars. After all, everyone needs a home, no matter where they live on the planet. A personal dwelling is all of the following: a safe zone, place of refuge and nourishment, work area, learning spot, recreation station, and setting for sleep. Basically, a personal residence protects you and meets your physical, social, and emotional needs. With that universal appeal, why not cultivate innovation in solar homes as a way to stimulate the push for clean energy and smart technology?

In Richard's words, "I had to do something, and it had to be vastly different, in order to make a big splash in some highly-visible spot where it would be immediately noticed. Here I was working at DOE's Forrestal Building in the shadow of the U.S. Capitol in the heart of Washington, DC. I couldn't just sit there and let policymakers stick with the same old story of fossil-fuel dependence. It was time for resolute action. I'm not usually a big risk-taker, but I was emboldened by the desire to make a difference. Lessons learned from the energy crisis of the 1970s were planted firmly in my brain, and I believed our nation had to chart a new course with renewables to achieve energy independence."

A competition for houses? What would that look like? Houses cannot race, so how would they compete? Houses are stationary, so how would they be transported and staged as part of one event? Who would design and build those houses? Where would the talent come from? How would recruitment for those designers and builders work? Who would fund this new idea? Who would provide logistical support? What time frame was realistic? What marketing strategies might attract the public? The list of questions seemed endless, but where there is sheer will and determination, there is a way.

As the idea for a competition of solar-powered dwellings took hold, Richard began formulating a conceptual blueprint for several contests. He wanted some performance-based contests, such as energy efficiency and indoor climate control, which could be evaluated with objective data collected during the competition. He also recognized the value of embracing professional expertise to assess certain design elements, such as architectural beauty, which called for onsite observation and evaluation. In addition, the competition had to attest to the capacity for solar as the primary source of power for residential housing units. That meant a variety of household tasks had to be carried out successfully during the competition. Model homes needed to show policymakers and consumers that solar power was a compelling substitute for fossil fuels that did not require compromise in functionality or aesthetics.

A vacation trip to the Caribbean Island of St. Martin in July 1999 offered a refreshing opportunity to contemplate this rather crazy idea. Could it take flight in the real world? Would it work as an innovation incubator? Would it accomplish its pivotal educational mission? Would it propel renewable energy to the forefront? There is nothing quite like the tranquility of snorkeling near colorful coral reefs to invigorate thinking, and that memorable vacation proved to be a turning point for Richard. Lively conversations with Melissa (his spouse and partner in solar adventures) centered around what contests might be included in this new competition and how they might be judged. One evening while enjoying dinner on a terrace overlooking a moonlit cove, Richard held up his fingers one-by-one as he described the contest possibilities. When he got to number ten, Melissa gave him a two-handed high-five and responded spontaneously with, "Hey, call it a Solar Decathlon!"

Hence, in one serendipitous moment, the clever name that would eventually become the world-famous brand for this unique competition was crafted. That moment galvanized Richard's determination.

Settling on a clever name for the competition energized Richard, and he decided to press on full speed ahead. He was certain he could transform his novel idea from dreams to reality. As Team Leader of PV R&D in the Office of Solar Energy, his role included responsibility for the PV Program at the National Renewable Energy Laboratory (NREL) in Boulder, Colorado. For the purpose of strategic planning and review, Richard was in frequent contact with Cecile Warner, an engineer who was the manager of NREL's University Participation Program at the time. In the past, Richard and Cecile had worked together on cross-country solar car races, where they established great rapport and mutual understanding. He was eager to pitch his wild idea for a competition with solar houses and decided to share it with Cecile in a phone conversation. Without hesitation, she said, "I love it!" She immediately grasped the potential value of the Solar Decathlon to achieve promising outcomes. Cecile's enthusiastic response was exactly what Richard needed, and he fortified his efforts to make it happen.

During a visit to NREL in August 1999, Richard and Cecile sat down with five other staff members to present the concept of a collegiate competition of solar-powered homes on the National Mall in Washington, DC. The challenge? Student teams would design and build net-zero-energy homes to demonstrate that solar power was capable of providing all the energy needs of a standard household, from cooking and laundry to lighting, showering, and running appliances. The public would be invited to tour these model homes, in order to advance the overall educational value. The goals? University students would gain valuable knowledge and hands-on experience, while policymakers and consumers would benefit from first-hand demonstrations of how solar actually worked when applied in residential dwellings. Richard explained that nothing like this had ever been attempted before, but he believed the vision would succeed in generating increased awareness and support for renewable energy.

The 10,000-foot conceptual view was well received, but the nuts-and-bolts perspectives of the participants at that first meeting led to considerable doubt. For some of them, this idea represented a quantum leap from concept to reality. Sure, the new competition would be ingenious enough to inspire resourceful innovation, but those analytical minds could not quite figure out how the complexities of that type of competition might get started or operate successfully from the ground up.

Reflections from Dr. Lawrence L. Kazmerski, Emeritus Member of NREL Staff, Fellow at the Renewable and Sustainable Energy Institute (RASEI), University of Colorado at Boulder

Remembrances: Early Beginnings of the Solar Decathlon

Solar Decathlon has reached worldwide acclamations for its contributions to bringing the understanding of energy importance in the built environment to multi-millions of our populations. The organization of SD events around the world has evolved to include thousands of experts and contributors. The rules have developed to meet or exceed technology and participant advancements and expectations. What has not changed are the enthusiasm and energy of the students and advisors who have devoted their time and effort to make this incredible event a success.

However, the beginnings were a bit more-humble, with less massive buy-in from a variety of decision makers, and some having other vested interests. The initial organization brings into focus the visionary behind this event—Richard King of the U.S. Department of Energy. This short contribution is meant to relive those beginnings from a different viewpoint than usual—that of a researcher at a National Laboratory accountable to Richard King and his colleagues in the Solar Program in Washington, DC for meeting the development of solar technologies in the U.S.

"In the beginning was the Word, and the Word" . . . was a Revolutionary Idea

Periodic reviews of our work at National Laboratories by the U.S. DOE was just a normal and expected way of life. The consolidation of the U.S. Solar Program under the collaborative umbrella of the National Center for Photovoltaics (NCPV) had just gotten underway in 1998-1999. We were meeting with Richard King, and I still remember where it was: in the NREL Director's conference room in Building 17. It was late in the morning of the second day of the review. The coverage of advancements in CdTe and CIGS thin-film research and record devices was completed, accomplishments in system analysis and reliability were highlighted, a summary of the status of U.S. industry had been described, and plans for R&D in the coming months were reasoned out.

It looked like everything had wrapped up, and we could get back to the security of our laboratories, equipment, computers, and operations. But there was another agenda—a surprise. Richard King stood up and said, "I have an idea." Richard noted that we had partially engaged the public with solar car races (that he had also headed), and that they were successful—to a point. Richard announced that he wanted to have a new event involving students and the public. One that was perhaps more meaningful to citizens—and one that could help educate government officials and increase visibility of new technologies to U.S. senators, representatives, the White House cabinet, "and yes, possibly the President himself." Richard explained, "I envision a competition of houses designed and constructed by university competitors, bringing in the best energy efficiency and renewable energy technologies, and judged on the basis of ten events: A SOLAR DECATHLON!"

Initial Skepticism

At this time, Atlanta had just hosted the Summer Olympics with a pretty good presence of PV (thank you U.S. DOE, Industry, and Georgia Tech). The next Olympics in Australia were being organized. Richard had modeled his vision after an Olympic Decathlon with ten different competitive events (running, jumping, throwing) over a short period of time.

The reaction from our group of scientists and engineers was mixed, to say the most. As I recall, the first question was, "Richard, we at the Laboratories have limited and decreasing funds for research. How much will this cost, and where will the funding come from?" Actually, I think I asked that question! The reply was quick and truthful. "Yes, funding is always an issue for our programs, but we have to be not so conventional or short-sighted. This is an opportunity — something that can lead to expanded budgets because the decision-makers will be involved and understand the importance and usefulness of these technologies for regular people. I do not know how much it will cost; that is for us to find out and plan."

The second question was, "When do you think this new event will take place?" Richard noted that the athletes for an Olympic Decathlon train for a long period before they perform, and that is what he expected. We would have to start planning immediately and expect to hold an event in the coming 2- to 4-year timeframe.

Someone then asked, "Where will this event be held?" Richard said that he envisioned this would take place where it had maximum visibility and access to all those Washington, DC government decision-makers: on the NATIONAL MALL, for example. Stunning and aggressive! Of course, this turned out to be the place, and one that caused Richard and his team many headaches, from gaining initial access, through delivering the site back to the Park Service in perfect condition afterwards.

The next question was, what are the 10 events? Richard talked around this a bit because he knew that the response needed some expert thinking and input to ensure a good and meaningful competition. After some hesitation, he said, "We will do this . . . "

The final question was, "Richard, who is this *we*?" His answer was that U.S. DOE, along with the National Labs (NREL, Sandia, others), would take responsibility. Richard was quite honest and said: "I need your help to pull this off." And so, the directive was given, and the seeds for the Solar Decathlon were planted.

Tom Surek, Manager of the PV Program at the time, listened carefully to this proposal. He seemed rather skeptical of this wild idea that suddenly emerged. As a realist, he could not fathom how an entirely new type of competition at such a high level with so many unanswered questions might come together and end up being a success story.

What Happens with Great Ideas?

This narrative about launching the Solar Decathlon reflects a beautiful *New York Times* best-seller book written by Kobi Yamada and illustrated by Mae Besom titled *What do you do with an idea?* (2013, Compendium, Inc.). This inspirational story with compelling illustrations considers multiple challenges and possible pathways for new and unique ideas that are daring or "just a little wild." In this marvelous book, a child ponders, struggles with, and befriends an unusual idea, gradually gaining more and more confidence and love for the idea as it grows and thrives. This idea demands lots of attention, and by accepting, nurturing, and playing with it, the child discovers that the idea brings joy. To the child's amazement, the idea eventually spreads its wings and soars above.

So, *"What do you do with an idea?"* Believe in it, hold on to it, then let it go. It just might help change the world. That was Richard's dream.

Chapter 3

Moving Mountains to Carry the Torch

Initial resistance often accompanies radically different ideas, simply because human nature tends to reject revolutionary pathways that go in unpredictable directions. An inner voice may tell us to slow down and be more careful before leaping ahead. Our sense of logic may warn us that uncharted territory is fraught with risk and possible failure. Despite this, bold leaders march on in pursuit of their vision. They hold fast to the firm belief that their idea has the potential to boost necessary or significant change. Visionary leaders who refuse to give up set the stage to empower remarkable transformations.

Reflections from Dr. Lawrence L. Kazmerski, Emeritus Member of NREL Staff, Fellow at the Renewable and Sustainable Energy Institute (RASEI), University of Colorado at Boulder

"Thinking back to that meeting at NREL in August 1999, I remember how Tom Surek's hesitancy gradually moved toward greater interest as he considered Richard's vision for this new endeavor. Tom realized that such a novel competition could be a big boost for the entire solar program, and he advised us that this should be done. Tom tended to be a high-level

---

Solar Decathlon: Building a Renewable Future
Melissa DiGennaro King and Richard James King
Copyright © 2024 Jenny Stanford Publishing Pte. Ltd.
ISBN 978-981-5129-47-2 (Paperback), 978-981-5129-13-7 (Hardcover), 978-1-003-47759-4 (eBook)
www.jennystanford.com

thinker, so his opinion was important; after all, he did control the funds for us at NREL. The spark was ignited, and Solar Decathlon was off and running."

The ensuing discussion was productive, and the positive aspects of Richard's new idea prevailed. The initial hesitation nudged toward cautious support, and a small team of believers emerged. That meeting established the groundwork for Richard's idea, but it was clear that plenty of work was on the horizon. Richard realized that transforming his novel concept into reality would require significant planning, undaunted perseverance, sheer determination, and a boatload of help. Looking back on this formative phase to launch the Solar Decathlon, Richard admits that he relied on *blind faith*, reflecting on that experience with these words: "I was rather overwhelmed with all the details necessary for an undertaking of this magnitude. Tons of questions were swirling around in my head as I tried to figure out what to attend to first. How would I recruit student teams? How would we select teams? How could we prepare teams for this complex design-and-build project? Where on the National Mall in Washington, DC could we place the solar houses? How would I get permission to use that piece of public land? How would the houses get to the competition site? What rules and regulations would be appropriate? What would the ten contests be? How would we judge and score each of those contests? How would we get sponsors to support the competition? How would we attract public visitation? How would we get policymakers to attend the on-site event? What did we need to do to get media attention? How would we announce winners? What would happen to the solar houses after the competition ended?"

"Despite the endless stream of questions, I held tightly to the belief that this crazy idea would become a real-world success story. I'm not one to show much anxiety, so few of my colleagues knew how stressful those early stages were for me. I'd wake up at night, trying to figure out which issues to tackle in what order and wondering how I could possibly make all this work out. Truth be told, I simply dug in my heels and put one foot in front of the other without allowing doubt or fear to creep in. I embraced the idea wholeheartedly and ran with it—full speed ahead—never looking back for a single moment. Purely *blind faith*."

This quote from Wilma Rudolph (1940–1994), an Olympic runner who holds three gold medals and several track-and-field world records, is an apt explanation for what kept Richard moving forward during this unpredictable time filled with extraordinary challenge: "Never underestimate the power of dreams and the influence of the human spirit. We are all the same in this notion: The potential for greatness lives within each of us."

Richard's steadfast pursuit of his dream never wavered, but from the get-go he was keenly aware of the need for staunch support. A key player was Cecile Warner, his trusted, reliable ally from solar car racing days who became the Competition Manager for the inaugural Solar Decathlon. Cecile's rock-solid commitment to Richard's vision, coupled with her nose-to-the-grindstone attitude and keen detail-orientation, made her a perfect choice to keep the dream alive and well. As an engineer with analytical prowess, Cecile's skillset was an excellent complement to Richard's luminous big-picture thinking. Together, they became a remarkable *tour-de-force*, catapulting an unorthodox idea into a masterful achievement.

Another bit of magic occurred in the fall of 1999. A cadre of talented individuals at NREL emerged as dedicated eager-beavers for Solar Decathlon. Among those hard-core believers who gave 200% to this endeavor were Ruby Nahan, Byron Stafford, Sheila Hayter, Amy Vaughn, and Mark Eastment. It is not stretching the truth to say that this core team delivered far beyond what needed to be done to lay the foundation for a grand-slam event. With a divide-and-conquer mentality, these folks organized efforts into distinct areas of need and provided strong leadership in various aspects of the competition. No stone was unturned as these highly qualified professionals searched for ways to answer most of the important questions that had been bothering Richard for months on end.

With Cecile Warner leading the charge, this team developed definitive blueprints for various aspects of the competition within six months. It was clear that teamwork and collaboration were necessary elements of success for this complex assignment. Solar Decathlon was a novel concept. As a bold, uncharted attempt to accelerate education and R&D for cleaner energy, while increasing public awareness of solar technologies for the residential market, the core team was united in a common goal: transform this idea into reality and make it happen—big time. They were fiercely loyal to the Solar Decathlon brand, and because of their shared belief that this crazy idea could make a difference for our energy future, this group was able to rise above conflict and disagreement during the planning process. They leaned on each other. They listened to each other. They applied problem-solving strategies. They remained open minded. They expressed opinions respectfully. They managed tasks efficiently. They worked tirelessly. And, they clung to hope.

This team of organizers embodied the principles expressed by Andrew Carnegie, the well-known steel magnate industrialist who made a fortune and then turned the tables to become a major philanthropist: "Teamwork is the ability to work together toward a common vision, the ability to direct individual accomplishments toward organizational objectives. It is the fuel that allows common people to attain uncommon results."

The results of their heroic efforts reflect these wise words by Henry Ford, the famous founder of the Ford Motor Company: "If everyone is moving forward together, then success takes care of itself."

Ruby Nahan was in that team of organizers, and her words speak volumes about behind-the-scenes work to prepare for the inaugural Solar Decathlon.

Reflections from Ruby Theresa Nahan, Director at Namaste Solar, was a Project Leader at NREL for 12 years and an independent communications consultant for 10 years. She served as Communications Project Manager for U.S. DOE Solar Decathlon in 2002–2008 and 2015–2017.

Creating the Solar Decathlon Brand

"As Communications Manager for the first U.S. DOE Solar Decathlon in 2002, it was my job to lead the initial branding effort. I continued as Communications Manager for the 2005, 2007, and 2009 events. I've had other honorary roles in several SD competitions since 2009, including VIP tour guide in 2011 and jury member for Solar Decathlon China and Solar Decathlon Europe. I returned as the Communications Manager for SD 2017. Once you start drinking the Solar Decathlon Kool-Aid™, it's difficult to stop.

The Solar Decathlon brand is part inspiration, part technical information, and part fun. Richard King was our visionary leader, and his clever idea was not hard to articulate: a student competition to design and build attractive, energy-efficient, solar-powered houses that included 10 contests reflecting how we use energy in everyday life. Basically, the SD brand aims to change consumer behavior in an inspired, informed, responsible way. The SD brand encourages the building industry to create a new business model that incorporates energy efficiency and renewable energy design and technologies. The SD brand aspires to show policymakers that public investment in endeavors that support clean energy are worthwhile. Along with those lofty aspirations, my team attempted to infuse the brand with a sense of fun. I'm happy to say that the original branding remains in place, except for minor tweaks over a 20-year period. We must have done something right from the very beginning!

Solar Decathlon had three major goals, and the brand was developed to align with these:

- Educate college and university students about renewable energy and energy efficiency through a hands-on design-and-build competition and ensure that students of architecture and engineering work together on the project. In addition, require that educational institutions create and/or augment curriculum in support of this goal.
- Educate consumers about energy efficiency and renewable energy technologies.
- Demonstrate to elected officials and industry leaders that today's consumers and tomorrow's leaders are hungry for energy efficiency and renewable energy technologies and solutions.

To support competition goals, the Solar Decathlon brand had to encompass the following:

- All communications and outreach had to relate to "average" consumers.
- The competition had to be like a "horse race," and that race had to be visible to people, no matter how they attended the event (in-person or virtually).
- Student work had to take center stage.

Pat Decathlete

Organizing a Solar Decathlon, be it the first one or any competition that has occurred since 2002, is a huge task. Participating in a Solar Decathlon—designing and building the competition house, taking it apart, shipping it to the competition site (sometimes overseas), putting it back together again, competing with other teams, providing tours to thousands of visitors, taking the house apart again, and shipping it to its final home—is a massive two-year undertaking of Herculean proportions. In designing the logo, we wanted something relatively simple and somewhat Olympian to convey the weight of the task, as well as the enthusiasm of the competitors.

Al Hicks, a skilled graphic designer at NREL, developed the first logo: a happy figure holding up the earth in orbit around the sun. Early on we named the figure "Pat Decathlete" (a gender-neutral reference to an old Saturday Night Live character skit). "Pat" is vaguely reminiscent of the Greek god Atlas, but instead of being weighed down by its burden, the figure appears to be floating and hugging the earth. After the inaugural event, Marjorie Schott (NREL staff) and I did a brand review. After minor tweaks to Pat Decathlete, we ensured that the SD brand would be applied consistently to different products. To date, the brand continues to appear on the U.S. DOE website and in many other materials related to the Solar Decathlon.

Solar Decathlon brand with Pat Decathlete, trademarked by the U.S. Department of Energy.

Inflatable Pat Decathlete in the 2002 Solar Village on the National Mall.

The Competition

Crafting a logo was probably the easiest part of creating the Solar Decathlon brand. Efforts to develop a competition of net-zero-energy homes that was compelling to both architecture and engineering students, as well as interesting and comprehensible to the general public was a much bigger challenge. Fortunately for this English major, that wasn't my primary job.

In my opinion, Richard King truly understood the importance of communications. He was clear from the very beginning that at least one of the ten contests would focus on communications. Richard and Cecile Warner, the first Solar Decathlon Project Manager from NREL, gave me a seat at the table from the outset. I was empowered to speak (on behalf of my profession) about what I thought would or would not work. We created an imaginary soap opera called "As the Paint Peels" to remind the team not to create a snooze fest. Ideas about contests, scoring, and the on-site event had to pass the "As the Paint Peels" test. Perhaps a suggestion was a good one from a technical perspective, but if we couldn't make it compelling or comprehensible to the public, we kept working on it.

I was part of a hard-working, passionate group charged with making Richard's visionary idea a reality. Solar Decathlon was the career opportunity of a lifetime, and I was given this role during my first year at NREL. I look back and marvel at how lucky I was. I owe a debt of gratitude to Eric Nelsen, my manager at the time. I vaguely remember how he described the job: "We have a client that I think you'd really work well with who needs a brochure." It certainly turned out to be a lot more than that! Nevertheless, I'm grateful that Eric saw something in me and gave me that great opportunity. He had worked with Cecile Warner on GM Sunrayce USA, and he thought we would click. He was right. Twenty years later, I'm lucky enough to call Cecile a dear friend.

I also worked with many other wonderful people on the first Solar Decathlon team at NREL: Mark Eastment and Mike Wassmer, the first and second competition managers; Ed Hancock and Greg Barker, who developed the objective contests and the scoring methods; Bryon Stafford, Robi Robichaud, and Dan Eberle, who managed the logistics of creating a small village in the middle of Washington, DC; and Sheila Hayter, who took the lead on subjective contests. I know I'm neglecting many more talented and worthy people (too numerous to mention here) who supported that core team, so we could attend meeting after meeting to debate what the competition and event would be like. We argued. We laughed. We worried. We solved problems. We invented all kinds of things. We worked incredibly hard. It was marvelous!

The Solar Decathlon story is a strong argument for the importance of communications and marketing folks working closely with others who are tasked with developing the products that the brand represents. Doing so creates a system of checks and balances. Do the products deliver on the promise implied in the brand? If not, something needs to change. Perhaps product developers should work harder to deliver on the aspiration of the brand. Or, perhaps the brand needs an integrity check.

Educating Consumers

Richard's vision to include architecture and design in the SD competition meant that every team made a significant effort to make sure the houses looked as good as they performed. We hoped that the public was lured in by the countertops, but left talking about appliances, lighting, space heating and cooling, water heaters, and batteries—mission accomplished! Although not every house was to everyone's liking, every visitor to the Solar Decathlon village learned that energy efficiency and renewable energy can be beautiful, comfortable, abundant, and within everyone's reach.

Looking back, the consumer education piece was easy to deliver, but ended up representing the bulk of hours that the communications team spent on the project. The SD brand relied on the ten contests and the fundamentals of energy-efficient design as a structure for developing consumer education. We provided basic information about the importance of building orientation, building envelope, insulation, windows, lighting, heating, cooling, energy production, storage, and management. We also included the impact of living and working at home and powering transportation from home. We shared information in print, online, through tours, and with workshops and displays. The idea was to convey general concepts, so teams could focus more on the specifics of their unique houses. This helped to underscore the reality that you don't have to be a Solar Decathlete to apply the principles on display in the competition—the same strategies and technologies are available to just about everyone.

Horse Race

Several people on the original Solar Decathlon team were veteran organizers from GM Sunrayce USA, Sunrayce 93, and Sunrayce 95. They understood from experience that the public needed something to watch. Solar Decathlon competitions (2002–2017) are comprised of a solar village with houses that don't move. Watching the temperatures of indoor air, water, refrigerators, and freezers rise and fall is like watching a slow turtle race. To add an element of a "horse race," the houses were wired and carefully monitored; Greg Barker huddled around a computer screen with an incredibly elaborate spreadsheet that ran calculations for contest scoring.

The Solar Decathlon web server frequently requested an update from that spreadsheet to display real-time scoring data online. That online scoreboard was also displayed in solar village welcome tents. The scoreboard was by far the most visited page of the Solar Decathlon website. The fact that each team had a person hunched over a monitor in their own house, refreshing and refreshing that page, likely skewed the stats a bit. Based on emails we received every time there was an error in the online scoring, I can confidently say it wasn't only students and faculty who were glued to the race. I recall a middle-of-the-night email in 2005 from someone in Japan politely pointing out an error (he was right), and there wasn't even a Japanese team in the competition!

Encouraging Students to Speak

Every competition has included a Communications Contest. Although details have changed over the years, the intent of this contest is largely unchanged. The contest requires each team to develop a "communications strategy" to help them as they seek sponsors and support for their project. Contest deliverables feed into the communications and media strategies of SD organizers. We had to keep interest up to grow sponsorship and build attendance; to do that, we needed content from SD teams. Meeting requirements of the Communications Contest also prepared teams for sharing information with the public during the competition.

SD teams were charged with creating a unique brand, creating a website (and beginning in 2009, creating a social media strategy and signature video), and implementing house tours with tour guides, displays, and handouts. Teams were evaluated on these efforts, plus leading the Communications Jurors on house tours.

The SD communications group mentored every participating team over the years. We emphasized the following:

- Having a communications plan with a clear purpose, goals, and strategies.
- Knowing how to measure the success of strategies to determine whether or not they were working, and what to do if they weren't.
- Creating a special story about the house and the team, to create a unique brand.
- The use of clear language, free of jargon.
- Speaking to visitors on their level. Sometimes students had a hard time speaking to consumers as consumers, and not as engineers or architects.
- Making people feel welcome and keeping them engaged, in-person and online.

In the end, the students, their hard work, beautiful homes, and compelling stories are the essence of the Solar Decathlon brand. Our job as event organizers was to offer compelling activities in an attractive venue to draw people in, and then provide context to help make the students' work comprehensible and transferable to a non-competition setting. Above all, it's the enthusiasm, energy, and excitement of student teams that shines through to make every Solar Decathlon a success.

Ultimate Inspiration

For me, being part of the Solar Decathlon was the opportunity and honor of a lifetime. I worked incredibly hard, and I loved nearly every minute. The students inspired me and filled me with hope and enthusiasm for the future. My colleagues at NREL and DOE worked equally hard. They came with sleeves rolled up, ready to get to work every day. Solar Decathlon is proof that each one of us can do great things. Given appropriate support and guidance, college students—tomorrow's leaders—can do even more than we might imagine.

Reflecting on Solar Decathlon 20 years later: Technologies have improved. Energy efficiency has gotten much better. Solar technologies are more productive. Battery storage has grown by leaps and bounds. And the cost has come down. In the meantime, the earth is getting warmer, while storms, floods, fires, and droughts have intensified. Alas, the predictions of climate scientists are coming true.

We have the technology and know-how to address the challenges of converting from a fossil-fuel world to a renewable-energy world. Thousands of college students from around the globe have learned how to design and build a house, which they knew little-to-nothing about before embarking on a Solar Decathlon. They've raised money, worked as teams to build houses, then taken them apart to ship across the country or the world to a competition site. They've put their houses back together and operated them through intense competition. Then, when the competition was over, they've taken the houses apart again and shipped them back to wherever they came from. If they can do all that and share it—in person at a competition site, on the web, and across social media—then we can certainly take the next steps toward a healthier planet. Surely, together, we can do more."

As the NREL team immersed themselves in the nuts and bolts of the Solar Decathlon, Richard focused on garnering support from external organizations. He realized that creating a new type of competition, implementing a huge onsite event in the nation's capital, and managing the logistics of a rigorous 2-year educational program would require substantial assistance. He wasted no time in reaching out to potential sponsors with cold calls and letters describing his vision and need for help. Richard also posted a notice in the *Federal Register*, the daily journal of the United States Government (June 12, 2000), asking for support for this new competition. With firm conviction and unfettered optimism, he singlehandedly tackled this task and was astonished but reassured with the results.

Richard's message on U.S. DOE letterhead reached the desk of Sir John Browne, CEO of British Petroleum (BP) in July 2000. Browne's spirited response confirmed Richard's belief that his vision would be accepted on a larger scale. Under Sir John Browne's directive, BP entered the U.S. solar business through its purchase of Solarex Corporation in 1999 and renamed the company BP Solar. At the time, BP Solar was one of the largest manufacturers of photovoltaic modules. Sir John Browne offered BP Solar's support as a major sponsor of the Solar Decathlon, and they offered to provide PV panels from any of their product lines to all participating teams "at cost." That offer translated to $2.50–3.00 per watt. BP Solar also offered assistance with local and national advertising, public outreach during the competition, technical support for the solar village, collateral material, and hordes of volunteers at the onsite event. Attaining the first major sponsor was a giant leap forward. BP Solar turned out to be a strong and steadfast ally for the inaugural Solar Decathlon held in 2002.

Knowing he needed backing from the architecture field, Richard picked up the phone that spring and called Dr. Ed Jackson, Jr., Director of Research at the American Institute of Architects (AIA) in Washington, DC, and Executive Architect for the Martin Luther King Jr. National Memorial. After learning about the goals and proposed objectives of Solar Decathlon, Jackson immediately recognized the potential power of this new competition and said, "How can I help?" Richard was invited to meet with Dr. Jackson at his office, and their conversation led to several tangible ways for the AIA to serve as a valuable resource and sponsor of the Solar Decathlon. Jackson and his team of experts provided indispensable assistance in planning the physical layout and design of the solar village on the National Mall. As a major sponsor, the AIA also supported nationwide marketing efforts, delivered significant help with jury member selection, offered office space for SD meetings, and assisted with logistics for the competition. Richard says he would not have succeeded without them.

Realizing that cellphones and internet service were necessary for the onsite event, Richard sought help for that infrastructure. Believe it or not, in those days no cell or internet accessibility was available on the National Mall in Washington, DC. Richard made a cold call to Electronic Data Systems (EDS), the world's leading provider of information technology services in the 1990s who prided themselves on delivering secure access to communications "anywhere and anytime." EDS invited Richard to visit their offices in Ashburn, Virginia, so they could learn more about his idea. After Richard's brief introduction to Solar Decathlon, they were intrigued.

EDS then formed a team of partners to meet the telecommunications needs of this unique competition. Cisco Systems had to get permission to set up the equipment (hardware, switches, antennae) that would relay cellular signals and a wireless local area network (LAN). Then EDS and its partners prepared to provide and monitor the infrastructure during the competition. As part of this sponsorship arrangement, Nextel provided wireless telephones free of charge to SD officials and competitors (student teams). In addition, WorldCom provided internet access with wireless broadband technology for the solar village; this was essential for day-to-day communications, as well as for data reporting related to competition scoring. As a major sponsor, EDS led the way to infrastructure developments of enormous importance for the Solar Decathlon. Looking back at the year 2000, cellphones, Wi-Fi service, and broadband technology were in the early stages of development, but reliable equipment for these communications were essential for the competition.

As a homeowner himself, Richard thought that securing support from a building supply company was of primary importance. He looked up the number for a local Home Depot store and was connected with Doug Zacker, who was quite excited when he heard about the Solar Decathlon. Zacker then reached out to his supervisor to heighten awareness at the corporate level, and soon afterward Richard was invited to make a trip to corporate headquarters in Atlanta, Georgia. There, Richard met a team of supervisors who responded enthusiastically to Richard's idea for a village of solar-powered homes open to the public. Home Depot generously offered in-kind assistance, professional advice, and various energy-efficient products to participating SD teams. As a major sponsor, Home Depot provided other valuable materials, such as signage, pop-up tents, solar walk-lights, and safety cones for the solar village on the National Mall. During the onsite competition, they also set up exhibits to help homebuilders and the public learn about energy-efficient products.

Reserving space on the National Mall was an essential step toward making Richard's dream a reality. The right location was crucial for overall success, and the National Mall was the perfect spot. First of all, it was a large, flat area with few obstructions. Second, it had east–west orientation, making it possible for solar houses to mount PV panels on south-facing roofs. Third, it was flanked by the United States Capitol Building and not far from the White House, the seats of power. Fourth, the Mall was across the street from the U.S. Department of Energy, the host institution for this competition. Fifth, it was easily accessible to the general public and public transportation services, including a Metro station. Finally, the National Mall is a beautiful iconic setting, well-known the world over. It turns out that gaining access to this space, managed by the National Park Service (NPS), for a 3-week period was no easy feat.

Richard remembers this challenge as one fraught with complexities. "On my first visit to the NPS Building in Washington, DC, Rick Merryman and his onsite office staff rejected my proposed plan immediately; they thought that putting temporary houses on the National Mall would destroy the grass. Furthermore, they didn't understand how an educational project about clean energy would fit their goals and purpose as an official agency of the federal government. I was a bit surprised but undeterred, so I wrote a letter to the Superintendent of the NPS, explaining the primary objectives of the Solar Decathlon. Once I clarified how this collegiate competition would showcase student efforts to demonstrate clean energy technologies to the visiting public, skepticism waned, and the superintendent granted permission for the Solar Decathlon to be held on the National Mall. But there was yet another step: I had to reserve specific calendar dates for this event. I mistakenly thought that part would be simple, but discovered it wasn't."

The National Mall in Washington, DC was an ideal spot for the inaugural Solar Decathlon.

Any group that wants to use the National Mall must first obtain a permit from the NPS, and this permit can be granted only one year in advance. Many factors were at play for determining the best dates to hold the inaugural Solar Decathlon in 2002, including local weather, annual college calendars, the government's schedule, national events, and federal holidays. After much discussion and research, Richard and his team from NREL decided that mid-to-late September would be an ideal time frame. They needed a three-week period to set up the village, construct solar houses, hold the competition, showcase these houses for the visiting public, and deconstruct the solar village. That turned out to be the maximum amount of time that any group was allowed to occupy the National Mall. That open space is quite special, and the Department of the Interior (responsible for the NPS) is rather choosy in terms of who gets permission to use this superior piece of land for public venues.

The following year on September 19, 2001, Richard stood in line at the NPS with a completed application in hand. After submitting that required document, there was a short window of time before he received an acceptance letter. Upon securing the dates requested, Richard met with NPS officials several more times to plan for the event, including: the exact location of the solar village, how to protect the grass with pedestrian walkways, when and where trucks could unload materials, how to manage the flow of visitors, and the need for security officers.

Solar Decathlon would mark the first time ever that a community of homes would be built on the National Mall in the shadow of the U.S. Capitol and in between several Smithsonian Institution buildings. The collegiate competition would be a historic development worthy of display on a national stage, but navigating the logistics was nothing short of assembling a complicated puzzle. The NPS has a set of strict regulations designed to protect the National Mall, including limits on building size and height, no excavation, accessibility for individuals with disabilities, and a limit of 21 consecutive days for any event taking place in that space. The merits of developing strong rapport with various government agencies, partners, and sponsors of the Solar Decathlon were crystal clear, so Richard carved out time for frequent communications with these groups.

Richard King revisits the site of the inaugural Solar Decathlon, 20 years later.

Reflecting on Solar Decathlons held on the National Mall, Melissa King stands in that historic location.

Starting with the 2002 event and continuing into successive Solar Decathlon events, location has proven to be a critical element. Putting a "Solar Village" on display in a prominent place that is easily accessible to the visiting public is part of the secret sauce for success. Richard's vision for this program encompassed public education and growing awareness of solar technologies, as well as gaining media attention to make a memorable mark with government leaders, legislators, and key decision-makers. When his dream became reality on the National Mall in Washington, DC, it was definitive proof of an ideal location for this competition. Additional SD events took place there in 2005, 2007, and 2009, and each one was clear testimony of the primary value of a great location. During that period, the volume of team proposals submitted to request participation and the overall number of public visitors to the Solar Village continued to climb, and the event became a popular, well-recognized biennial venue for the DC metropolitan area.

The first Solar Decathlon on the National Mall with the U.S. Capitol in the background was a strategic location for the design-and-build competition.

The primary ingredient of a Solar Decathlon is a set of participating collegiate teams. To that end, a formal Request for

Proposals (RFP) was needed. Richard crafted a letter to pitch his idea to colleges, universities, and institutions of higher learning to encourage them to consider entering the competition. He secured a list of U.S. schools of architecture and then sent out an invitation to the deans at those institutions on October 19, 2000. Since this was a completely new concept, that was a quantum leap of faith. Richard hoped a few of those leaders would recognize the educational value of this unique "hands-on learning adventure" for students. The RFP presented three guiding principles for solar houses that would participate in the SD competition:

1. Supply the energy requirements necessary to live and work using only renewable energy incident on the home during the event;
2. Exemplify design principles to increase public awareness of the benefits of solar energy; and
3. Stimulate the acceleration of research and development (R&D) of renewable energy.

A comprehensive set of rules and regulations for the inaugural competition was posted on the U.S. Department of Energy's Solar Decathlon website. Proposed structures for the competition had to be modular, so they could be transported from their respective colleges and universities and quickly reassembled on the National Mall before the official start of the Solar Decathlon. In accordance with the RFP, interested schools had to complete and send in an online registration form by December 29, 2000, to indicate their intention to submit a proposal, which was due February 16, 2001. The RFP provided complete instructions for proposal preparation, as well as a description of how the documentation would be evaluated.

By the end of that year, 11 colleges and universities had responded. Richard wanted a few more teams, so he made a phone call to William McDonough, FAIA, who was Dean of the School of Architecture at the University of Virginia at the time. After listening to Richard's overview of the Solar Decathlon, McDonough, a well-known leader of sustainable development and "green" architecture, gave an enthusiastic thumb's up. He offered to help by reaching out to colleagues and fellow deans at other architecture schools to share this educational opportunity and encourage them to enter. His influence proved fruitful, and subsequently, three more institutions sent in SD registration forms. That made a total of 14 teams, enough for a healthy collegiate competition.

Once again, Richard's perseverance proved to be an essential ingredient for forward motion. His actions echoed the words of famous poet and author Henry Wadsworth Longfellow: "Perseverance is a great element of success. If you only knock long enough and loud enough at the gate, you are sure to wake up somebody."

Early in 2001, Richard King, Cecile Warner, and several others at NREL reviewed proposals from 14 colleges and universities. They evaluated submissions according to the following criteria: technical innovation and content, organization and project planning, curriculum integration, and fundraising. All the proposals were deemed acceptable. These universities became the first group of Solar Decathlon teams, which were slated for the 2002 competition:

- Auburn University
- Carnegie Mellon University
- Crowder College
- Texas A&M University
- Tuskegee University
- University of Colorado at Boulder
- University of Delaware
- University of Maryland
- University of Missouri-Rolla and The Rolla Technical Institute
- University of North Carolina at Charlotte
- University of Puerto Rico-Mayaguez
- University of Texas at Austin
- University of Virginia
- Virginia Polytechnic Institute and State University

Those teams were excited about participating, and over the course of the next 12 months, Richard visited many of these institutions to meet students, professors, and college administrators. Enthusiasm for this design-and-build project was evident. Teams were abuzz with earnest commitment for the challenging task of creating an innovative solar-powered home for public display in the nation's capital.

The select group of teams set the stage for the inaugural Solar Decathlon. They were motivated, clever, dedicated, hard-working, knowledgeable, passionate, determined, and exuberant. As they worked together to plan and construct beautiful, energy-efficient homes for the competition, students discovered that teamwork and collaboration were crucial elements of success. Diverse talents and divergent ideas became prized assets as students made steady progress with their projects. For many, it was the most difficult project they had ever undertaken, but their thirst to master the tough assignment never waned. In the words of Pelé, Brazilian soccer star considered one of the greatest professional players of all time, Solar Decathlon teams found out firsthand that, "Success is no accident. It is hard work, perseverance, learning, studying, sacrifice, and most of all, love of what you are doing or learning to do."

Competition Manager Cecile Warner in the Solar Village at the inaugural event in Washington, DC in September 2002.

With 14 collegiate teams committed, SD organizers turned their attention to details of the 10 contests. A team of experts in building science, solar technologies, architecture, and engineering developed a set of priorities to guide their decisions: dwelling livability, aesthetics of structure and components, and integration of dwelling with energy systems. From an engineering perspective, homes created for the Solar Decathlon had to be structurally sound and in compliance with all relevant codes and standards. Each house was required to use only solar energy to enable the various activities and modern conveniences of everyday living, such as cooking, washing, doing laundry, running a home business, and operating an electric vehicle.

Ten distinct contests that would be evaluated and scored independently of each other were established: nine contests worth 100 points, and one contest (Design and Livability) worth 200 points, for a total of 1100. The main objective was to ensure that participating teams focused on all important aspects of designing and constructing solar-powered homes that were energy efficient, aesthetically pleasing, structurally sound, and functionally reliable. In addition, to honor the public education aspect of the competition, teams were required to "tell the story of their unique dwelling" with the public, sharing and communicating innovative ideas, key concepts, and clever technology applications that set the foundation for their design plans. The 10 contests for the 2002 Solar Decathlon were:

1. Design and Livability
    - Are design, innovation, aesthetics, and renewable energy technologies integrated into a pleasing domestic environment?
2. Design Presentation and Simulation
    - Do drawings, scale models, and computer-generated models effectively illustrate house construction and energy performance?
3. Graphics and Communication
    - Are the team's website, newsletters, and outreach materials effective?
4. The Comfort Zone
    - Does the house maintain interior comfort (heating, cooling, humidity, ventilation) with a minimum amount of energy?
5. Refrigeration
    - Does the refrigerator (with freezer) maintain appropriate and consistent temperatures with minimum energy use?

6. Hot Water
   - Does the house supply an ample amount of hot water for bathing, dishwashing, and laundry?
7. Energy Balance
   - Can the team complete all tasks for the competition using only the sun's energy?
   - Energy used = Solar energy supplied
8. Lighting
   - Is lighting for the house sufficient, energy efficient, and high quality?
9. Home Business
   - Is the house able to produce enough energy to power a home business?
10. Getting Around
    - Does the house generate sufficient solar power to operate an electric car for local transportation?

As with any endeavor, the first attempt at doing something new is fraught with risk, uncertainty, and unpredictable challenge. Figuring out which 10 facets of solar home design and construction were the most appropriate for this novel competition was far from easy. Determining fair, accurate, and reasonable methods for scoring the 10 contests was even more difficult. These decisions called for many heads together, lots of discussion, and plenty of compromise. In the end, SD organizers realized that "a perfect system" to satisfy everyone's perspectives and opinions was impossible. It turns out that SD planners have learned volumes with successive Solar Decathlon competitions held in different locations over the years. As a result, the original 10 contests have been modified to reflect new developments in engineering, materials science, "green" architecture, photovoltaics, technology, and increased understanding of sustainable construction. Powerful testimony about the Solar Decathlon legacy as an enduring competition is that Richard King's visionary concept remains intact, even though the nuts and bolts of the program have morphed over time. The fluid, dynamic nature of Solar Decathlon allows it to continue as a transcendent force with unparalleled effectiveness for workforce development, education, emergent leadership, and valuable R&D.

The next step for SD organizers was to devise an evaluation system for the 10 contests. How would these contests be scored? Who and what would manage the scoring? How would points be awarded and winners determined? From the outset, Richard and his team recognized that the 10 different contests would require different methods of assessment. Just as architects and engineers create from different perspectives with divergent goals in mind, evaluation of Solar Decathlon homes needed to reflect those differences. They decided that some contests would be judged subjectively by experts from the field, while other contests would be evaluated with objective measurements. Keeping careful balance between subjective judgment and metric-based performance was crucial, in order to demonstrate that elegant design and engineering excellence are both important aspects of high-quality dwellings.

After lively discussion and expert input about what mattered most and how to judge those elements, contest scoring methods were designated as follows:

- Subjective evaluation: Design and Livability, Design Presentation and Simulation, Graphics and Communications
- Objective measurement: Energy Balance, Getting Around
- Combination of subjective and objective (hybrid): Comfort Zone, Hot Water, Refrigeration, Lighting, Home Business

For any competitive endeavor, overall fairness and unquestioned accuracy are very important in the determination of winners. For the inaugural Solar Decathlon, participating teams were ranked from best to worst (1–14) in each of the 10 contests; points were then awarded to each team based on subjective merit or performance-based metrics. This was rather complex, and SD organizers quickly realized that to accomplish effective assessment of these prototype models in a limited time frame (10–14 days), scoring methods had to be highly efficient. Therefore, in subsequent competitions, the participant ranking system was eliminated, and points were awarded based on specific subjective and objective criteria delineated for each of the 10 contests.

For performance-based contests, a variety of monitoring devices plus a data logger were placed in every SD house. Real-time data were collected from these instruments and made available to all teams via the LAN setup for the event. These data were also used in populating electronic scoreboards that were part of the Solar Decathlon website available to teams and the public. A special team of officials managed the scoring systems. They installed monitoring equipment in every house and developed a data system to keep track of contest scoring. This group was also responsible for house inspections to ensure physical compliance with current building codes and the Americans with Disabilities Act (ADA). For the first event, the following trailblazer individuals spearheaded scoring-related tasks: Greg Barker, Zahra Chaudhry, Michael Deru, Ed Hancock, Mark Eastment, Sheila Hayter, Charles Newcomb, Paul Norton, Shanti Pless, Paul Torcelli, and Norm Weaver. During the onsite competition, team members were easily identified by their bright red T-shirts.

To evaluate the subjective contests and subjective components of hybrid contests, experts from the field were invited to serve as "jurors" for Solar Decathlon. A talented group of jury members for each subjective contest was assembled for the competition. They were charged with (1) reviewing the required documentation that teams submitted, (2) visiting SD homes onsite for a "team tour," and (3) meeting with teams to ask questions and learn more about each house. Jurors were provided with scoring criteria and detailed guidelines for evaluation in advance. After the house visits, jury members for each contest met to deliberate, determine how many points to award to each house in specific scoring categories, and make decisions about winners.

Jurors play a vital role in the competition. SD jury members are highly esteemed in their respective fields, and their participation raises the stature of the competition. In addition, collegiate students are honored to share their creative efforts with world-famous talent. Students have opportunities to present their innovative homes to accomplished professionals with prestigious backgrounds, and this strengthens student motivation to go above and beyond. It is notable that every Solar Decathlon juror has commented on how impressed they are with student-generated prototypes for the competition.

Jury members for the inaugural Solar Decathlon included the following experts:
- Design and Livability Jury
    - Glenn Murcutt (winner of the 2002 Pritzker Architecture Prize)

- o Edward Mazria (author of *The Passive Solar Energy Book*, a classic volume)
- o Steven Paul Badanes (designer/researcher of sustainable building construction)
- o Dr. Edward Jackson, Jr. (Research and Program Director at AIA)
- o Dr. Douglas Balcomb (specialist in passive solar design systems)
- o Stephanie Vierra (Executive Director of the Association of Collegiate Schools of Architecture)
- Engineering Design Panel (support for Design and Livability Jury)
    - o Dr. Hunter Fanney (Team Leader, National Institute of Standards and Technology)
    - o Dr. Dick Hayter (Associate Dean of Engineering for External Affairs, Kansas State University)
    - o Ron Judkoff (Director of the Center for Buildings and Thermal Systems, NREL)
- Graphics and Communications Jury
    - o Jill Anderson
    - o Susan Moon
    - o Paula Pitchford, Nancy Wells, Shauna Fjeld, Kristine McInvaille, and Jim Snyder from NREL
    - o René Howard from WordProse
    - o Jill Dixon from the National Building Museum
    - o Ben Finzel from Fleishman-Hillard Communications
    - o Lani Macrae from U.S. DOE
- Building Energy Analysis Experts (support for Design and Simulation Contest)
    - o Dr. Douglas Balcomb (specialist in passive solar design systems)
    - o Greg Barker (Mountain Energy Partnership)
    - o Michael Deru (NREL)
    - o Russ Taylor (Steven Winter Associates)
    - o Norm Weaver (Interweaver)

Members of the Design and Livability Jury for SD 2002.

Sheila Hayter deserves abundant credit for masterminding the organization of the first set of jury members. She worked closely with Richard King and Cecile Warner to ensure that selected jurors were top-notch leaders in their fields, and she helped with communications, travel plans, and accommodations for jury members. At the onsite event in 2002, Sheila orchestrated a complex schedule to ensure that jury members and teams met at designated times for "show and tell," ushering them from house to house and answering any questions with grace and finesse. Through her stellar example, Sheila set the bar for excellence. Ever since the inaugural competition, jurors have been eager to support and participate in Solar Decathlon events, thanks to Sheila's impressive standards.

To maintain the integrity of the competition, a team of volunteer "observers" was formed to observe team activities at the onsite event. On each day of the competition, one official observer was stationed in every SD house as an impartial, third-party official. The observer kept a written log and completed a checklist to record what teams were doing throughout the day. This documentation helped the organizers know whether or not teams were following the rules. Volunteer observers have been used in every Solar Decathlon (nationally and internationally), and their added value to the competition cannot be overstated. The observer role as an impartial witness ensures that the competition is fair, honest, and squeaky clean.

Reflections from Susan Piedmont-Palladino, an architect, writer, curator, and professor based in Washington, DC who is currently the Director of Virginia Tech's Washington-Alexandria Architecture Center (WAAC).

2000 — Engineers and Architects

It sounded like a certifiably crazy idea: turn over the National Mall in Washington, DC to college students, let them build a neighborhood of little off-the-grid houses powered only by the sun, and have them run through a set of real-life tests. Who could say no to that? My Virginia Tech colleagues on the main campus in Blacksburg, Virginia, with ample space to work and an abundance of design and engineering talent, planned to answer the call for teams to compete. The Alexandria campus, wedged into the snug fabric of Old Town, took a different path and instead got involved planning the neighborhood, or the "Solar Village" as it came to be known. Like any town-planning exercise, this one had to have some rules: each house had site boundaries, lot coverage and size limitations, and prescribed set-backs from each other. The Solar Village would be organized along a "main street" with assembly spaces in large tents, like a town hall. That fall, students from fourteen schools from across the United States would move in, as neighbors and competitors, to model a future of solar-powered living on the nation's front lawn.

What interested me, though, as an architect and educator who had been involved in the behind the scenes work of designing rules for several major design competitions, was how to frame the problem to yield high performance in both energy and aesthetics, engineering and design. How to pitch this competition to schools so that both architecture and engineering students, who often occupy separate buildings across campus from each other, would be equally attracted to enter? If the tone were pitched to an engineering frequency, the architects would think, "They're not interested in *design*!" If it sounded the right notes for architects, the engineers would think, "Why bother; this is just a beauty contest." And they would both be right.

It was crucial to set the context to attract the best talent. From my perspective, that was an easy task. For millennia, humans have designed and constructed buildings as if their lives depended on it, because they did. It is not a new problem, but it became a bigger, more complex problem as humans evolved and expanded our buildings and survival technologies. So, while my colleagues on the main campus in Blacksburg, Virginia were designing their house, and my colleagues at the WAAC in Alexandria, Virginia were designing the Solar Village, I nudged my way into helping write the inaugural competition brief. My opening paragraph for the "Introduction" to the *2002 Rules* put it simply: "The Solar Decathlon is a new intercollegiate, interdisciplinary design and construction competition that takes up a persistent and age-old question: How do we integrate architecture and technology with a dwelling? In other words, what makes a good house?" Implied in that statement is that "good"-ness demands integration of disciplines and systems, which became a foundational principle for all Solar Decathlon competitions.

In the design world, there is no such thing as a problem with only one solution. That's why a design studio with 15 students can yield 15 completely different solutions to the same design brief. Give the same brief out the following year to anther 15 students, and you will get another 15 different solutions. If a design professor is really lucky, they may even get a solution that challenges the brief, that questions the question. Questioning the question is a particularly valued skill in design education; it propels innovation, yet at the same time it is the bane of a rule-writer's existence. Few designers, and fewer design students, are inclined to accept a problem statement as given, particularly in a competition.

Engineers see problems a bit differently. In *Thinking through Technology: The Path between Engineering and Philosophy*, Carl Mitcham suggests that all engineering activities are "efforts to save effort."[1] Engineering students are also adept at questioning the question, but where designers tend to open and expand a question, solving an engineering problem is a process of narrowing the possibilities to reach the optimal result as measured against the relevant criteria. Put a design student and an engineering student in a room together, and the engineer will set about figuring the best way to get from here to there, while the design student might begin by asking where "there" really is. Yet, the two disciplines share a pedagogical foundation that both disciplines' problems are "wicked problems," that is, problems that entangle many different disciplines which can only be solved by experimenting, or learning by doing.

---

[1]Mitcham, Carl, *Thinking through Technology: The Path between Engineering and Philosophy*, University of Chicago Press, 1994, p 221.

It was already clear in the first Solar Decathlon that the full range of design disciplines, including interior, industrial, architecture, and landscape; and engineering, including computer, electrical, mechanical, and structural, would be needed to compete successfully. They all would have to bring toolboxes, metaphorical and actual, full of skills for the task at hand.

2005 — Designing Rules

The 14 houses entered in the 2002 inaugural Solar Decathlon were an eclectic group, yet each one was an 800-square-foot answer to the question. Some were never completely finished, as student teams underestimated the work involved in actually building the houses on the National Mall.[2] There were ordinary-looking houses that performed extraordinarily well in quantitative competitions, and extraordinary-looking houses that didn't. And there was plenty of questioning of questions on the part of the competitors and the organizers.

I officially joined the Solar Decathlon team after the close of the 2002 competition to help assess and revise the rules and regulations. All inaugural events have a few things in common: first, only when the event is over can you see whether your predictions and expectations were fulfilled; and second, for the first event to be "inaugural," there needs to be a second. Joining the team to plan for 2005 was an opportunity for me to exercise some of my earliest research as an educator on the nature of rules and their consequences in competitions.

I had become interested in rule-setting, the invisible design work that precedes design, decades earlier. The task of setting and framing of competition briefs is to set a table of more or less likely solutions. If organizers want to keep certain things off the table, or if they want more of other things, they need to back up and rethink the parameters of the prompt. Competitions are different from commissions: the goal of a competition is to get results you can't entirely predict. The language of rules and regulations can nudge, entice, discourage, and forbid. They can be productively ambiguous, but they can never cover all possibilities. Like many architects, I had entered my share of competitions and been regularly perplexed by the winners, which

---

[2]The Virginia Tech (VT) team was fortunate to have a supply of student and alumni labor just across the river, which is how Mr. Solar Decathlon, Joe Wheeler, first got involved. VT would go on to compete under Joe's direction in 2005 and 2009. The 2009 entry, the well-travelled *LumenHAUS*, went on to compete in Madrid in Solar Decathlon Europe in 2010, where it won first prize. Under Joe Wheeler's leadership, VT then entered the *FutureHAUS* in the 2018 Solar Decathlon Mideast in Dubai, winning first prize there as well.

sometimes seemed as if they had ignored the rules. So, I turned to the rarified world of yacht racing to look at a vivid example of rules and their unintended consequences, a quirky research path which ended with the publication of my first academic paper, "The America's Cup: A Design Parable," in 1993.[3]

My conclusions in the paper shaped how I saw the task for 2005: every set of rules offers three possible responses. First, one can choose to follow the rules closely and aim for the best result within those boundaries. Or, a competitor can interpret the rules as a "necessary but incomplete description of the possible," as I wrote in 1993. Those competitors are the "question questioners," the ones who read the rules and say, "It doesn't say we *can't* do it." The third response is to challenge the validity or relevance of the rules. This can risk disqualification, but it can also lead to radical leaps in the competition itself. Ultimately, the rule-setting depends on the competition organizers deciding what they want to get out of the effort.

Each successive Decathlon has faced the task of sorting the space and activities of everyday life into actionable contests and responding to larger social contexts to determine which contests to elevate and which to elide into others. Between 2002 and 2005 the contests evolved in subtle ways. The Appliance Contest was new in 2005, broadening the more restrictive Refrigeration contest to better reflect the energy demands of daily life. The Appliance contest rules, though, seemed to have bias toward energy use, albeit efficient use. A clever student might question the laundry question, for example: it's much more energy efficient to simply not use a clothes dryer. So, the question arose: Can we just hang up the laundry to dry? Not until the 2011 Decathlon does the Appliance contest explicitly mention passive drying, allowing competitors to use "active or passive drying methods." Other questions would come up: What if a few neighboring houses formed their own micro-grid? Or shared water? And what about the buildable envelope? What if a house transforms, expanding or unfolding?

The Design and Livability contest from 2002 split into two contests for 2005: Architecture and Dwelling. We described the distinction between the two in the Rules as "The Dwelling contest begins, in a sense, where the Architecture contest ends: after the design and construction of the houses are complete and living in them begins." As with Design

---

[3]The brief paper can be found in the *Journal of Architectural Education*, Vol. 46, No 4 (May, 1993) pp 266–269. One of the peer reviewers of the paper was not impressed; he questioned the relevance of my argument to architecture. Architecture is not a contest, he commented; there aren't "winners."

and Livability, though, Architecture and Dwelling stayed at the top of the list. Although technically the ten contests are not listed in order of importance, putting Architecture and Dwelling first and second in 2005 conveyed that for true sustainability, "We have to lead with beauty," as the late Lance Hosey wrote in *The Shape of Green*.[4] Later Solar Decathlons move "Market Appeal" to Contest #2, to emphasize that beauty counts.

To illustrate the challenges of trying to set the "rule-scape," consider that the 2002 Rules and Regulations document is just 43 pages long. The same document for 2005 is 177 pages; 2007 weighs in at 143 pages. Interestingly, by 2009 the Rules and Regulations were slimmed down to 78 pages, a weight that SD organizers have been able to maintain. Why did the Rules and Regulations balloon for the second and third SD events? Deciding exactly what to describe in a set of regulations as permitted or prohibited, prescribed, or proscribed, is itself a challenging design problem, particularly when one of the primary goals is *innovation*, which actually becomes its own contest in 2017. The Solar Decathlon challenges student teams to bring something new into the world. How do you set limits on that?

One of my favorite rule-challengers was the Rhode Island School of Design house in 2005. This spunky team upended expectations about building orientation, positioning it like a rowhouse, and challenging the competition to include denser, more urban contexts. That kind of change—there was nothing in the rules to prevent that orientation—came from "questioning the questions." Later Solar Decathlon competitions added considerations of density and scale to the prompts, recognizing that the challenge of changing how we power our buildings can't be solved one house at a time. Nor can we make meaningful change without keeping the public engaged and informed.

One powerful but unanticipated outcome of the Solar Decathlon since the beginning has been the emergence and accelerated growth of student leaders on collegiate teams. Creating and constructing unique, net-zero energy, solar-powered homes for public display to compete in and win on a national or international stage is a sizable challenge for young adults who are in the throes of learning about architecture, engineering, and sustainability. Fortunately, this first group—and every successive group of Decathletes since then—has been highly motivated, ambitious, eager, industrious, passionate, and driven to excel. Decathletes believe in the Solar Decathlon

---

[4]Hosey, Lance, *The Shape of Green: Aesthetics, Ecology, and Design,* Island Press, 2012.

mission and want to lead the way forward. As students work together, they discover that collaboration is essential for successful innovation in the design-and-build process. It also became clear that capable, effective leadership is a foundational pillar for cooperative engagement in any group. As a result, some students with the appropriate skillset have assumed the mantle of leadership, often without prompting.

For professors and mentors working with SD teams, it has been a stunning example of impactful work-force development. Typically, student leaders enact this quote from John Quincy Adams, the sixth president of the United States: "If your actions inspire others to dream more, learn more, do more and become more, you are a leader." Reflections from Mike Wassmer, a student at the time, capture the essence of how Solar Decathlon influenced many young adults actively engaged with the competition.

Reflections from Mike Wassmer, a licensed professional engineer and independent consultant who designs, implements, and manages technical aspects of collegiate, design/build, zero-energy home competitions and the evaluation of prototype zero-energy homes.

I graduated with a Mechanical Engineering degree from Rutgers University in 1999 and subsequently landed an entry-level job in Philadelphia, PA as a staff engineer for the Amtrak High-Speed Trainset project, now known as Acela Express. I had a casual interest in environmental issues in my teens and early twenties, so I felt I was making a positive contribution by helping advance U.S. rail transportation into the 21st Century. One day, after about 16 months on the job, I began reading Al Gore's book *Earth in the Balance* during my lunch break. The book shook me to my core, and I suddenly felt a personal obligation to make a bigger contribution than I would ever be able to make in the rail transportation sector.

As a kid, I had always wanted to be an architect, but I grew up in a family of engineers and had much more aptitude in math and science than I did in art and design. Nevertheless, I tickled my architecture bug in freshman year of college by taking a "Drawing for Architects" course, in addition to some core engineering courses. I struggled in the drawing course and subsequently abandoned my dalliance with an architecture degree. I resigned myself to the fact that I wouldn't be an architect, but I never lost my interest in architecture.

After I finished reading *Earth in the Balance*, I recalled my previous aspirations of working in the architecture field and decided to spend more lunch breaks visiting a few large architecture firms in the city. My intent was to stumble upon a firm that needed an engineer to assist in growing its sustainability program. I knew it was a long-shot, and I didn't find a job opening, but a couple firms suggested I would be useful to them if I supplemented my mechanical engineering degree with a buildings-related master's degree. Right away I began researching and applying to graduate programs for architectural engineering. I chose the University of Colorado at Boulder (CU-Boulder) because of the heavy emphasis on sustainability in their "Building Systems Program" within the Department of Civil, Environmental, and Architectural Engineering. I was set to start at CU-Boulder in Fall 2001, but I decided to get a head start by taking some courses at Penn State University in Spring 2001.

One of the course offerings for my first semester at CU-Boulder that fall was called "Green Building Design." The course description was vague, but the title was compelling, so I enrolled. On the first day of class, Dr. Mike Brandemuehl, the professor, announced that the ultimate objective of the course was to design and build an ultra-efficient residential building in collaboration with a design studio in the College of Environmental Design. I couldn't believe my good luck! I would be working day-to-day with architecture students on a real-world design-build project. My excitement was raised further when Dr. Brandemuehl informed us that we would transport the house to Washington, DC in Fall 2002 to participate in an event called the Solar Decathlon.

Most people remember where they were when they became aware of what was transpiring on 9/11/2001. On that day I was sitting in the Green Building Design classroom watching the initial news coverage on a projector screen. Having grown up in New Jersey with countless NYC visits during my formative years, I was really shaken and glued to nonstop news coverage for the next week or so. However, once I got out of my funk, it was full-steam ahead on the Solar Decathlon project. Considering the state of the world at that time, my sense of mission was stronger than ever.

The next 12 months of the design-build process leading up to the competition was an unforgettable experience. I met many amazing people who shared my passion and were intensely motivated to create a building that would demonstrate our shared vision for the future. Our team was somewhat unique in that most of us were grad students, so we came into the project with significant life and work experiences (unusual for Solar Decathlon teams). We were keenly aware of and motivated by the bigger-picture importance of the Solar Decathlon. It meant much, much more to us than a typical "cool science project."

The 3–4 weeks in Washington, DC were a once-in-a-lifetime experience. Our team camaraderie, collective sense of purpose, and the exuberance we felt after winning the competition made all the sleep deprivation, heat exhaustion, and frustrations along the way well worth it. I also value two personal memories from that time in DC. First was talking to guests who were waiting in long lines to walk through our house. I wasn't sure what to expect during this experience in DC, but it turned out that my favorite part was explaining our project to visitors who had little or no knowledge of the technologies and strategies that were on display in our house. Most visitors entered the Solar Village with a little curiosity and virtually no knowledge of photovoltaics or green energy principles. All those sunbaked hours that I spent educating people in our house queue were immensely gratifying. I was excited by the prospect that some of them might take serious action steps because of something I had explained to them or opened their eyes about. It was the "multiplier effect" of the Solar Decathlon in action.

The second particularly meaningful personal memory was being recognized for producing the best building energy model and simulation among all teams and earning a perfect score in that sub-contest. Early in the project, I volunteered to lead the modeling and simulation effort for our team. It was a challenging, painstaking process that involved me working countless lonely hours in the office on a computer, while the rest of the team was outside working and having fun at the construction site. So, when the CU-Boulder team earned a perfect 50 points for modeling and simulation, it was personally gratifying after all that effort to be recognized by expert jurors as well as my teammates.

After the competition ended, I had a tough time coming down from the experience. In a sense, I felt the highlight of my career was already in the rear-view mirror at the age of 26. What could possibly be a worthy encore to what I had just experienced? As I was finishing my master's thesis and beginning the job search, I became disillusioned with the prospect of sitting in a cubicle working for a large engineering firm on projects with few or no sustainability features. At that time, jobs in the sustainable buildings sector were tough to come by. I applied for and was accepted into a PhD program, but I wasn't particularly excited about getting another degree. I think I was just buying time until the job market improved, and then I could hopefully continue pursuing my passion in the workforce.

During the summer before I was to begin the PhD program, a few of my Solar Decathlon teammates from CU-Boulder and I attended an American Solar Energy Society (ASES) conference in Kansas City to make our last-ditch attempt to find jobs that satisfied our desire to make real positive change in the world. After one conference session, I had a chance encounter with Cecile Warner, who at the time worked at NREL as the Solar Decathlon Project Manager. I told her why I was at the conference and asked her if she knew of any opportunities at NREL. She told me that the previous Solar Decathlon Competition Manager had recently left the job, and she was scrambling to hire a replacement. I couldn't believe my luck. I enthusiastically expressed my interest in the job, and within a few weeks I was hired. I spent the next eight years in my dream job as the Solar Decathlon Competition Manager.

Then in 2011, I started working as an independent consultant to help the Sultanate of Oman run their own collegiate design-build competition. Today, twenty years after that memorable first day in the Green Building Design course at CU-Boulder, I am still working on collegiate design-build competitions. I sure hope this is what I will be doing until the day I retire!"

Solar Decathlon has an uncanny way of affecting anyone who is involved with this special program, shining a light on what's vitally important to advance the path toward renewable energy and sustainability: by taking positive action. In the words of Leonardo da Vinci, well-known scientist, artist, astronomer, and mathematician: "I have been impressed with the urgency of doing. Knowing is not enough; we must apply. Being willing is not enough; we must do." This is the essence of Richard's vision.

Faculty advisors and mentors of the collegiate teams are unsung heroes of Solar Decathlon. Dedicating two years or more to redesign courses that will prepare students for the heavy lifting of planning and constructing innovative net-zero-energy homes that rely solely on sun power is no small feat. Providing guidance but demonstrating restraint in terms of how to accomplish this is certainly not easy, especially for educators who are more accustomed to leading the way. The educators who engage with Solar Decathlon strive to energize their students, giving tirelessly to a complex hands-on project of Herculean proportions. This quote from Oprah Winfrey sums up what these faculty members have achieved through their selfless example of leadership: "Leadership is about empathy. It is about having the ability to relate to and connect with people for the purpose of inspiring and empowering their lives."

Richard's voice reflects on the significant contributions of university faculty. "I have been totally impressed with the unwavering commitment of college professors in every Solar Decathlon, here in the U.S. and around the world. They offer deep knowledge, rich expertise, and can-do attitudes as mentors. They remain open minded and supportive as students investigate various design solutions. They are strong and steadfast leaders who give 200% to this effort. They personify what education is all about."

Reflections from Michael J. Brandemuehl, Professor Emeritus of Civil, Environmental, and Architectural Engineering at the University of Colorado Boulder, where he served as faculty advisor for the University of Colorado Solar Decathlon teams in 2002, 2005, and 2007. He is a licensed Professional Engineer who performs teaching, research, and consulting related to the design, operation, and analysis of building energy systems, with emphasis on the modeling and simulation of HVAC&R systems and their controls, smart building systems, and application of renewable energy technologies.

For me, Solar Decathlon was a fifteen-year adventure. It began in December 2000 when a small group of students at the University of Colorado approached me about joining this new competition. By the time I retired, I had guided three teams as faculty advisor, served as proposal evaluator or engineering juror for four competitions in the U.S. and China, and spoke with thousands of industry professionals, policymakers, school children, homeowners, and school children about the opportunities for sustainable building design.

One of the great challenges of engineering education is providing the fundamental and technical foundation to solve engineering problems while also preparing students for real-world applications. As an engineering educator, teaching the fundamentals always seemed easier than the task of exposing students to meaningful engineering experiences and projects. It is especially difficult in the civil and architectural engineering disciplines, where the real-world projects are so large and do not lend themselves to the laboratory or benchtop. In addition, the ABET accreditation criteria for all engineering programs expect students to demonstrate broad non-technical abilities, and include terms like *multidisciplinary teams*, *communication*, *contemporary issues*, *societal context*, and *professional responsibilities*.

Along comes Solar Decathlon, checking all the boxes. Admittedly, we couldn't make all our engineering students commit two years of their lives to develop an architecturally compelling and technologically advanced home with the latest sustainable design features, build that home, and then prove it all works on the most public of stages. For the bright and motivated students who immersed themselves in the projects, though, Solar Decathlon exposed them to the real-world of architectural engineering like no course ever could. Solar Decathlon defined their academic careers and launched their professional futures. As word got out, it also attracted the best students.

Curiously, Solar Decathlon emerged at a time that project-based learning was growing in other programs in our Department of Civil, Environmental, and Architectural Engineering. At the same time that I was mentoring Solar Decathlon students, my colleague, Bernard Amadei, was leading other students to work on civil and environmental engineering projects in developing communities in Central America. Within the year, he formed Engineers Without Borders. In both cases, students were offered extracurricular opportunities to participate in meaningful projects aligned with their professional callings.

While I have been a keen Solar Decathlon observer over most of its lifetime, the most fascinating years were the early ones when teams (and organizers) were still trying to figure out what the competition was all about, and the public was less informed about the opportunities for solar buildings. The first three competitions were also the ones in which I was most immersed as faculty advisor.

The 2002 Solar Decathlon was the most fun, largely because it was so completely different. College students building solar houses and taking them to Washington, DC for a public competition on the National Mall? Unheard of. There was an elaborate set of competition rules, but it was not obvious what it would take to win. It was a design competition, where elegance and innovation would be expected, but the house also had to work. There was no budget, little knowledge of the other competitors, and only a vague understanding of juror expectations. No precedent, no template for success. And wait—the Department of Energy is only giving us $5000? Where do we get the rest of the money? Nevertheless, the uniqueness of the Solar Decathlon and the nobility of the cause were powerful motivators for a passionate group of students in our engineering and architecture programs. The novelty and naiveite' of the first competition were liberating.

The 2005 and 2007 competitions offered a very different experience, for both students and faculty, especially after our team won the 2002 and 2005 events. Student interest exploded. Like the first competitions, subsequent Solar Decathlon projects began with the same blank sheet of paper and a new batch of wide-eyed and inexperienced students, but expectations had been set.

The core 2002 team responsible for delivering the project comprised only 16 students—four from architecture and twelve from engineering—eight undergraduate and eight graduate students. It was a small, tight-knit group, but a team that eschewed formal job titles. There was no hierarchy, but everyone knew their respective roles. Even the obvious team leader demurred to an egalitarian structure. One student was the public face of the team on the National Mall, but the whole team chafed when an interviewer suggested that he might make an executive decision. By comparison, the 2005 and 2007 teams were much larger, each with an organized and formal structure, and with strong student leaders.

Solar Decathlon is described as a *student competition*, but I'm not sure that people understand how truly student-driven these projects can be. Among the three University of Colorado teams, there were never more than two or three faculty actively involved—the students had to do almost everything.

The work on design, construction, and operation of the house is well-documented, and the benefits of working in multidisciplinary teams and developing oral and written communications are well-known. I think one of the less appreciated benefits came from the fundraising efforts required of most teams. The University of Colorado administration typically provided only about 10% of the project costs. Each of our teams sold their houses, but there remained large fundraising needs. Students had to write proposals to potential funding agencies, contact manufacturers for product donations, reach out to local and national professionals for construction assistance, and pursue donations from alumni and the public. Even our original 2002 proposal to the U.S. Department of Energy was co-authored by three undergraduate students and four graduate students. Perhaps the most inventive fundraising idea was when leaders of the 2005 team organized a campaign on the CU-Boulder campus in which the student body voted to assess a fee of $0.45 per student per semester, from 2005 through 2009, for support of the Solar Decathlon. Those fees provided over $100,000 for the 2005 and 2007 teams.

The job of fundraising required that students put themselves out there and sell themselves—to convince a stranger to support their ideas and share their passions. It's an empowering experience, especially for young engineers or architects who will spend their professional careers defending their designs to clients and peers. For me, one of the most satisfying experiences of each Solar Decathlon was the first day that homes were open for tours on the National Mall. Inevitably, all the students had been up till dawn putting finishing touches on their houses. A few hours later, they would return, transformed from exhausted construction workers to fresh-faced tour guides, beaming with pride, and flush with confidence as they showed off their designs and fielded questions from visitors. The transformation was stunning, but they were also exceptional ambassadors.

The students who participated in Solar Decathlon got an invaluable introduction to their upcoming professional careers, leveraging their academic training with hands-on experience in the industry. They were working in multidisciplinary teams to solve complex problems. They saw their designs through to operating the finished product. They raised funds to fulfill their ambitions. They learned to communicate their ideas to their peers, professionals in the industry, policymakers, and the public at large. They were part of something bigger than themselves, having a ton of fun, doing something noble, and doing their part to try to save the world.

While students at every competing university gained valuable skills and experience through participation in the Solar Decathlon, the uncommon success of the University of Colorado teams also brought priceless public exposure to the University. Our teams' achievements were highlighted by the University in most outreach and recruiting material. Enrollment in our graduate Building Systems Engineering program during the 2003–2004 academic year more than quadrupled over the previous year, and enrollment in our undergraduate Architectural Engineering program rose steadily through 2008.

The success of the Colorado teams was embraced by the University community. At the time, the University was weathering some less favorable publicity. The Princeton Review cited CU Boulder as the top party school of 2003. The University was also struggling with backlash from the controversial writings of Prof. Ward Churchill in the wake of the 9/11 attacks and a football recruiting scandal involving alcohol, drugs, and sex. The issues resulted in the firing of Ward Churchill and the resignations of the University president, athletic director, and football coach. By contrast, the Solar Decathlon team offered a wholesome counterbalance of bright and articulate students demonstrating global leadership in sustainable building design.

Finally, Solar Decathlon had a profound effect on me. While the professional success of an engineering professor is measured largely by research productivity, we differentiate and define ourselves as professors by our service to our students. In my case, Solar Decathlon allowed me to earn my job title. There was something special about working with those three teams of students on challenging and high-stakes projects, driven by the societal imperatives of energy responsibility. It was invigorating to see students so motivated, so excited, and so proud of what they had accomplished. In the process, they were an inspiration to me and to everyone around them."

One more essential facet of Solar Decathlon planning is the schedule of events and the completion of the required tasks for the 10 contests. As SD organizers outlined myriad details of the competition for the participating teams, volunteers, and officials, they realized that everyone needed a clear road map for the onsite event. The NPS allowed only a 3-week period of occupancy for the National Mall. Team arrival at the site, assembly of houses, judging and scoring, showcase of prototype models for the public, and disassembly of the solar village had to take place with a 21-day window. Success required a finely tuned calendar made available to everyone involved. It also meant honoring the official schedule with activities that ran on time and in an orderly fashion. Richard King and his team of SD organizers pulled it off with sheer grit, composure, and dignity, and the impact was felt far and wide. The inaugural Solar Decathlon was a big hit, worthy of a Grammy Award, Academy Award, and Nobel Prize for excellence on so many levels. In the years to come, the competition would be repeated at regular intervals in the United States and replicated in eight other countries on five continents.

On September 19, 2002, 14 collegiate teams began to arrive on the National Mall—in the dark. Their solar homes had been disassembled to travel in trucks and tractor trailers to Washington,

DC for the Solar Decathlon. Once there, heavy-duty machinery, cranes, and forklifts were needed for the unloading process. However, that equipment is not allowed to disturb regular traffic patterns between Constitution and Independence Avenues during the daytime. The permit to occupy the National Mall started at midnight, so at exactly 12:01 am, big rigs and other transport vehicles showed up. Like a stealth operation, they unloaded their precious cargo. It was quite a sight, but only teams and organizers got to witness this unusual scene unfold in the wee hours of the morning.

In the dark of night, trucks and cranes deliver modular sections of SD houses to the Solar Village site in September 2002.

Thanks to Dan Eberle's expertise and ability to lead any group whatsoever, this mega-movement of huge equipment onto the National Mall came off without a hitch. As Director of the American Solar Challenge (car race) and road navigation expert, Dan understood how to coordinate the passage of extra-large vehicles into specified team lots that were extra-small within a designated time limit—no small feat! Mission accomplished.

The Trojan Goat from the University of Virginia arrives on the National Mall at night.

Once they "landed" onsite, these teams became part of a massive construction zone. Decked out in hardhats and work boots, students started reconstructing the houses they had originally built at home campuses in various regions of the country. From sun-up to sundown, the National Mall was a whirlwind of activity as over 500 students rebuilt solar houses, piece by piece, from bottom to top. Most were incredibly organized, and voluminous tool boxes and supply cabinets were clear evidence of meticulous advance planning.

Richard admits that he was concerned about all this activity. "Ultimately, I was responsible for this buzzing beehive, and I sure didn't want to see anyone get hurt. I'm immensely grateful to our fantastic safety team, who developed important safety protocols and requirements. They made sure that all teams and every team member were in compliance 100% of the time. Thanks to their expertise and due diligence, there were no serious injuries at the inaugural event and very few in any of the competitions since then."

As teams reassembled their prototype houses, the National Mall became a huge construction site.

Decathletes from Tuskegee University, a land-grant institution in Alabama founded by Booker T. Washington in 1881, planned and built a dwelling that linked the past with the future. Their dog-trot design featured a central breezeway to help cool the house.

A Tuskegee team member applies paint to an exterior wall of their house.

Early in the morning on September 26, Richard King stood on the National Mall gazing out at the Solar Village. He wondered whether anyone would come see this amazing display of innovative solar-powered homes equipped with the latest and greatest technology. He crossed his fingers, hoping the public might pay attention and show up.

Chapter 4

The Dream Unfolds

Solar Decathlon 2002

The Opening Ceremony for Solar Decathlon was set for 10 am on Thursday, September 26, 2002. More than 300 invited guests arrived to hear the opening remarks, followed by a moving rendition of the national anthem performed by "The President's Own" United States Marine Band. After words of welcome from the U.S. Secretary of Energy and major sponsors of the event, the 14 teams were introduced. An official ribbon-cutting ceremony declared the Solar Village open to the public. Richard reflects on that memorable moment: "I could hardly believe this was happening. Focusing on all the details required to pull this off called for my full attention. Then all of a sudden here I was, watching my far-fetched dream come true on the National Mall. Teams were there, houses were built, exhibition tents were open, staff and volunteers were ready, and dignitaries showed up. Everything was in place, but I was still worried about one fundamental aspect of my vision: Would the Solar Village attract visitors to fulfill the mission for public education?" The answer to that question was soon to come.

Bright blue skies bestowed untold blessings on the opening weekend of the inaugural Solar Decathlon. *The Washington Post* offered readers in the Washington, DC metro area a wonderful glimpse of what was in store for those who ventured out to the National Mall. On Friday, September 27, 2002, the cover of the

---

Solar Decathlon: Building a Renewable Future
Melissa DiGennaro King and Richard James King
Copyright © 2024 Jenny Stanford Publishing Pte. Ltd.
ISBN 978-981-5129-47-2 (Paperback), 978-981-5129-13-7 (Hardcover), 978-1-003-47759-4 (eBook)
www.jennystanford.com

"Weekend Section" of that newspaper included a stunning photo of students from the University of Maryland installing solar panels on the roof of their Solar Decathlon house. This was an appetizer for the captivating, well-researched feature article called "Here Comes the Sun" by Caroline Kettlewell. The compelling human-interest story about why and how the competition got started appeared in a four-page spread, brimming with colorful, upbeat photos taken by Mark Finkenstaedt.

The next day, on Saturday, September 28, *Washington Post* staff writer Benjamin Forgey penned another article on the front page of the Style section that bolstered the public's attention. Titled "Bright Idea: Solar Village Lights Up Mall," Forgey stated: "With the Capitol dome and Washington Monument as background decoration, it's probably the best advertisement solar housing has ever had in this country." The writer's powerful testimony came after his own visit to the Solar Decathlon, where he observed first-hand the awe-inspiring "solar subdivision" on the National Mall. A wire version of Forgey's article was picked up by the *Los Angeles Times,* the *Modesto (California) Bee,* and the *Juneau (Alaska) Empire.* Back then, printed newspapers were a primary source of information for current events, so it is safe to say that securing these prime-time hits was a tipping point for public exposure to Richard King's bold vision.

University of Maryland Decathletes assembling their solar system on the rooftop.

University of Maryland team members engaged in construction work for their Solar Decathlon house.

During two weeks of intense construction activity, teams of dedicated Decathletes had worked diligently for countless hours to assemble their unique dwellings on a national stage. Students were exhausted, but they were also excited about the prospect of sharing the fruits of their labor. They hardly had time to relish their accomplishments, because suddenly the Solar Village became a "must-see" exhibition in Washington, DC. Fortuitously, the Solar Decathlon was situated next to the escalator exit from the Smithsonian Metro Station (subway). After the village opened at 9 am on September 28, a steady stream of visitors poured out of that exit to take a peek at this unusual collection of net-zero-energy homes, launching an unanticipated predicament. With 14 small houses open for public viewing, what would happen when hordes of curious onlookers showed up to tour them?

Well, thousands of eager visitors of all ages arrived in full force that first weekend. Decathletes were thrilled to have an audience, and house tours commenced with wonderful show-and-tell. However, long lines formed outside of each house as people waited to get inside. Students did not really understand how to talk about their creative dwellings "in brief," and SD organizers were concerned about the negative impact of long wait times. Richard King recalls his reaction: "With every passing hour, more enthusiastic visitors appeared and wait times to view the houses increased. Of course, we wanted to make a positive impression, so we had to take quick action. I asked a few of the Organizers to help me by going to every team house to offer tips on crowd control. We suggested that they: 1) station a strong student communicator outside each house to share talking points with the public before they entered, 2) shorten the length of time for house tours, 3) answer any questions from visitors as they waited, and 4) pass out a printed overview of their house when visitors showed up, so people could read it while waiting (each team had created a handout for this purpose)."

"Although teams were caught off guard with throngs of inquisitive attendees, they rose to the occasion and adjusted accordingly. It was fun to see which Decathletes enjoyed being on the front lines with the public, but a few lost their speaking voice as a result of such lively interaction. In subsequent competitions, we arranged to have a student at the entrance to each house to communicate with inquisitive spectators. In all honesty, crowd control has been a thorny issue for most Solar Decathlon events. It's a *good problem* because it indicates we're reaching multitudes."

As soon as the Solar Village opened to the public, crowds appeared. Here, they wait to see the prototype from the University of Colorado at Boulder.

Outside the Virginia Tech house, enthusiastic visitors wait patiently to get a glimpse of this innovative dwelling.

Richard continued to be a persistent advocate for strong internal and external communications about his "big idea." To succeed with public education, Solar Decathlon messaging had to be loud and clear. No one could guarantee appropriate media attention for the competition, but Richard knew it was a necessary ingredient to spark interest and instigate gradual changes with public perception and behavior. For the inaugural SD event, local news stories and photos captured thousands of avid fans in the Solar Village, which motivated journalists to see what was going on. Reporters and photographers arrived on the site for a glimpse of the action, and the competition was abuzz with media attention.

Over the next few weeks, 507 stories about Solar Decathlon were published in newspapers, magazines, and internet sites across the country. In addition, 45 television and radio stories about the 2002 event were reported by the Video Monitoring Service (VMS). NBC's *Today* show ran a segment for 4 minutes and 28 seconds, and National Public Radio's *Weekend Edition* show included coverage of the event. Spanish language media, including Orlando's *Latino International*, and African-American news organizations, such as Black Entertainment Network (BET), reported on the Solar Decathlon. Voice of America also had crews at the Solar Village, so they could share stories in several different languages for other countries. More coverage of the Solar Decathlon appeared in home magazines such as *This Old House* and *Fine Homebuilding*. Of course, the hometown of every competing team ran informative stories to applaud their participation in the competition.

For each day of the inaugural event, the Solar Decathlon website recorded an average of 400,000 hits and 20,000 unique visitors. Team scores and standings were updated every 15 minutes, so interested virtual viewers could get the latest and greatest news about the competition. In addition, each participating team had its own website that included details and photos of the competing solar-powered homes.

Producers of the *Do It Yourself* (DIY) network were so enthralled with Solar Decathlon that they developed two shows for cable television and several shorter programs about the competition on their website. A DIY series called "Solar Solutions" introduced viewers to some of the innovative products and energy-saving technologies used in SD homes, encouraging people to adopt some of these ideas in their own dwellings. Clearly, many more consumers were being exposed to renewable energy strategies as a result of student-powered ingenuity.

In the words of Gary Schmitz, the Outreach and Public Affairs office at the National Renewable Energy Laboratory, "The Solar Decathlon captured the imagination of the media and the public alike. The contest managed to put a national spotlight on new and environmentally beneficial concepts and technologies in a way rarely—if ever—seen before."

Reflections from Dr. Lawrence L. Kazmerski, Emeritus Member of NREL Staff, Fellow at the Renewable and Sustainable Energy Institute (RASEI), University of Colorado at Boulder.

The "Idea" Becomes Reality, and a Success Story

After working diligently on designs and construction for two years, 14 university and college teams descended on the National Mall in Washington, DC on September 19, 2002 to begin the on-site realization of their solar homes. Then, on September 26, the opening ceremony of the inaugural Solar Decathlon took place. After the ribbon cutting with the U.S. Secretary of Energy for the new Solar Village, teams conducted public tours through October 6, dazzling throngs of visitors. Then the village was disassembled in three days, and teams took their solar houses back to their respective schools.

I had the privilege to visit the first three Solar Decathlon events and was always impressed with the enthusiasm and knowledge of the students involved with these complex projects. And, I was equally impressed with Richard King's high level of energy and enthusiasm. He gave every bit of his strength to make this "new idea" a success.

The first two Solar Decathlons had an overall winner that was special for me: the University of Colorado at Boulder. By the rules, NREL was excluded from working with any single team, but we were all pulling for the local University to do well. Eventually, a former NREL staff member, Ron Larsen, purchased the winning house from the University of Colorado at the inaugural event and turned it into his own residence. He still occupies that home today!

My role with the Decathlon was not a major one, but I enjoyed designing special commemorative Solar Decathlon scarves and ties, which I presented to each participant in the first two SD events (shown in photo).

Commemorative ties and scarves designed by Larry Kazmerski and presented to student teams at the Solar Decathlon (2002 is multicolored brand and 2005 is gold and blue).

In closing, I include a photo (courtesy of NREL) of Richard King and Ray Sutula, U.S. DOE Solar Program Manager at the time, receiving the Paul Rappaport Award for special contributions to solar energy.

Paul Rappaport was the first Director of SERI/NREL. Here's a quote from me that shows how the Solar Decathlon and Richard King were successful at mentoring an initial skeptic of this idea: "Kazmerski lauded the Solar Decathlon as *"an event that was key to elevating PV and solar technology to a bigger audience."*

It's true, thanks to Richard King and all those who believed in the Solar Decathlon. You succeeded in helping to make solar energy real for the world. You were right!"

DOE PV Team Leader Richard King (right), who conceived and directed the Solar Decathlon, and DOE Solar Program Manager Ray Sutula (center) accept the 5th Paul Rappaport Award for the Solar Decathlon and the organizer team that made it possible from National Center for Photovoltaics (NCPV) Director Larry Kazmerski (left). Kazmerski lauded the Solar Decathlon as "an event that was key to elevating PV and solar technology to a bigger audience."

As the inaugural competition unfolded, Richard King and SD Organizers kept watch on daily performance data of the solar-powered homes. The big question: Would they work as designed? They also watched in amazement as an endless stream of spectators continued to pour into the Solar Village. As luck would have it, the 2002 event boasted delightfully warm weather with plentiful, uninterrupted sunshine for three solid weeks. This was great for the Energy Balance Contest, which required teams to demonstrate that their PV systems could provide enough power for daily household needs, a home business, and an electric vehicle. The goal: finish the competition with the same amount of stored energy (or more) in the battery as measured at the start of the competition. In other words, "net-zero energy." Perhaps this was Mother Nature's nod of approval for Richard's vision, along with applause for the ingenuity and perseverance of those bold Decathletes.

Sunny skies during the 2002 competition were optimal for rooftop photovoltaic systems in the Solar Village, and five of the 14 teams had "positive energy balance" when monitoring ended (e.g., they had produced more than enough power). The other nine teams were unable to meet Energy Balance criteria, primarily because their PV systems were not powerful enough to supply the energy needed or their PV systems malfunctioned during the contest. This leads one to believe that installing a larger, more reliable PV system might ensure success with the Energy Balance Contest in future competitions, but it is not that simple. More on that topic to come later.

Richard reflects on formulating plans for the 10 contests as an important part of his "big idea." "A dwelling designed for everyday life in today's modern world is inherently a complex structure. In order for a home to adequately meet the various needs of its inhabitants, such as shelter, safety, sustenance, work, recreation, and relaxation, it must be well-balanced. Distinguishing which Solar Decathlon entries deserved recognition as "winners" required comprehensive and fair assessment of many components and contributing factors.

"Including ten distinct contests helped ensure that these innovative homes represented excellence in different domains, from architectural beauty to high-quality engineering, from efficient lighting and hot water systems to comfortable humidity levels and temperatures. My goal was for students to demonstrate appreciation for the necessary trade-offs between aesthetics and performance. Sure, these state-of-the-art homes had to be visually appealing, but they also had to be models of seamless execution; they had to function as designed.

"To win, teams needed to prove that their houses could outshine competitors in more than one or two contests. To accomplish that, Decathletes needed in-depth understanding of how to eek out the most points in all ten contests. Top-tier teams in every Solar Decathlon have developed ingenious strategies to reap scoring benefits. As noted, figuring out how to deal with adverse sky conditions (e.g., cloudy days, intermittent sunshine) and planning for energy management is crucial for the Energy Balance Contest. For the Comfort Zone Contest, teams utilize natural ventilation, clever shading devices, and interior arrangements that maximize airflow to maintain indoor thermal comfort and decrease energy loads. For the Lighting Contest, taking ample advantage of daylighting and incorporating low-energy fixtures (e.g., light-emitting diodes, or LEDs, were new to the market in early SD competitions) can help teams reduce power consumption. Just as homeowners want dwellings that function well, Decathletes want to show the public that their net-zero-energy homes have the same capacity to meet daily needs without sacrificing comfort or modern conveniences."

Wendy Butler Burt, U.S. DOE staff, was an organizer for Solar Decathlon 2002, 2005, and 2007 in Washington, DC and communications contractor for the 2009 and 2011 competitions. She will forever be a Solar Decathlon "groupie."

It caught my attention in early 2002 during a brief hallway conversation with Richard King. I learned about the Solar Decathlon scheduled for that Fall; once I'd heard about it, I couldn't get it out of my mind.

The U.S. DOE Office of Energy Efficiency and Renewable Energy is engaged with interesting projects featuring technologies designed to move our economy toward more environmentally-friendly practices. These technologies, such as electric cars, "smart" windows, energy efficient appliances, new industrial materials, superconductors, and wind turbines, are important. They may even be transformational, but few move the needle far enough to excite the general public. And yet, public excitement is an important step toward acceptance and changed behavior, which can help mitigate climate change. Solar Decathlon, on the other hand, had ALL the hallmarks of an event that could energize and motivate everyone: students, journalists, politicians, bureaucrats, and the general population.

Honestly, Solar Decathlon not only caught my attention; it completely captured my imagination. I'd spent my entire environmental communications career trying to convince people that individual actions could help change the world, and that their own actions could help improve human health and environmental quality. Yet here, wrapped up in one program, was a competition that could accomplish just that.

Soon after discovering Solar Decathlon, the entire focus of my job shifted to promoting it, within and beyond the U.S. Department of Energy. I characterized my new Solar Decathlon communications job as the "department of everything else." Richard King and his remarkable team from the National Renewable Energy Laboratory managed the technical and logistical elements of the competition. On the other hand, I tried to manage SD-related, nontechnical activities within DOE. It was truly amazing how well everyone worked together to pull off such a mini-miracle: a functioning solar village on the National Mall.

Of course, Solar Decathlon morphed over time—contests changed, technologies evolved, teams and SD Organizers gained new understanding. But one thing didn't change: enthusiasm for the event. Here's why it worked so well:

Elements of a Successful Solar Decathlon

- *Human interest.* Competing students were of varied ages, from all over the U.S. and abroad. They also represented a wide range of academic disciplines, with a strong focus on engineering and architecture. Due to this extraordinary diversity, there was something familiar to every person who read about or visited the solar village. Such overwhelming enthusiasm for the Solar Decathletes turned all of us into believers. The collegiate students helped everyone believe that the world could change for the better.
- *Location.* The National Mall is beloved by Washingtonians, as well as millions of Americans who have visited that coveted spot over the years. Some have participated in the Smithsonian's National Folklife Festival, fourth of July festivities; political demonstrations, and inaugural celebrations. Many have simply walked on the National Mall while visiting the city's iconic monuments and the U.S. Capitol.
- *Technology.* Every Solar Village is filled with clever and "smart" technologies that contribute to a brighter environmental future. Visitors have the opportunity to imagine using these technologies in their own homes.
- *The competition.* Everyone loves a contest. Everyone likes to see how well "their" team is doing against competitors; how well teams represent "their" state, territory, or country; and how beautiful and efficient solar homes can be.

Leveraging Communications Resources

My "department of everything else" for Solar Decathlon flourished by adapting to whatever needs developed. Flexibility was essential. I was fortunate to work closely with talented NREL staff. They built the Solar Decathlon website and created media that became the core of communications for the competition. I augmented their role within DOE, orchestrating formal press releases from the Secretary of Energy's office and reaching out to staff at supporting associations and private sector partners. I also navigated the political forces that such a highly visible public event can bring to bear upon the event itself. As it was, Solar Decathlon had such a positive effect on everyone that raw political concerns diminished over time. Truth be told, getting good publicity for Solar Decathlon was a relatively easy sell. After all, the concept is compelling, team accomplishments are outstanding, the Solar Village is appealing, and the National Mall setting was truly spectacular.

Inspiring the Troops

Solar Decathlon Organizers needed hundreds of volunteers to support the event. Within the government, the challenge was not so much finding people who wanted to get outside for several hours on beautiful fall days, but rather, getting management to approve. I encouraged managers to allow staff to actively participate in a government-sponsored event in the great outdoors. Well, we got our volunteers and assembled an enthusiastic group of Federal employees, representatives from sponsoring corporations, local teachers, members of environmental organizations, and Solar Decathlon "groupies." Payment consisted of highly coveted swag, such as t-shirts, hats, bags, and umbrellas. These volunteers escorted visitors through the solar village and helped everything run smoothly. From start to finish, they were a valuable asset for the competition.

Energizing Visitors

For me, the most exhilarating aspect of Solar Decathlon participation was my interaction with visitors. I enjoyed expanding their understanding of renewable energy and how to incorporate energy efficient technology into their daily lives. I watched as they wandered in wonderment through the solar village, joined tours of SD homes, checked out displays and exhibits about solar power, attended workshops presented by researchers and corporate sponsors, and discovered new products and online resources. I was amazed and thrilled at how captivated they were.

Over the years, I have met builders who had flown across the country to see a remarkable village of innovative, solar-powered homes. I met young people who were motivated for further study. I met political appointees so awestruck that they were determined to incorporate renewable energy strategies into department activities. I met families so intrigued that they decided to change their lifestyle. I met lots of people who were over-the-top inspired by the talent, sophistication, and determination that it took to create a Solar Village. It seemed that not a single person could remain unaffected after experiencing a Solar Decathlon event. For me, it had an unforgettable impact.

Many others have shared similar sentiments after experiencing Solar Decathlon. For a variety of reasons, this competition has a way of "getting under your skin" to leave a lasting impression. For those who had doubts about this crazy idea, watching young Decathletes work diligently to transform the National Mall into a stunning Solar Village jam-packed with innovation was not only convincing, but also truly inspirational.

Innovation is a crucial element of a Solar Decathlon competition. Collegiate teams are encouraged to challenge the status quo by integrating clever solutions, novel ideas, and new materials in their homes. By pushing the envelope, Decathletes are pioneers who open doors for unexplored opportunities. However, such bold action is fraught with risk. For example, when teams experiment with new products in SD homes, there is no guarantee that everything will work as planned. If equipment or materials function as designed, great, but if they do not, teams may lose precious points for operational failures.

Over time, Decathletes have personified inventive thinking while demonstrating dramatic leaps forward with technology, proving that open-minded R&D can be fruitful. Richard King smiles as he talks about the "wow factor" in Solar Decathlon homes. Innovative design and high-tech applications on display in these dwellings continue to impress visitors and are a big draw for the media as well. In addition, new ideas can motivate practicing designers and builders to rethink conventional construction methods. This subtle push forward encourages the homebuilding industry, typically rather slow to change, to incorporate novel approaches and products in residential construction. Richard King's vision for Solar Decathlon is for students to lead the way, "disrupting" the status quo to foster change.

> Keep in mind that the first Solar Decathlon took place in 2002, when advanced equipment for telecommunications was just getting off the ground. That year pre-dates the arrival of the iPhone (2007) and the iPad (2012), and cell phones at that time were mainly for voice calls. This also came before the now-ubiquitous presence of SMS (short message service) for text messaging. Furthermore, "Smart Home" technology was not yet on the market, yet whiz-bang Decathletes were way ahead of the curve in masterminding and applying whole-house control systems in their SD entries. Some teams integrated "Smart Home" features in the 2002 event, and viewers were in awe as they observed this futuristic technology in action.
>
> In preparing for the inaugural event, Richard chuckles as he remembers communicating with Jury Member Glenn Murcutt in Australia, on the other side of the globe. Their mechanism for staying in touch? Believe it or not, via FAX (Facsimile Transmission). For those who may be too young to have had experience with that technology, the user loads a printed piece of paper into a copier equipped for FAX delivery, that machine sends a copy of the material to a recipient whose FAX number has been entered into the machine, and their printer then spits out a printed copy of the material that was sent. Voila'! (That sure seems like light years ago.)

The 2002 competition featured teams who dared to dream about "possibilities." Their model homes soared above and beyond to showcase the aesthetic integration of solar technology with state-of-the-art solutions in energy-efficient dwellings that dazzled, inside and out. Richard King believes that "Nothing motivates creativity like a good contest." He refers to the inaugural event as "a blueprint for the future." Here are some examples of the stunning ingenuity on display at the first Solar Decathlon.

The University of Virginia (UVA) team designed a dwelling to echo the "rebellious nature" of the younger generation. A "Smart Wall" near the entrance was a large-scale light-emitting diode (LED) with touch screen that functioned as a control center for the whole house, including all appliances. As the interior temperature changed, the wall also changed color: pink for warm and blue for cool. In addition, students installed translucent ceramic wall tiles to create an unusual, appealing "glow" inside the house. To support their theme of sustainability, the UVA team used birch and bamboo, as well as other reclaimed materials to build their house. The house boasted a green roof with living plants and reclaimed tires for outdoor landscaping planters.

The exterior siding of the UVA *Trojan Goat* was made of reclaimed copper cladding from the rooftop of an old building. The University of Virginia team used wood from scrapped shipping pallets to make movable window louvers for the large window on the south wall of the house, which controlled the incoming sunlight (and temperature).

The UVA *Trojan Goat* had super-insulated walls, radiant floor heating, and valance cooling that consisted of a pipe with liquid coolant pumped through it at the top of the south wall.

Jury member Glenn Murcutt, world-famous architect from Australia and winner of the distinguished Pritzker Prize, was very impressed. In his words, "Virginia had a design that showed a solar home is not just plunking a panel on top. It must be as poetic as it is rational. It must be whole-building sustainable, and all those components must be integrated in an elegant way." The University of Virginia team took first place for Design and Livability in the 2002 competition, earning all 200 points possible for that contest. The Jury's overall comment about this entry was "absolutely fantastic."

Glenn Murcutt announces winners of the Design and Livability Contest for SD 2002: first place—University of Virginia; second place—University of Puerto Rico-Mayaguez; third place—University of Texas at Austin. Teams were honored to be in the winner's circle for this highly coveted contest.

Auburn University, located in Alabama, created a house that represented architectural elements typical of southern states. Student Lesley Hoke said their goal was to create a contemporary dwelling that "blended the best of southern living with the best of modern convenience." The house featured large indoor solar water columns on the south wall that moderated indoor temperature (practical) and contributed to the ambience (beautiful). To brighten the interior, the team created "solar megaphones," skylights filled with prisms to enhance daylighting effects. Their use of structurally insulated panels (SIPS) resulted in a tight exterior envelope for an energy-efficient house that showcased marvelous engineering and delightful architecture.

Auburn University Decathletes install solar panels on the south-facing roof of their house.

The Auburn University house reflected a traditional "dog-trot" design with a central walkway dividing the house into two parts.

Decathletes from the University of Puerto Rico (UPR) brought much more than their house to the inaugural competition. Determined to showcase the bright, joyful spirit of their island home, this team brought abundant effervescence to the Solar Village. From the moment of their arrival on the National Mall, UPR students dressed in colorful t-shirts and carried their flag while singing, dancing, and playing musical instruments. During every phase of the competition (Assembly, Contest Week, Disassembly), this team found moments to parade down the "Main Street" of the Solar Village with a lively band, encouraging others to join the impromptu entertainment. Decathletes, organizers, staff, volunteers, and visitors appreciated those opportunities to lighten up and be happy.

Visitors characterized the UPR house as "light and breezy," which was accurate because the dwelling remained astonishingly comfortable, even when outdoor temperatures hovered above 85°F during Contest Week. The Design and Livability Jury commended the team for creating an attractive model home that appropriately reflected local culture and climate. The Decathletes said they aimed to build "a home to honor who we are." The UPR house earned second place in the Design and Livability Contest, and jury members called it a stunning example of architectural design that fit superbly with its environment. Their comment about this entry: "elegant in its simplicity."

Decathletes from the University of Puerto Rico assembling the louvered shades on the south wall of their house.

The University of Puerto Rico house was designed for a tropical climate, featuring cross-ventilation in a rectangular shape with metal shading devices on the entire south wall.

A few more examples of creative solutions on display at the 2002 competition prove this point. Carnegie Mellon University used shredded blue jeans for denim batt insulation, plus wheatboard (made of straw) and Plyboo® (made of bamboo) for flooring and room dividers. Virginia Tech created and installed transparent, translucent Skywall panels with high insulation value that functioned as thermal walls, active solar collectors, and hot water heaters. The aerogel with high thermal resistance inside those panels also enabled daylighting. The University of Missouri and Rolla Technical Institute built a sunroom with floor-to-ceiling windows and added porcelain tiles on the floor for passive heating and cooling. In a clever twist, those tiles were engraved with sponsors' names. This house comprised three distinct modules (for easier transport) and claimed a strict "no nails policy"; only glue and screws were used to assemble the structure.

The house from Carnegie Mellon University featured a green roof with living plants on top of the modular pod in the northwest corner that included the kitchen and hot water heater.

The south side of Carnegie Mellon University house demonstrated passive solar design with large windows and overhangs to provide ample shading from the hot summer sun.

Active solar collectors on the south-facing Skywall panel of the Virginia Tech house made it possible for radiant floor heating inside the dwelling.

The Skywall panel on the south side of the Virginia Tech house was transparent and translucent. Photovoltaic panels on the roof provided power for the house, as well as shading the structure, and the Virginia Tech team, called that PV system a "benevolent umbrella."

In the evening and at night, the luminosity of the Virginia Tech house was especially beautiful.

The University of Missouri-Rolla house was made of three distinct modules and had a separate storage area for the batteries.

All the houses in SD 2002 were standalone structures (not grid-tied), so every house needed battery storage. This photo shows the University of Missouri-Rolla battery storage unit.

In a special nod to exceptional people, jury members play a significant role in every Solar Decathlon, and professionals are chosen carefully for their deep expertise. Decathletes appreciate the chance to share their unique creations with well-known scholars and masters from the field, and everyone learns in the process. Richard King believes that engaging top-notch experts for the inaugural competition enhances the stature of this collegiate competition and elevates the status of the onsite event. In the beginning, what he did not anticipate was how inspired jurors would be after they reviewed documentation from participating teams and experienced onsite house tours. The jurors applauded the ingenious solutions, creative designs, and exemplary engineering innovations on display. Many were surprised at the advanced level of knowledge of the Decathletes. The jury members left the competition exclaiming that they had gained valuable new understanding from these young college students.

As an example, Design and Livability Jury Member Glenn Murcutt, Australia's Architect Laureate and 2002 Pritzker Prize winner who has taught at leading universities on every continent, was literally gushing with positive comments when he presented awards in the 2002 Solar Village. Murcutt has won numerous prestigious awards and is well known across the globe for designing buildings that are in "harmony with the environment." He believes that "a building should be able to open up and say, *I am alive and looking after my people.*"

Edward Mazria, a 2002 Design and Livability Jury Member who is a transformative figure in architecture, was also impressed. Mazria has received many awards, including the Gold Medal from the American Institute of Architects in 2021. He is the author of a timeless classic, "The Passive Solar Energy Book" (c. 1979), and founder of Architecture 2030, an organization that supports real-world solutions to global issues by reshaping the dialogue on energy, emissions, and climate change. He shares these comments.

At the time I was asked to judge the first Solar Decathlon in 2002, we had just discovered that the building sector was responsible for about half of all CO_2 emissions in the U.S. It was a startling discovery, so I was eager to see zero energy homes powered entirely by passive and active renewable energy systems tested and showcased on the National Mall in Washington, DC that year.

I arrived the day before the "Solar Village" was open to the public. Late that day, I toured the village with two of my fellow judges, unannounced. As we walked down the main street you could sense the excitement as student and faculty teams worked feverishly through the night to complete their homes. It was an impressive sight.

The following day we were given a tour and explanation of each dwelling's design and solar system by members of student teams. After each tour, we gathered with the team in their home's living room and began to ask very tough questions. It was the conversations with the students that I remember vividly after all these years. What surprised me most was how much the students knew about siting their buildings and integrating passive heating/cooling and active systems in their designs – sizing shading devices, solar glazing, PV, solar hot water, natural ventilation, daylighting, and thermal mass—at a time when those subjects were not widely taught.

In 2003, Metropolis Magazine published the "Architects Pollute" issue and alerted the design and planning professions that they were complicit in fueling climate change. It quickly became clear that we needed to design and build differently, which is why I was thrilled to judge the second 2005 Solar Decathlon. Again, I wasn't disappointed. Submissions had grown from 14 to 18 teams, and the quality, ingenuity, and sheer variety of systems and design strategies were impressive.

It's evident that the Solar Decathlons held to date have played a role in the public's acceptance of solar applications, as well as the propagation of sustainable design principles and strategies in professional education which has led to the explosive growth of renewable energy generation and integration in the building sector today.

For the closing ceremony at the inaugural Solar Decathlon on Saturday, October 5, each team drove their electric vehicle across an imaginary finish line on the Main Street of the Solar Village. The NREL team responsible for data analytics and scoring was still at work until noon, as students, officials, and media teams waited in suspense for the final results. Fortunately, the University of Puerto Rico kept everyone entertained with lively music and dancing! When official winners of the 2002 Solar Decathlon were finally announced, it was University of Colorado Boulder in first place, University of Virginia in second place, and Auburn University in third place. Loud cheering and high-fives were definitely in order to celebrate the achievement of those overall champions who proved that creativity, hard work, and team effort would prevail.

The University of Colorado Boulder was honored to win first place. Ironically, one of their main goals was proving that solar could work on "just about any house." They thought that public acceptance was essential, so they set out to build a house that might be found in any neighborhood. Perhaps that would coax visitors to consider PV or energy-efficient features as add-ons to existing homes. The team intentionally chose a corner lot in the Solar Village to maximize public exposure. Their house showcased environmentally friendly bamboo flooring, recycled plastic sheeting, and sunflower board cabinets. The team made a special louvered door that slid into place to shade the door and large windows, effectively controlling indoor temperature.

The University of Colorado Boulder team delivered three prefabricated modules to the Solar Village, making it relatively easy to assemble the entire house in 6 days on the National Mall.

The University of Colorado Boulder dwelling won first place in the inaugural Solar Decathlon competition.

Faculty Advisor Michael Brandemuehl holds the first-place trophy surrounded by the University of Colorado Boulder team at the entrance to their solar house.

Although the CU-Boulder entry did not earn the top prize for Design and Livability, it did demonstrate strong, consistent performance in all other contests during the onsite competition. That team was quite proud to earn special recognition for "Engineering Excellence" from a distinguished panel of experts. In the words of Secretary of Energy Spencer Abraham, the Solar Decathlon was "a real test of their abilities and their willingness to pit their talents against some of the best schools in the nation, and they proved themselves worthy of this honor."

The University of Virginia's *Trojan Goat* won second place overall at the 2002 Solar Decathlon. UVA student Josh Dannenberg noted that the team's main goal was to integrate "future design technologies with a lifestyle that's both ecologically aware and comfortable." They certainly succeeded with this architectural beauty that embodied sustainability.

Auburn University won third place in the competition. Faculty Advisor Henry Brandhorst noted that the team aimed to encourage homeowners to incorporate some of the advanced technologies showcased at SD 2002 into their own homes.

Auburn University won third place with their house that radiated Southern charm.

Richard reflects on the final awards ceremony, held in the evening on that last day of the 2002 competition. "I was elated that the first Solar Decathlon was such a spectacular success. In fact, the outcome surpassed my expectations by far. Witnessing the enthusiasm at the awards ceremony brought tears to my eyes. All participants had invested incredible effort to create an amazing Solar Village on the National Mall, and I worked closely with SD Organizers to ensure that every team received at least one award. In addition to the top three awards for each of the ten contests, special awards, such as Herculean Effort (University of Puerto Rico), Perseverance (University of Delaware), Open Door (Tuskegee University), and Best Logistics Plan (University of Texas at Austin) were announced, so that every team member had a chance to be center stage."

The People's Choice Award was a unique way to engage the public at the inaugural competition. Visitors at the event could vote for their favorite Solar Decathlon house, and the team that received the most votes won. Crowder College Decathletes from Neosho, Missouri, along with Faculty Advisor Art Boyt (former solar car racer), were elated that their house was chosen for this award at SD 2002.

The house from Crowder College, the only community college in the event, won the People's Choice Award.

The Crowder College team designed and built their own special rooftop hybrid system with photovoltaic panels on top of solar thermal hot water tubes.

Hot water from the unique handmade system was used for a hot tub on the south porch of the Crowder College dwelling. The Decathlete pictured here was an older student who had gone back to school to learn about solar energy.

In addition to team recognition, many others deserved credit for their commitment and support of the first Solar Decathlon. That included: (a) dedicated NREL staff, such as Project Manager Cecile Warner, who gave their all every single day at the Solar Village and were at-the-ready 24/7 without fail; (b) individual superstars like Pamela Gray-Hann, who handled event registration, distribution of Solar Decathlon attire (for volunteers, observers, rules officials), and dissemination of SD programs; and Byron Stafford, who managed onsite logistics such as tent and walkway setup, signage, and basic equipment for the event; and (c) enthusiastic volunteers (including DOE employees) who freely offered their time and energy to keep the Solar Village alive and well.

Another group that deserved heaps of gratitude was the sponsors. The U.S. Department of Energy, the National Renewable Energy Laboratory, and other Solar Decathlon sponsors provided incredible support that led to resounding success. For the first competition, the U.S. DOE gave each participating team $5000 as "seed money" to begin planning their solar-powered houses, but a design-and-build program of this magnitude requires more capital and logistical assistance. For the 2002 competition, collegiate teams literally "pounded the pavement" to raise enough money to create and transport their solar-powered houses to the National Mall. So, for success with any future competitions, Richard quickly realized that funding would need to be increased.

The National Renewable Energy Laboratory, a primary sponsor of the Solar Decathlon, was celebrating its 25th anniversary in 2002. This competition proved to be an ideal platform for NREL, which maintains the National Center for Photovoltaics. On the National Mall, the results of leading-edge research in renewable energy and energy efficiency were showcased via those solar-powered houses. NREL provided talented staff and key leadership to help make Richard's vision shine. They also contributed critical expertise for data analytics, scoring, software, and digital communications. Fast forward to this day, and NREL continues to support Solar Decathlon competitions as they proceed with world-class R&D to advance renewables.

Additional sponsorship for the 2002 competition included a wide range of contributions, from photovoltaic panels and exhibit tents (BP Solar) to gift cards for building materials and portable flooring for walkways in the Solar Village (Home Depot) to mobile communications on the National Mall (EDS, WorldCom, Cisco). Timely efforts of public affairs staff from SD sponsors boosted publicity for the competition, which helped get the word out. A special call-out to some dedicated sponsors who have offered their support for every Solar Decathlon in the United States since the inaugural event. Notably, the American Institute of Architects (AIA) has been a remarkable champion over the years, and their membership of 80,000+ has extensive influence nationwide. Richard King says, "Honestly, I wouldn't have succeeded without the valuable professional expertise and logistical support of our fabulous sponsors."

Following the 2002 awards ceremony came the disassembly phase of the onsite event. Teams had 3 days to deconstruct their houses and prepare them for transport back to home campuses. The National Mall was once again abuzz with activity, but with a distinctively different set of emotions. Students were upbeat, knowing they had accomplished the main objectives of the competition. Their innovative solar-powered houses were a big hit, capturing the imagination of 100,000 visitors in the shadow of the U.S. Capitol in Washington, DC. Most of the houses returned to their respective colleges or universities, where they would serve as living laboratories for continuing research or temporary housing for visiting faculty. Some, like the UVA and CU-Boulder entries, were sold to new homeowners in their respective states to help defray the costs of Solar Decathlon participation. Even today, those two dwellings still serve faithfully as local residential homes.

For SD Organizers, there was finally a chance to breathe a sigh of relief and catch up on much-needed rest after several weeks of nonstop action. Richard King marveled at what had just happened and immediately began planning for another competition. Following the undeniable triumph of the inaugural event, it did not take much to convince the U.S. DOE to make a commitment to sponsor another Solar Decathlon. Knowing that it would be more worthwhile for student efforts to focus primarily on design and construction, Richard concentrated his efforts on securing DOE buy-in to increase the "seed money" for team entries. It worked. DOE agreed to offer $100,000 (sizable increase) to each team selected for the next competition. This funding makes a huge difference for teams getting their projects off the ground and securing their own sponsors.

Solar Decathlon 2005

On December 12, 2002, the U.S. DOE announced plans for Solar Decathlon 2005. The formal Request for Proposals (RFP) was issued 2 months later. SD projects required two years of preparation, so the timing of this announcement was critical. The Solar Decathlon website, filled with engaging photos and videos from the inaugural competition, delivered important information about the upcoming competition. Participants from the 2002 event also shared photos and stories of their experiences on their home school websites. Sponsors posted media and summary reports about their involvement on their websites, and hometown newspapers continued to carry stories about the achievements of 2002 Decathletes. This was prior to the current era of ever-present social media, so website traffic, word of mouth, printed media, and television were the predominant means of spreading news.

Fortunately, most SD Organizers, including the NREL team, were more than happy for a repeat performance, so Richard had the advantage of working alongside capable colleagues with Solar Decathlon experience. In December 2004, another champion boosted Richard's spirits: Dr. Samuel Bodman was named the new U.S. Secretary of Energy. With his doctorate in chemical engineering and background as a professor at MIT, Bodman felt an immediate kinship with this collegiate competition focused on engineering and architecture. From 2005 to 2009, his whole-hearted support and multiple in-person visits to the National Mall "to learn from Decathletes" helped to grow SD sponsorships and earned greater respectability for this "crazy idea" that had developed into a national phenomenon.

Act Two begins! A second Solar Village rose up on the National Mall from September 29 to October 16, 2005. Eager Decathletes from 18 colleges and universities (chosen from an applicant pool of 23 teams) came to Washington, DC to show their prowess with technological innovation, energy efficiency, photovoltaics, green building, and creative design. The selected teams came to present bold, creative solutions. They arrived onsite with truckloads of confidence and tools.

House from CU-Boulder, defending champions from the 2002 event, arrives at the Solar Village.

Modular section from the Pittsburgh Synergy house is dropped onto the National Mall with a crane.

Two modules of the Pittsburgh Synergy house are placed on top of each other in the Solar Village.

Some modules, like this one, arrived on the National Mall fully equipped with appliances and cupboards already in place.

Houses being assembled on Decathlete Way in the Solar Village, with the Washington Monument in the background.

Student representation for this event was different. Alongside the 16 teams from 13 states across the United States were two teams from other countries: Canada and Spain.

The *Magic Box* from the Universidad Politecnica de Madrid (UPM) being assembled on the National Mall.

How did news of the competition reach the other side of the Atlantic? Apparently, a Spanish manufacturer of PV systems had visited the 2002 event and talked it up with professors from the Universidad Politecnica de Madrid. They were excited, and faculty members from the UPM School of Architecture decided to enter the Solar Decathlon. Richard King was thrilled, and he explains why. "Climate change and fossil-fuel dependence are global problems, so we all need to work together for viable solutions. By assembling creative minds from across the country and around the world in one location to demonstrate innovative ideas, we create valuable synergy. Every participant in a Solar Decathlon gains new knowledge and experience, which in turn, generates more innovation. Including teams from other nations will accelerate these beneficial outcomes, and I'm all for it!"

The participating teams for Solar Decathlon 2005 were as follows:

- California Polytechnic State University (Cal Poly)
- Rhode Island School of Design
- Concordia University and University of Montreal
- University of Texas at Austin
- Washington State University
- University of Colorado, Denver and Boulder
- University of Puerto Rico (UPR)
- University of Michigan
- University of Missouri-Rolla and Rolla Technical Institute
- Cornell University
- Crowder College
- University of Massachusetts Dartmouth
- Pittsburgh Synergy (Carnegie Mellon, University of Pittsburgh, Art Institute of Pittsburgh)
- Virginia Polytechnic Institute and State University (Va Tech)
- Florida International University
- Universidad Politecnica de Madrid (UPM)
- New York Institute of Technology (NYIT)
- University of Maryland

U.S. DOE Secretary Samuel Bodman greets officials from the Spanish government and UPM in the Solar Village.

The team from Cornell University in New York State included beautiful vegetable gardens around their house. They had grown the plants on campus in advance and then transported them to the Solar Village.

In a nod to careful use of resources and sustainability, the Cornell team fashioned outdoor seating from the sod around their house, so visitors could relax while waiting for an inside tour.

What was different in the second competition? First of all, there was more diverse geographic representation. Second, architecture was front and center, and house designs were notably more sophisticated. Knowing that the Architecture Contest was worth 200 points, Decathletes zeroed in on architectural elements that would delight viewers, while still honoring principles of energy efficiency. Some teams extended their artistry to beautify the outdoor area around their homes, and several utilized rooftop areas as living spaces. Another noticeable, positive change was more substantial focus on sustainability. Houses showcased environmentally friendly materials, rainwater collection and reuse, and gardens with herbs and vegetables for human consumption.

The 10 contests for 2005 differed slightly from the 2002 competition, but the main goal remained the same: demonstrate that attractive, energy-efficient homes powered solely by the sun can produce enough energy for daily living functions, including work, play, sustenance, and rest. Evaluation of some contests was based on task completion and performance measurement (objective); for others, assessment relied on the expertise and judgement of jury members (subjective). Each contest was worth 100 points, except for architecture, which was worth 200 points. The 10 contests for the second Solar Decathlon were as follows:

- Architecture (juried)
    - How well and how beautifully does the house integrate solar technology and energy efficiency into its high-performance design?
- Dwelling (new contest, juried)
    - Is the house designed well for everyday living? (livability)
    - Are construction methods sound? (buildability)
    - What is the "curb appeal" of this house? Does it appeal to potential homebuyers?
- Documentation (juried)
    - How well does the team document house design and energy analysis of system performance?
    - In addition to as-built drawings and energy analysis, teams must provide a cost report. What will it cost to build this house?
- Communications (juried)
    - Does the team effectively communicate information about the house and their experiences with the project?

- Comfort zone (monitored)
  - Can the house maintain comfortable temperature and humidity levels during the contest week?
- Appliances (monitored)
  - Do appliances function as designed to enable teams to complete required daily tasks in their house?
- Hot water (monitored)
  - Is there ample hot water provided by systems in the house for teams to complete a daily shower test?
- Lighting (juried and monitored)
  - Is lighting in all areas of the house attractive and sufficient (based on required levels for day and night)?
- Energy balance (monitored)
  - Is the house net-zero energy (it produces as much energy as it uses) during the contest week?
- Getting around (electric cars, monitored)
  - Does the PV system on the house provide enough energy to power an electric car during the contest week?

As in the first Solar Decathlon, jury members played an important role in the competition. Jurors were carefully selected for their knowledge, experience, and leadership in their respective fields. Some were winners of prestigious awards and highly acclaimed for noteworthy accomplishments. Decathletes were honored to share their house projects with jury members, and in most cases, the experience was mutually beneficial. Members of the 2005 Solar Decathlon Jury Panels were as follows:

- **Architecture:** Steve Badanes, Ed Mazria, Sarah Susanka, Ken Wilson
- **Dwelling:** Dennis Askins, Eric Borsting, Steve Easley, Sam Grawe, Katherine Salant
- **Architectural documentation:** Phil Bernstein, Katherine Prigmore, Grant Simpson
- **Energy analysis:** Doug Balcomb, Mike Deru, Pete Jacobs, Russ Tayler, Norm Weaver
- **Engineering:** Steve Emmerich, John Mitchell, Terry Townsend
- **Lighting:** Howard Brandston, Sandra Stashik, Gary Steffy

- **Communications:** Ethan Goldman, Kim Master, Alan Wickstrom, Ben Finzel, Jaime Van Mourik, Craig Savage

Architecture Jury at the 2005 Solar Village: (left to right) Susan Piedmont-Palladino (coordinator), Sarah Susanka, Ed Mazria, Steve Badanes, and Ken Wilson.

Jury Member Katharine Salant announces winners of the Dwelling Contest; Richard King (center of photo) is wearing one of Larry Kazmerski's special solar ties.

On Thursday, October 6, 2005, Secretary of Energy Samuel W. Bodman introduced the 18 teams and remarked, "These homes are helping to bring the promise of solar power to reality. It's inspiring to see these young people work together on these next-generation homes. I want to congratulate all the teams who are competing and note that the awards they win here won't compare to the prize of knowing, down the road, that their work helped strengthen our world's energy supply with more available use of solar power." Secretary Bodman joined the ribbon-cutting ceremony and officially declared the Solar Village open to the public.

The animated team from the University of Puerto Rico proudly carried their flag while singing as they marched together to the opening ceremony on October 6, 2005.

Richard King's smile was a mile wide as he watched exuberant teams assemble at their houses on "Main Street." Their enthusiasm never waned, and over the next 11 days, more than 120,000 eager visitors of all ages and backgrounds toured these unique model homes. Richard recalls feeling more confident in 2005. "We had learned from the first event, improved the competition, and expanded the venue. This time, we were all more than ready for broad public exposure and plenty of media attention. Of course,

Decathletes took center stage with their phenomenal showcase of ingenuity, advanced technologies, and 'Smart' solar houses. It didn't matter that most were surviving on very little sleep during Contest Week because they were bursting with excitement and abuzz on adrenaline."

The second Solar Decathlon featured more than onsite tours and real-life demonstrations in the Solar Village to educate the public. There were also interactive educational exhibits, such as "The Anatomy of a House," that offered tips on saving energy and showed how grid-connected PV systems worked. There were free workshops about solar energy, power technologies, and sustainability for consumers and industry professionals. There was a Product Expo near the National Mall where solar-related companies displayed their current products and services.

Reflections from Carolynne Harris, a museum planner specializing in implementation of new museums, renovations, expansions, and large exhibitions. After working at the Smithsonian Institution and Fernbank Museum of Natural History, Carolynne started her own firm. She teamed up with Solar Decathlon organizers in 2005, continued through 2011, and served as a jury member in 2015.

When the first Solar Decathlon was launched in 2002, I was working at the Smithsonian Institution and had recently completed a special project on the National Mall—a semi-permanent outdoor scale model of the Solar System that stretched the length of one side of the Mall. I was impressed with the varied designs of Solar Decathlon houses and the intention of what I saw as a 'festival' feeling, a combination of model homes and educational programs for the public. When the U.S. DOE contacted me in early 2005, I was thrilled at the opportunity to work on this important project. It was right up my alley, as I specialized in unique and large exhibitions and capital projects. I also have a fondness for architecture, sustainability, and renewables. Plus, I had a good idea of what it took to deliver projects on the National Mall, and the many entities around DC that could potentially contribute to the event.

The first Decathlon was extraordinary, but the DOE team realized that visitor information, signage, and educational communications could be vastly improved. My role was multi-faceted: 1) Create signage to draw people into the Solar Village, showcase both DOE and event branding and messaging, and guide visitors to the various program areas over what was essentially a 10-block area with multiple entrance opportunities; 2) Identify each school's location on various maps and signs; create a signage program that described the 10 Decathlon contests in simple terms, and showed the contest scores, judging status, and school rankings; 3) Guide and facilitate the development of this "first-of-its-kind" outdoor exhibition to engage tens of thousands visitors and demonstrate the choices they had for how to build and use their homes, including more energy-efficient materials and strategies; 4) Connect the Organizers with like-minded museums and institutions in Washington, DC who might develop programming or cross-promote the event, including special signage for bus stops and Metro Stations leading to the National Mall.

Let's not forget that all this had to withstand gale-force winds, potential heat waves, rain, tens of thousands of visitors over a two-week period, the stringent National Park Service rules for installation (e.g., no drilling past 10-12" into the soil for signs that were 10-30 feet tall), and Solar Village security measures. All this required permitting, approvals, and collaboration with various agencies, U.S. DOE, and Solar Decathlon sponsors. So many logistics, possibilities, and players! However, everyone had such passion for the competition and its purpose, so no matter what the obstacles or details, energy remained high and every participant was optimistic, which made it a fun challenge.

For the signage, branding, wayfinding, and educational information, we worked with Hargrove, Inc. in Maryland, which was very experienced with large events and trade shows. The DOE/NREL team provided content, while Hargrove designed, built, and installed the signage. Quatrefoil, a Maryland-based exhibit design and fabrication firm, was our partner to develop an outdoor exhibit called "The Anatomy of a House." This was a house separated into components with 'behind the walls' views to show interventions that could be utilized for energy efficiency. The exhibit also included a solar fountain and other moveable and tactile items, which kids loved.

Of all the vast details, planning, coordination, reviews, approvals, and agendas, what I remember most is the infectious energy and excitement of the Decathletes. I'll never forget standing on the National Mall at midnight the first day that semi-trucks flowed in like a caravan to their designated spots where they would offload the houses in various stages of construction. SD Organizers choreographed this carefully, but student teams were bursting with anticipation—after all, they had worked for two years for this moment. Watching them work literally 24/7 to build their houses during the Assembly Period was incredible. As we started our install, any time I felt tired, frustrated, or stressed (*who knew that soil on the National Mall baked into concrete in a matter of days, making drilling for sign poles a Sisyphean effort?*), I just walked over to one of the houses to see the progress and work of these amazing Decathletes. The unending pride in their projects and desire for top-notch performance was evident in the great tours they gave to thousands of visitors. They were some of the best docents I've ever seen!

Being on the National Mall every day to check for blown-over signs, wipe down exhibits, and look for bits and pieces that needed fixing was a lot of work. I loved seeing all the visitors flock to the Solar Decathlon. Observing them tour the Solar Village and participate in educational activities, I knew we had achieved our goals: People could find team houses or special areas they were interested in; they asked great questions; and they remarked that the Solar Village felt like "a neighborhood of cool houses." This showed that solar could be both attractive and energy efficient. The 2005 event made a big statement and was memorable for those who came from nearby, as well as far away, to attend this special competition.

Regrettably, weather during the 2005 Solar Decathlon was the exact opposite of the 2002 event. Skies were frequently dark and overcast, and it rained almost every day of the competition. In fact, more than seven inches of rain were recorded on one of the weekends that homes were open to the public. As a result, umbrellas and rubber boots were "must-have" accessories for visitors, rather than sunglasses and wide-brimmed hats. Oddly enough, inclement weather did not deter enthusiastic crowds; more than 120,000 viewers toured the Solar Village in 2005, despite drenching rain and messy mud puddles. Exuberant crowds confirmed that curiosity and fascination were compelling enough to overcome the inconvenience of soggy ground and waterlogged outerwear.

Meanwhile, a drastically different weather scenario meant that Decathletes had to revise their strategies for the Energy Balance Contest. A larger PV array wouldn't necessarily produce more power if sunlight was missing. Instead, would an increase in battery size make the difference? Contest data proved that neither one of those elements made the critical difference for energy balance. It turns out that less-than-ideal weather forced SD teams to be nimble, adapting strategies to accommodate for whatever outdoor conditions emerged. Lesson learned? Careful "energy management" was essential for securing the most points in that contest. In fact, close analyses of performance data led teams to discover that efficiency in just about every aspect of their dwellings would consistently yield the highest scores.

Rainy weather at SD 2005 did not deter visitors from showing up to tour the innovative houses on display in the Solar Village.

Like their predecessors at the inaugural event, Decathletes showcased all sorts of clever solutions to ensure energy efficiency, reliable functionality, sustainability, and consumer appeal. Some examples from the 2005 competition illustrate the awesome ingenuity of participating teams.

The entry from Virginia Tech was exemplary in many ways, illustrating how innovative applications can add delight and beauty, as well as practical value. That house aced the Architecture and Dwelling Contests, earning an impressive 299 of 300 possible points. Contained within the wood frame of the home were polycarbonate panels filled with translucent aerogel insulation that allowed light inside and warmed the house on cooler days. Motorized "MechoShades" (which could be absorptive or reflective) supported control of indoor temperature and provided privacy at night. Natural daylighting entered through clerestory windows above the walls, and colorful LEDs at the bottom of the walls created attractive nighttime displays. The pitched, adjustable PV system that curved upward and appeared to "float" over the rooftop collected plenty of sunlight in all seasons. In presenting Virginia Tech with the first-place award for the Architecture Contest, Sarah Susanka, distinguished designer and member of the Architecture Jury, expressed heaps of praise and shared these complimentary words, "Everything about this house is wonderful. It took my breath away."

Virginia Tech Decathletes work together to ease the supporting trusses for their exterior deck into place on the National Mall.

The Virginia Tech house, which won the Architecture and Dwelling Contests, was absolutely beautiful, inside and out. The team designed a rainwater collection system to water the plants that surrounded the decking.

The team from the Rhode Island School of Design (RISD) envisioned their Solar Decathlon house in an urban setting with a rooftop garden. The north and south sides were open, but the east and west sides had "sliver" windows that could be eliminated in a row house configuration. A metallic exterior with vertical heliotropic louvers to track the sun simulated a chameleon look, and the use of phase-change material in "bricks" under the floor absorbed or released heat as needed to control interior temperatures (each brick could store 4000 Btu). This house was the only one in the 2005 competition with a radiant heated ceiling. By maximizing energy management and efficiency (e.g., solar tracking louvers were less expensive than PV panels), the RISD team was able to minimize the size of their solar system while still providing ample power for daily functions. Jury members praised the collaborative efforts of students of art and design to create such an aesthetically pleasing, transformative, and futuristic dwelling.

The team from the Rhode Island School of Design created a transformative urban dwelling that would function well as a rowhouse. Insulated, vertical aluminum louvers on the east and west sides of the house tracked the sun, and this metallic skin helped block unwanted heat.

The south side of the Rhode Island School of Design house featured a green roof with living plants and large glass doors that opened up completely for fresh air.

The team of trailblazers from the New York Institute of Technology (NYIT) took innovation to new heights. Electricity from their PV system was used to separate hydrogen from water (through electrolysis), and then the stored hydrogen was later used in a fuel cell to produce electricity and heat "on demand." Team leader David Schieren referred to this process as a "clean, efficient, and elegant cycle." The team used a shipping container for mechanical systems, as well as the kitchen, bathroom, and roof garden (the Green Machine), and created another section of the house (the Blue Space) for sleeping, relaxing, and working. The NYIT team also designed indoor micro-climates, where lighting, heat, and furnishings were independently controlled for appropriate comfort levels. In the congratulatory words of Jury Member Sarah Susanka, esteemed architect, teacher, public speaker, and celebrated author of "The Not So Big House" series, "I'm so impressed. Everything about this house is an invention!"

A highly innovative prototype (from an engineering point of view) was the New York Institute of Technology (NYIT) house powered by a hydrogen fuel cell. Electricity from their 10-kilowatt PV system was used to split water into hydrogen, which then powered a fuel cell. This system made storage batteries unnecessary. All equipment for the fuel cell was contained in the shipping container next to the house.

The NYIT fuel cell was a huge attraction for curious visitors who wanted to learn more.

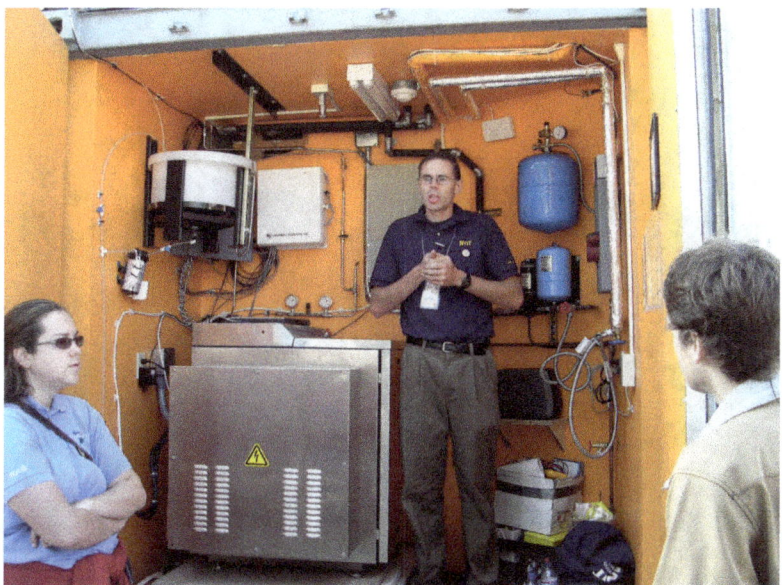

Gregory Sachs, NYIT engineering team leader, enjoyed describing how the NYIT hydrogen fuel cell worked.

Decathletes from Crowder, a small community college in Neosho, Missouri, chose to emulate the "Arts & Crafts" style championed by Gustav Stickley and reflected in their midwestern hometown. They created an open-flow bungalow of detachable pods in a U-shape with an outdoor courtyard. Their home featured a unique hybrid PV/solar thermal system capable of handling higher voltage at higher temperatures to produce abundant electricity and hot water for daily needs. The team used hardwood from the Ozarks Pioneer Forest, where renewable forestry has been practiced for over 50 years, for the cabinets, trim, and flooring. Faculty advisor Art Boyt, who also came to Solar Decathlon 2002 with a team, explained, "The greatest asset of a two-year school is that there are always fresh ideas. We ask a lot of questions and scramble harder for the information. We reach out into the community for people with expertise."

The entire rooftop of the Crowder College entry showcased their hybrid PV/solar thermal system.

The Fibonacci Sequence, or the Golden Ratio, was the focus of the design for the prototype from the University of Missouri-Rolla and the Rolla Technical Institute. This sequence, which represents many shapes and patterns in nature, is displayed in some elements of the house.

The curved kitchen island inside the University of Missouri-Rolla house demonstrates the application of the Fibonacci Sequence in interior design.

Over the years, innovative technologies used in Solar Decathlon houses have influenced research, funding, and commercial development nationwide. Participating universities have become testing sites and proving grounds for new strategies, materials, and equipment. Numerous innovations showcased for the first time in SD houses were considered "leading edge" at the time, but many have since moved into mainstream use. Some examples are modular design and construction, highly efficient residential appliances and induction cooktops, advanced electronics and "Smart" technology for whole-house monitoring and control, structural insulated panels (SIPS), phase-change materials, bamboo wall framing, ultra-efficient air sealing, and incorporation of green, recycled, upcycled, and reused materials in homes. The competition also involves careful code review and approval, which provides an appropriate, expedited avenue for commercialization. In general, Solar Decathlon houses are about 5 years ahead of market adoption and 10 years ahead of code adoption.

The team from Cal Poly Pomona traveled the farthest of all—from California to Washington, DC; so they fit everything for their prototype onto one truck. They embraced simplicity as the overarching concept for their design, including less reliance on mechanical equipment to operate the house.

Reflections from David Schieren, Decathlete from the New York Institute of Technology team, is co-founder and CEO of EmPower Solar, a New York-based residential and commercial solar and battery development, engineering, installation, and operation firm. David is Board President of the New York State Solar Energy Industry Association. His wife, Cristina Zancani, a sustainable architect and urban designer, was a 2005 Decathlete with the RISD team.

First learning about the Solar Decathlon in 2003 remains one of the most vivid, striking, thrilling memories of my life. NYIT faculty advisor Michele Bertomen and lead architect Heather Korb were pitching the engineering department via video conference to team up with the School of Architecture on an off-grid, sustainable, solar home project. I can see her now, in slow motion, almost dreamlike. The words were magical to me. Independently and seemingly all alone (in the NYC

Metro area), I had become obsessed with building homes powered entirely by sunlight. The Solar Decathlon project and the vision that Michele presented got my heart racing so fast that I had to step out of the room. So, a special, purposeful journey began.

NYIT's team was architecture-led. Dynamite Dean Judy DiMaio and renowned faculty allowed the students to shine, and struggle, in every way. As a graduate student in the Energy Management Program with entry level engineering knowledge, respectable clean-tech knowledge, and business experience, I took responsibility for the solar-hydrogen-fuel cell power plant. I had never had the opportunity to work with architecture students, but found those incredibly talented people excited to have me on board. After our electric utility company, the Long Island Power Authority, introduced me to Greg Sachs, an electrical engineering professor at the U.S. Merchant Marine Academy and now a business partner), the team was fully established. We were on our way.

The classical fault lines were established early on. Form versus function. Solar Decathlon competition scoring is careful to reward design and engineering features, but energy was dominant, including energy balance: consume only what you produce. At that time most homes used lead acid battery technology, but we decided to experiment with a hydrogen fuel cell energy storage system. There are pros and cons to hydrogen storage. A major con is the efficiency loss going from solar kilowatt hours in, to Fuel Cell kilowatt hours out, with a roundtrip efficiency of less than 20% without using the heat. Standard battery systems had closer to 80% efficiency. While the energy loss only applied to energy stored—not used immediately by the operating loads during sunlight hours, we knew the efficiency hit meant that the solar power system had to be bigger than any other system at the competition. Indeed, it was. However, this involved tense battles in countless meetings where Greg and I would show calculations and beg for roof area; every single inch of roof area. We got a lot of what we asked for, including some panels installed on the ground using valuable footprint space.

The design development phase from summer 2004 to summer 2005 was intense. When on-campus construction began, this phase was equally intense, time consuming, and emotionally exhausting. The acute stress was related to the ability to deliver a functional, workable house in a very short time frame. Fundraising was a continuous effort. The energy equipment budget alone was over $250,000.

Remarkably, Greg had experience assembling hydrogen piping, which was wild to me. So, we constructed, commissioned, and tested the entire thing at the NYIT mechanical lab. To secure approval to install a hydrogen storage system on the National Mall, Greg and I presented the system to the Washington, DC Fire Marshall. Receiving the approval letter was perhaps the greatest sub-milestone for our team.

"Team energy" was the constant that powered us through. Such strong faith in each other and belief in the mission I had never experienced before and never thought possible. The Solar Decathlon was much bigger than us and much bigger than our academic pursuits.

While a blur, because of the incredible effort without sleep, there are countless indelible memories from deconstructing the house, shipping it down to DC, and then reconstructing it on the National Mall. For example, we hired a fine art moving company to transport the sacred hydrogen electrolyzer.

The experience in Washington was utterly exhausting, but it easily ranks as the most important and memorable in my life. Many Decathletes I know say the same. I think my wife, a member of the Solar Decathlon team from the Rhode Island School of Design, would also agree.

The October 2005 competition suffered from continuous rain. Every day Richard would forecast better, sunnier days to come. This was true, but he wasn't really speaking about the weather. What he knew deep down was that we were all dedicated to a transformative project that would become our life's work. We must collaborate, team up, give our everything, even when it's hard, to create a world powered by clean energy. We must stop relying on fossil fuels as quickly as possible.

At a dinner event hosted by BP Solar, a Solar Decathlon sponsor, I was speaking with a representative from the American Petroleum Institute. He was courteous, but highly dismissive of the possibility that solar and sustainable construction could supplant the need for fossil fuels. He asked, "What should we [the fossil fuel industry] be afraid of?" I looked around the room, gestured to all the Decathletes in attendance, and said, "**Us** - what's now inside of us, now part of our DNA."

The 2005 event definitely raised the bar, and media attention was profuse. There were 864 stories on television and cable, on the radio, in print, and online, with a calculation of 884 million impressions. Solar Decathlon was featured on the *CBS Evening News*,

DIY Network, HGTV, This Old House, and the Discovery Channel. Richard King soon discovered that responding to media inquiries was a major part of his job as Competition Director, and he was often surrounded by video cameras as he talked about the competition. With SD houses from Spain and Canada on the National Mall, plenty of international visitors and diplomats were also in attendance.

Several Congressional delegations from Capitol Hill visited the Solar Village, and the U.S. DOE sponsored special tours for members of the House Science Committee and Renewable Energy Caucus. The White House took notice, and Jeff Lyng, the Team Leader of the CU-Boulder team, was later invited to sit with First Lady Laura Bush during the January 2006 State of the Union address, when President Bush announced increased funding for solar technology R&D. In sharing their inspirational solar-powered houses with multitudes, Decathletes had become "solar ambassadors."

The *Magic Box* from the Universidad Politecnica de Madrid got its name from movable walls that divided the interior space into a bedroom, office, kitchen, dining, and living areas; those walls could also be moved to allow for a totally open area.

Reflections from Betsy Black, a communications and energy consultant based in North Carolina. She was hired in 2009 by the U.S. DOE Office of Energy Efficiency and Renewable Energy as the SD Organizer responsible for fundraising, communications, and sponsorships.

My devotion to Solar Decathlon began when I discovered the event as a volunteer in 2005 while working for NBC Washington to promote environmental programs within the DC metro marketplace. I decided right then and there to become part of the Solar Decathlon family. Four years later I landed a job at DOE and had the good fortune to work closely with Richard King. One of my tasks as an Organizer of the competition was to convince potential partners and corporations to contribute to the building of a Solar Village on the National Mall. On that spot, down the street from the White House and across from the U.S. Capitol, students would construct solar homes in just ten days. My role included a focus on sponsorships.

Our team of Organizers was charged with educating the public about building technologies focused on renewables. The Solar Decathlon was not only a student competition; it was also a living laboratory to demonstrate energy efficiency and develop consumer understanding of how to save money at home with clean energy solutions. Well, if major TV networks could market the Olympics, surely, I could explain the need to provide real-world training for college students about to start their careers in the energy sector. Yet, unlike the Olympics, this competition could not boast an established worldwide reputation, multi-million-dollar budget, or gold medals for winners. It did promise bragging rights for schools, buckets of pride, and a sizable professional network.

In figuring out my best approach to attract potential sponsors, I considered various options: Should I show them photos from days when it was so cold in Washington that I had ski pants and work boots on while setting up visitor tents? Send possible funders photos of on-site signage as a "welcome gateway" for more than 150,000 visitors? Show interested partners statistics for the 3.2 million visits to the SD web page? Share lists of visiting dignitaries, Capitol Hill staff, and members of Congressional Delegations who attended? Play a video clip of Virginia Tech's *LumenHAUS* in Times Square on *Good Morning America* (January 27-28, 2009) to encourage contributions?

Well, I discovered my best approach with potential sponsors was to begin my presentation with this: "Let me show you a picture of participating collegiate teams." The rest of the story wrote itself. After showing one photo of the Solar Village on the National Mall, sponsors were hooked. It's true that "a picture was worth a thousand words." At first, interested parties were intrigued. Then they became skeptical that such a massive undertaking could be completed in such a short time frame. Eventually, they jumped in to make a commitment and be part of the "Solar Decathlon family."

Most sponsors continued their support for subsequent Solar Decathlon events, but the next time they made larger commitments for greater involvement, plus longer hotel reservations, often bringing their families. Sponsor contributions grew, and unique programs were created to enhance the competition, such as: an employee community service team, a clean energy exposition, electric vehicle ride-and-drive experiences, electric bike test drives, direct access to top students for workforce opportunities, worldwide recognition through media coverage, and the creation of onsite STEM educational workshops for 3,000+ middle school students.

Like the students, sponsors first thought the Solar Decathlon was just another event, but it soon turned into a relationship built on firm belief in the mission and solid determination to educate the public. One memorable benefit for sponsors was earning a coveted ticket to dine with Decathletes in their solar-powered homes. This was part of the "Home Entertainment Contest." Students hosted two dinner parties for Solar Village neighbors, and they awarded the host team points based on ambience, quality of the meal, and the overall experience. I remember how excited sponsors became when given special access to students in their houses. After all, many Decathletes were seen as potential new hires. Most wrap-up conversations with sponsors ended like this: "How else can we support these students?" Companies were eager to get involved, and their engagement was contagious.

Early on (in 2005), "going green" was a desirable trend for sponsors. They would comment, "Isn't home energy efficiency what they do out West?" Certain corporations also wanted to check the box for energy-efficient policies, mandates, regulations, legislation, public policy, or education. Others hoped to partner with the government for marketing purposes and potential access to senior officials at the U.S. Department of Energy. Some were confused about how the government offered energy incentives, grants, and loan programs, and they needed guidance.

I have great memories of sponsors committing in-kind and financial support. As the program grew and became more successful, they were happy to jump in professionally and personally. An executive from Wells Fargo lobbied her institution for increased grant money and helped secure long-term investments in multiple events. Another executive gathered his own sponsors because he believed so passionately in the mission and convinced Dow Corning to create an educational program for school kids. Schneider Electric created an employee community service team, which ended up challenging other sponsors to beat them in the number of on-site company volunteers that helped out at the Solar Decathlon. A government affairs executive from Bosch contributed a large investment towards village infrastructure; he even brought his son for the week, who ended up applying to schools that would be competing in upcoming Solar Decathlon competitions. Lowe's created and brought a trailer with almost every power tool available to assist teams on-site with their construction efforts, which was immensely helpful to teams from other countries who were not able to bring all their own tools along.

Big picture: The Solar Decathlon simulated virtual reality glasses for visitors. The event brought to life a collegiate competition that exemplified energy-efficient strategies for everyday use, such as LED lightbulbs, solar panels, rainwater collection systems, and recycled materials like denim for wall insulation. As energy efficiency came alive on the National Mall, sponsor support for individual teams and for the event as a whole proved to be incredibly valuable.

Solar Decathlon engaged college students who wholeheartedly believed in what they had designed and built. Despite sleepless nights, traveling thousands of miles, and overcoming real life challenges such as tractor trailers stuck under bridges or power tools failing during construction, Decathletes shined on. The event also created lasting networks for everyone involved, including students, faculty mentors, attendees, sponsors, and volunteers. Sponsors grew their involvement over time and "traveled" with Solar Decathlon as it moved to new locations because they understood how effective this program was for work-force development.

To wrap up, one of my favorite memories was a student, who said to me at the beginning of this two-year project, "I'll never get a job." A year later I heard from that same student, who told me he had just been offered a job as a Manager of New Business Development at Lowes. In other words, this program creates opportunities for success. It works like a charm!"

Collectively and scientifically speaking, the 18 houses and teams that participated in the 2005 Solar Decathlon showcased a robust, comprehensive, onsite study of residential solar applications in real time. Every team set its sights on winning, and 10 separate contests motivated Decathletes to "invest their best" in order to outwit and surpass competitors. They accepted this daunting challenge with intensity and passion. They demonstrated incredible determination and resourcefulness. They designed, redesigned, built, and rebuilt with a collaborative mindset. They dared to dream and refused to give up. In the end, *everyone* was a winner. In fact, Jury Member Howard Brandston said just that, "First, let me tell you how impressed I am with all of you. We saw a range of products and expense, but in the end, it didn't matter. It was a job superbly done. You should all celebrate, because *you are all winners*."

On October 14, the last day of the competition, Decathletes were anxious to find out which teams had captured top honors. Given the persistent rain and cloudy conditions during the Contest Week (not ideal for solar-powered houses), teams had to come up with clever strategies to succeed with "Energy Balance" and find various ways to earn as many points as possible in every other contest. Teams were aware of trade-offs in choosing where to invest their greatest efforts: objective (performance-based) contests versus subjective (juried) contests. Ultimately, houses that reflected strength in just about every aspect (both architectural design and engineering) earned a place in the coveted winner's circle. The overall winners for the 2005 Solar Decathlon were: University of Colorado at Boulder in first place, Cornell University in second place, and California Polytechnic State University in third place.

The University of Colorado at Boulder successfully defended their title with their first-place finish at SD 2005.

Cornell University's attractive house with impressive landscaping won second place overall in the competition.

Decathletes from Cal Poly Pomona who trekked all the way across the country to appear in the competition were thrilled to win third place overall.

Throughout the onsite event, SD organizers kept up their nonstop schedule of activities, despite pouring rain and unavoidable mud puddles, but dark days did not dampen their vibrant presence on the National Mall. A steady stream of lively volunteers, many from sponsoring organizations, contributed to a variety of tasks, from guided tours to educational workshops. They provided the needed support for the dedicated organizers who were running low on their supply of physical energy. Richard was amazed at the volume of requests for interviews with journalists, reporters, and news groups and pleased that most of the media attention zeroed in on student teams. After all, beaming Decathletes were the best "advertisement" anyone could ask for!

As in the first event, student teams "shined on" every single day of the competition, from start to finish. They greeted visitors, gave tours, monitored how well their houses performed, and fixed whatever components needed attention. During the course of the onsite event, Decathletes appeared transformed from college students into articulate, responsible, well-informed professionals. Leadership blossomed, friendships grew, and a delightful community developed on the National Mall. In fact, as a democratic gesture, students decided to choose a ceremonial "mayor" for their Solar Village.

At the end of the disassembly period on October 18, when the last truck carrying a "deconstructed" solar house left Washington, DC, Richard looked around at the huge empty space and felt a strange sense of exhilaration. Twice, Solar Decathlon had been a smashing success. The second competition had expanded, with teams from the United States, Puerto Rico, Canada, and Spain. The result? Greater impact on more lives. Still looking to build on his crazy idea, Richard believed that more international representation in the next event would raise the bar further. Including more geographic locations, countries, and cultural backgrounds would surely spawn even more innovation and technological progress.

The U.S. DOE issued the Request for Proposals for Solar Decathlon 2007 in the summer of 2005, and the applicant pool was bigger than ever. Quite a few teams that hoped to compete had toured the 2005 Solar Village to learn more about the event and the logistics of building and managing a solar-powered house. Teams were raring to go, eager to jump start the process, and itching to get ahead of competitors. Richard had a hunch that upcoming Solar Decathlons might be full of delightful surprises.

Chapter 5

Expectations Surpassed

As Solar Decathlon 2007 approached, leaders at the National Renewable Energy Laboratory (NREL) and the U.S. Department of Energy (DOE) realized that the competition had reached a new phase in its development. The program was deemed highly effective, and seasoned organizers had a strong grip on how to manage the competition. Solar Decathlon had earned a solid reputation as an outstanding educational experience. Responses to the request for proposals from collegiate teams indicated increasing interest across the country. The U.S. DOE was convinced that funding this special program on a biennial basis was worth the investment.

After heavy lifting to get the 2002 and 2005 competitions off the ground, what was in store for 2007 and beyond? As the Solar Decathlon story continues, compelling narratives and stunning images shine brightly on the ingenuity, artistry, skill, and technological expertise of dedicated Decathletes. As they forged ahead to demonstrate leadership capacity, their innovative, state-of-the-art model homes delivered sustainable, clean energy solutions that "wowed" everyone who toured them.

Solar Decathlon: Building a Renewable Future
Melissa DiGennaro King and Richard James King
Copyright © 2024 Jenny Stanford Publishing Pte. Ltd.
ISBN 978-981-5129-47-2 (Paperback), 978-981-5129-13-7 (Hardcover), 978-1-003-47759-4 (eBook)
www.jennystanford.com

Curious visitors flocked to the Solar Village for SD 2007 in Washington, DC.

Solar Decathlon 2007

On January 5–6, 2007, Decathletes for the upcoming competition gathered at the Crystal Gateway Marriott Hotel in Arlington, Virginia. SD organizers invited students from each participating team to attend a weekend workshop to get to know each other and meet the Solar Decathlon leadership team. Over the past 12 months, students and faculty mentors had joined monthly conference calls by phone with Richard King and SD competition managers, but this was their first opportunity to assemble face to face. Note that this was years before the advent of online Zooming where all attendees "see" each other in the digital space, so it was quite exciting. The Friday evening welcome included team introductions and a social hour. Decathletes savored the chance to chat informally with peers and hear about the trials and tribulations others had experienced in designing net-zero-energy houses for the competition.

On Saturday, students learned more about contest rules, National Park Service regulations, safety requirements, code compliance, instrumentation, performance monitoring, logistics, professional juries, and communications. After an action-packed weekend full of important information, students returned to their respective schools with renewed enthusiasm and motivation to excel. Richard understood the tremendous value of shared experiences for people who are invested in a common purpose. After all, that is the fundamental premise of Solar Decathlon. He also realized that building relationships ahead of the competition would reap tremendous benefits.

Assembling the Decathletes altogether before the onsite event was a significant milestone. The January workshop planted seeds for the strong sense of community that developed on the National Mall where Decathletes took center stage. From its inception, Solar Decathlon has been a superior example of a competitive challenge that fosters a cooperative spirit among participants. Without doubt, teams must soar above and beyond to win, but somehow, the Solar Village fosters exceptional camaraderie across competing teams.

That spring, as Richard King reviewed the required documentation and design plans from SD teams, he noticed definite improvement and made a note to himself: "This event is likely to dazzle everyone. The collaborative, iterative process of brainstorming, design, build, and evaluate, followed by redesign, revision, and reassessment of models leads to ongoing discovery and innovation. When you apply that process to the construction of full-sized houses for a competition with highly motivated students, you accelerate progress. Framing the purpose of Solar Decathlon as one way to address the global problem of climate change is a catalyst for continued advancement. We're on a roll."

The roster of participating teams for SD 2007 included 20 schools representing 13 states and four nations. This was a diverse set of universities from many different geographic locations.

- Carnegie Mellon University
- Cornell University
- Georgia Institute of Technology
- Kansas State University
- Lawrence Technological University
- Massachusetts Institute of Technology
- New York Institute of Technology

- Pennsylvania State University
- Santa Clara University
- Team Montreal (Ecole de Technologie Superiore, Universite' de Montreal, McGill University)
- Technische Universitat Darmstadt
- Texas A&M University
- Universidad de Puerto Rico, Rio Piedras and Mayaguez
- Universidad Politecnica de Madrid
- University of Cincinnati
- University of Colorado at Boulder
- University of Illinois at Urbana-Champaign
- University of Maryland
- University of Missouri-Rolla
- University of Texas at Austin

Whenever possible, Richard King visited teams at their respective home campuses to get a firsthand update on progress with house plans and construction. Those onsite visits established and deepened key relationships. They also provided excellent opportunities for the Competition Director to energize teams and spread excitement for the upcoming event. Richard always returned home feeling inspired with renewed hope for what these students might accomplish.

SD 2007 was the third SD event held on the National Mall in Washington, DC, and teams were well prepared. Many sent "scouts" to attend previous Solar Decathlons, and that helped them mastermind the competition. With better understanding of how to calibrate winning metrics, teams developed sophisticated scoring tactics to manage potential success in the 10 contests. Decathletes were learning more and more about the tricky dance steps required to balance architectural beauty and engineering prowess.

Clever innovation was on full display at the 2007 competition, with notable advancements in technology, materials, and design solutions. Teams enlisted greater multidisciplinary support from a broad spectrum of majors at their respective schools, including business, marketing, finance, and communications, and this led to increased sponsorship. With stronger financial backing and support, teams could focus on new ideas, creative designs, and state-of-the-art technology. The houses they built proved to be excellent examples of creativity, practicality, and resourcefulness.

The capacity of SD organizers had also reached new heights. The capable leadership of Cecile Warner and Mike Wassmer (Decathlete from the first-place CU-Boulder team in SD 2002) ensured a well-run competition. The management team from the National Renewable Energy Lab was second-to-none in their enthusiasm for Solar Decathlon. A savvy SD scoring team had sharpened their skills for monitoring performance and analyzing data for the 10 contests. Communications and media outreach expanded, thanks to the expertise of John Horst, Wendy Butler Burt, and others at the U.S. DOE, leading to increased press coverage and greater public awareness of the event.

To top it off, the weather in Washington, DC was absolutely delightful for most of the 2007 competition. With no measurable rainfall from October 1–18 and an average temperature of 67°F, Decathletes and the visiting public experienced the warmest October on record for the DC metro area (at the time). Conditions were ideal for Solar Village tours. Approximately 120,000 people enjoyed getting a firsthand look at the marvelous solar-powered houses on display during that event.

As designed, 10 Solar Decathlon contests reflect how people use energy in their daily lives: at home, for work, and at play. The contests and their assigned point values for SD 2007 were slightly different from the previous event. Five contests were evaluated by professional jury panels, and five contests were objectively measured for onsite performance during the competition. Teams could earn up to 1200 total points in the following contests:

Juried

- Architecture—200 points
    - Are energy efficiency technologies seamlessly integrated into an aesthetically pleasing dwelling that delights the senses?
- Engineering—150 points
    - Does the house represent high-quality engineering and craftsmanship, from the building envelope to the mechanical, electrical, and plumbing systems?

- Market Viability (new)—150 points
    - Can the house accommodate a variety of residents? Can it be readily built? Is it affordable? Does it appeal to the current market of potential homebuyers?
- Communications (new)—100 points
    - Does the team deliver clear and consistent messaging and graphics on its website, signage, collateral materials, and during onsite house tours?
- Lighting—100 points
    - Does house design incorporate both natural and electric light to create elegant, functional, and energy-efficient solutions?

Monitored

- Comfort Zone—100 points
- Appliances—100 points
- Hot Water—100 points
- Energy Balance—100 points
- Getting Around—100 points

The distinguished jurors dedicated considerable time and talent to SD 2007, reviewing team documentation in advance and participating in scheduled house tours during the October competition. The following esteemed professionals were a vital part of that event.

- **Architecture:** Gregory Kiss, Susan Maxman, Grant Simpson
- **Engineering:** Kent Peterson, Bill Rittleman, Miles Russell
- **Market appeal:** Bob Burt, Jim Ketter, Doug Lowe, Joyce Mason
- **Economic analysis:** Stuart Bernstein, David Kline, Miles Russell
- **Communications:** Jaime Van Mourik, Scott Shepherd, Alan Wickstrom
- **Lighting:** Nancy Clanton, Naomi Miller, Sandra Stashik
- **Energy analysis:** Brent Griffith, Sue Reilly, Norm Weaver

Find out more about these esteemed leaders at: https://www.solardecathlon.gov/past/2007/juries.html

Reflections from Joyce Mason, a marketing expert who worked at Pardee Homes and developed their Living Smart® brand to encourage energy efficiency, water conservation, and use of sustainable materials, which won an "Earth Award" from the State of California. Her efforts helped Pardee Homes earn NAHB's "Green Builder of the Year Award." Joyce also organized a special playhouse at the National Building Museum to educate children about green building.

I was invited to be part of the Market Viability Jury for Solar Decathlon 2007. At the time I was Vice President of Marketing and champion of the sustainable building program for Pardee Homes, a multi-regional production builder doing business in California and Nevada. Our company built 2–3,000 homes per year, and in 2004, Pardee Homes built the first *Zero Energy Show Home* for the International Builder Show in Las Vegas. As a company, we were committed to exceeding energy efficiency standards in all our homes. Our participation in the U.S. DOE's Energy Star™ program brought me, on occasion, to Washington, DC, and it was there that I first met Richard King.

I didn't know much about the Solar Decathlon competition, but I was totally enamored with the concept. For an industry driven by building costs, codes, and market dynamics, and where the tendency is to keep doing what you have always done, it was very exciting to think about what college students could bring to the housing sector.

I had been on judging panels for building industry awards, but nothing in my experience prepared me for the true magnitude of the Solar Decathlon. When I saw the Solar Village on the National Mall with the Washington Monument on one end and the U.S. Capitol on the other, I was completely swept away. I've been a believer ever since. Witnessing the intensity of the Decathletes, the quality of their projects, along with the dedication of Organizers who had high and fair standards for the competition, I immediately felt the "bigness" of the event. This was no ordinary competition, but rather, quite an extraordinary one. The unbridled enthusiasm of student teams, their desire to build "outside the box," their ability to bring creative academic excellence to the field, their efforts in all aspects of the process, and the respect for each person's contributions were inspiring.

The logistics required and the dedication of the Organizers to produce such a high-quality event were impressive, and that spoke to the leadership and vision of Richard King. His team had to set rules and standards for the competition, prepare the site where 20 houses would be constructed all at once, plan for the press and public to attend safely, and provide space for Organizers to maneuver and solve problems. The list goes on. It seemed as though they'd thought of everything, including managing a flock of professional jurors.

The judging process, I quickly learned, was serious, with rigorous time constraints. Jury panels in the Market Viability category were asked to evaluate SD entries for *Livability, Buildability, and Flexibility*. In short, we had to determine how well each house might do in the real marketplace. Student presentations described target markets and how their project was designed and built to reach that market. They also had to explain how technology could benefit and support the target market. After the presentations we were permitted to wander, take a closer look, and make notes. Each jury panel had an assigned coordinator to keep us on track; this helped ensure fairness and the development of constructive criticism in our deliberations.

As a building industry marketer, I appreciated the teams that understood their target market and built to appeal to that market. That meant attention to house location and climate, so students could showcase appropriate technology and cost impacts. Multidisciplinary teams made it possible for several departments or schools to work together, which was great. The Market Viability Contest "nudged" Decathletes to produce a full-package model home, as expected in the real world.

Solar Decathlon houses never failed to impress. Despite meeting challenges along the way, teams showed resilience and resolve to overcome and carry on. Not every house can win first prize, but every house can stand out in some way. Even now, many years after my involvement with SD competitions, some still resonate. Here are a few examples:

- SD 2007, University of Colorado Boulder: Using a shipping container as a mechanical core and creating living spaces around the core made this house relatable in a production environment. It was easy to see how the house could expand in size while achieving cost

advantages with a core that could be replicated. Shipping containers have lots of possibilities for solving housing challenges (such as at FEMA) in places where infrastructure is limited and for low-income projects.
- SD 2009, Santa Clara University: Home design that took advantage of adjacent market areas in Silicon Valley and the agricultural regions of California. Students conducted focus groups with their target market and, as a result, simplified the technology to make it easy to use.
- SD 2011, Team Canada, University of Calgary: This house addressed the housing needs of the First Nations of Canada and was designed with representatives of local tribes. Native American Customs and traditions were integrated into the design. I'll never forget the special spiritual quality of this house.

I was fortunate to be a Market Viability Juror for three Solar Decathlons: 2007, 2009, and 2011. In that role I became keenly aware that the competition showcased housing innovation in real-time. The Building Industry needed to see this. Solar Decathlon was an opportunity to see what might be possible in a future marketplace. These houses were incredibly energy efficient, and they used advanced technology that actually worked. They incorporated creative use of existing building materials, as well as sustainable building practices. Ingenious Decathletes want to do more than what is required, going above and beyond. This attitude shines brightly: "Let's try to do the *most* we can!"

As the event expanded into other locations across the U.S. and around the world, more people get a chance to see what Decathletes can do and the team spirit that emerges. The building industry is legitimately constrained in many ways, particularly in production housing. Costs, supply chain and distribution channels, trades, warranty issues, local restrictions, and market dynamics create obstacles that can seem insurmountable. Exposure to what's new is often limited to building shows, manufacturer improvements, or occasional break-out projects. The residential construction industry has benefited greatly from exposure to the Solar Decathlon. Given this opportunity, student teams have soared. Just think of all the Decathletes and the incredibly talented workforce being delivered to the housing industry!

Solar Decathlon competitions showcase advanced technologies and further building science in creative ways, illuminating new options for both custom and production housing. In reflecting on my time as a Solar Decathlon Juror, it was a privilege to participate. I've been inspired.

Maryland's *LEAFHouse* arrived in the dark of night on October 3, 2007.

After night-time delivery of housing modules for SD 2007, the construction of the Solar Village began.

Due to National Park Service regulations and requirements of Washington, DC, oversized trucks could enter the city for only a limited period of time and only at night. That said, the assembly of the Solar Village began in the dark on October 3, 2007. It was quite a sight to behold! Huge flatbed trailers and cranes swooped onto the National Mall to unload their precious cargo. Students were at-the-ready, watching carefully to be sure that every section of these unique structures arrived safely and were placed correctly on assigned "lots." On the morning of October 3, the entire area became a major construction zone.

As he walked around the emerging Solar Village, Richard King reflected on his observations: "I immediately noticed that teams had gotten bigger. Each lot was abuzz with energy and activity as Decathletes focused on a wide variety of tasks. There were large containers filled with tools and equipment alongside piles of various materials. Somehow, everyone seemed to know what they were supposed to do. Spirits were sky-high as parts were assembled and houses emerged. Watching all the students in construction boots and hard hats work tirelessly for hours on end gave me pause. Over the next six days I witnessed an incredible transformation, but I never heard one complaint or saw a single person slouch off. Instead, I saw teams help each other, lending a hand or sharing tools as needed. Competitors became friends, and a neighborly community took shape right there on the National Mall."

The sights and sounds of a construction zone snuggled in-between museums of the Smithsonian Institution attracted plenty of interest from the public. Curious onlookers strolled down "Energy Street" to get a glimpse of all this purposeful action. The Mall was particularly busy during lunchtime, when federal employees came out of nearby offices for a sunshine break and suddenly discovered that a collection of unusual structures had appeared.

Prototypes for the 2007 Solar Village taking shape on the National Mall. Teams had just eight days to complete assembly of their houses for this public showcase.

Aerial view of 2007 Solar Village on the National Mall in Washington, DC.

By Friday, October 12, every team was ready. As he welcomed everyone to the Opening Ceremony, U.S. Secretary of Energy Samuel Bodman expressed his own excitement. He congratulated Decathletes, event sponsors, supporters, and SD organizers for creating such an impressive living laboratory on the National Mall. Here are some of the clever innovations on display at Solar Decathlon 2007.

U.S. Secretary of Energy Samuel Bodman cutting the ribbon at the official opening ceremony for SD 2007.

Georgia Institute of Technology (Ga Tech)

The Georgia Institute of Technology team made it crystal clear that their house was on the leading edge of technology. They created a delightful, mood-enhancing space filled with natural light. Translucent walls made of polycarbonate sheets and aerogel filler made with ETFE (ethylene-tetrafluoraethylene) provided excellent insulation, corrosion resistance, and abundant filtered light. Aerogel is also called "solid smoke" because it is one of the lightest solid materials. Clerestory windows above the walls and a flat roof made with that same translucent film contributed to a bright interior. The Georgia Tech house was the very first residential application of this material in the United States.

The house from Georgia Tech featured tilted solar panels mounted above the roof that were positioned to follow the sun, in order to maximize solar gain. Posters mounted on the deck railing described house features for visitors.

The Georgia Tech house had translucent walls and roof filled with aerogel insulation and ETFE (ethylene-tetrafluoraethylene). Those same materials were used in the Aquatics Center for the 2008 Olympics in Beijing.

The east and south walls of the Georgia Tech house were translucent, thermally efficient cellular polycarbonate panels. The aerogel inside these panels is temperature resistant and very lightweight.

The Georgia Tech team installed an innovative PC-controlled system, so the house could operate according to a model programmed for efficient energy use. In today's world of "smart" homes, this might be a standard feature, but 2007 Decathletes were at the forefront in designing and using these systems. Team members said that close collaboration among students, faculty, and industry partners was a powerful educational experience. In their own words, "the greatest aspect of the SD project was learning by doing." Their elegant Georgia Tech house took fourth place in the highly coveted Architecture Contest. This forward-thinking team was also delighted to win third place in Getting Around and fifth place in the Communications Contest.

Carnegie Mellon University

Carnegie Mellon University (CMU) collaborated with the Art Institute of Pittsburgh and the University of Pittsburgh on the Solar Decathlon project. Their house represented the fusion of art and technology. CMU brought a "plug and play" house to the National Mall, offering a set of adaptable modules that prospective owners could rearrange or upgrade as desired when daily needs changed over time. Each room was a "living pod" with a junction box for electrical use. A central high-tech core with easy accessibility contained all mechanical systems for the dwelling. CMU students developed a unique lighting tool that informed visitors about electricity use in the house (how, where, when). This became a great educational asset during public tours.

Carnegie Mellon and the Technical University of Darmstadt (TUD) had developed close friendships at SD 2005, and those students stayed in touch over time. In May 2006, members of the CMU team visited TUD in Germany to brainstorm design ideas. As a result of their time together, they decided to create "shared community space" between the two houses at SD 2007. As the competition progressed, it was not unusual to see Decathletes from those two teams enjoying each other's company. What a great example of how Solar Decathlon creates connections among people with diverse backgrounds.

The Carnegie Mellon house showcased a set of "living pods" that arrived on the National Mall as prefabricated modular units.

Special lighting at the Carnegie Mellon house created a beautiful sight in the evening.

Attractive interior of the Carnegie Mellon house created a friendly, inviting space.

The Carnegie Mellon team added a "greenscape" of living plants on the walls and roof to provide extra insulation, rainwater purification, and natural beauty.

Pennsylvania State University

To test the concept of market viability, the Pennsylvania State University (PSU) team made a double commitment to SD 2007. They built two houses for two distinct communities: one for MorningStar, Pennsylvania, and the other for the sister city of MorningStar, Montana. The Pennsylvania site would become a research laboratory and educational facility. The Montana site would be an affordable version of the prototype and a residence for visiting faculty at the Chief Dull Knife College on the Northern Cheyenne Reservation. The team wanted to show how renewable technologies can be beneficial in different environmental and economic settings. They used a hybrid construction process, mixing pre-fab elements with site-built features, and incorporated "materials of opportunity" that were locally available. For example, attractive Pennsylvania bluestone and reclaimed slate shingles provided thermal mass for the house. That was a great application of Building Integrated Photovoltaics (BIPV). Photovoltaic solar cells were expensive, so one way to reduce the cost was to have PV cells serve "double duty." For example, if solar cells were integrated as water-tight roofing or siding material, then conventional roofing or siding was unnecessary, offsetting the cost of a PV system.

Expectations Surpassed | 161

Signage outside the Penn State house featured multiple languages to increase user-friendliness.

The exterior of the Penn State house showcased several different locally available materials to create strong contrast and visual appeal. Students called them "materials of opportunity."

PV shingles on the Penn State house were a great example of Building Integrated Photovoltaics (BIPV).

Deep, movable louver shades on the south wall of the Penn State house enabled control of direct sunlight to moderate heating and cooling effects.

Inside the PSU house, lumber from a recently fallen 100-year-old oak tree was used to make a table for dining, and wood from a fallen white oak tree became sun shades on the south wall of the house. Dedicated to spreading learning across many disciplines, the SD project engaged more than 900 students at PSU, as well as the Northern Cheyenne Tribe. The Penn State team had undeniable spunk. They could have won an award for their big smiles and unabashed friendliness, so their third-place finish in the Communications Contest was not surprising. They aced Market Viability and Lighting with third-place awards and were proud of their first-place finish in the Hot Water Contest and fifth-place award for Engineering.

Glass bottles on movable shelves inside a window on the south-facing wall of the Penn State house filtered incoming light.

Universidad Politecnica de Madrid

Eager Decathletes from the Universidad Politecnica de Madrid (UPM) traveled across the Atlantic to participate in SD 2007, to follow-up on their first appearance at SD 2005. This team from Spain included a diverse group of students from Chile, Ecuador, Peru, Brazil, Venezuela, Mexico, and Puerto Rico. Team member Maria Gomez believed that having many different points of view "brought out the best in everyone." Their main goal was to build a contemporary prototype for single-family or multi-family use with strong market appeal. Being situated prominently at one end of the Solar Village directly in front of the U.S. Capitol gave *Casa Solar* tremendous visibility. A distinctive exterior with prominent green wall and iconic sloped roof, coupled with a colorful, elegant interior left a memorable impression on every visitor. Electrochromic windows were capable of getting darker or lighter to block or allow sunlight indoors. Lightweight construction materials, water-saving features, and kitchen bins for direct disposal of compostable kitchen scraps, waste, and recyclable items were highlighted during house tours to promote greater awareness of sustainability.

The distinctive roofline of the prototype from the Universidad Politecnica de Madrid was a striking complement to the Washington Monument in the background.

Casa Solar from Madrid sat on a foundation containing phase-change gel that helped with temperature control inside the house. The bold color scheme of the exterior walls made it easy to identify this house.

PV panels commandeered rooftop space of the Madrid house. Sliding glass doors opened onto a spacious deck.

Team members from the Universidad Politecnica de Madrid represented eight different countries. Photo credit: Kaye Evans-Lutterodt, U.S. Dept of Energy Solar Decathlon.

The delightful interior of the UPM house displayed the same bright colors as the exterior walls. The sliding doors also featured BIPV, with solar cells mounted on the glass surface.

The warm, friendly Madrid team was full of sparkle and positive energy. It is not surprising that they had huge success with media outreach before, during, and after the Solar Decathlon. Students were amazed at how many people were interested in buying this type of dwelling after experiencing it firsthand. The Madrid house, which placed third in the Architecture Contest, was a masterful blend of stunning architecture and splendid interior design. They also won fourth place for Appliances and fifth place for the Getting Around Contest.

Santa Clara University

The story of the Santa Clara University (SCU) team is full of surprises. First of all, they were the only university west of Colorado to participate in the 2007 event, representing Silicon Valley, California. From the get-go, they were determined to showcase innovative technology. Santa Clara Decathletes, committed to building a more sustainable society, chose "design with purpose" as their theme. SCU does not have an architecture program, but students connected early on with the SD 2005 Cal Poly Pomona team for design support. Team manager James Bickford led a close-knit group to engineer a dynamically smart, energy-efficient house that performed consistently well during the competition. Their unique solar thermal collection system produced a high-temperature fluid for space and water heating, which also powered an absorption chiller for air conditioning. PV panels to charge batteries were incorporated into the siding, so the house could operate off the grid. Later, those panels could also be connected to an inverter, in order to deliver excess power to a utility company.

The SCU team was excited about their sustainability meter that gathered data on the home's power use and carbon emissions. The meter then made those data available to homeowners online, so they could view, make adjustments remotely, and manage the sale of carbon credits. Electrochromic windows in the house had special glass that could lighten or darken a room with a switch to block or allow sunlight. Students created hybrid lighting with compact fluorescent bulbs (CFLs) and light-emitting diodes (LEDs). Although commonplace today, LEDs were brand new at that time. To support sustainability, I-joist beams, floors, and other interior elements were made of bamboo. Fortuitously, SCU Decathletes built their house in modular units that could be dissembled and reassembled fairly quickly.

The house from Santa Clara University arrived three days late, but fortunately, the modular sections allowed for quick assembly on the National Mall.

The dedicated Santa Clara team worked diligently to get their house ready for the ribbon-cutting ceremony in less than a week. Many of the interior features had been prefabricated and installed in advance.

Santa Clara called their prototype the *Ripple House* because it aimed to help people understand solar energy and sustainability concepts and their importance in today's world.

As it turned out, the Santa Clara house arrived on the National Mall three days late. The reason? The team had designed their own low-bed truck to carry modular sections of the house from California to Washington, DC. Unfortunately, one of the axles bent enroute. They had to call a welder to cut and reweld the axle before they could get on the road again, and that cost them precious time. SD organizers figured this team might never catch up. Thanks to wise foresight, many interior elements of the house were finished before the trip. As a result, the entire modular structure was up and ready in short order. The SCU team started at the bottom of the pack, but surprised everyone by steadily moving upward in the rankings.

With astute strategizing, this team did remarkably well in all the performance-based contests: first place for Hot Water, first place for Energy Balance, second place for Appliances and Getting Around (e.g., They drove their electric vehicle 323 miles in just 5 days.), and fifth place for the Comfort Zone Contest. SCU also impressed the Communications Jury, taking second place in that contest. In the end, they placed third overall, a spectacular comeback for a team that arrived late. In his announcement of the winners, U.S. Secretary of Energy Samuel Bodman referred to SCU Decathletes as "the Cinderella team from California that never gave up."

University of Maryland

The University of Maryland (UMD), located just outside of Washington, DC, had the distinct honor of being the local favorite. Given substantial media attention, it did not take long for throngs of visitors to line up, eagerly waiting for a tour of the *LEAFHouse*. According to team member Brittany Williams, a leaf is one of nature's most efficient organisms, and they were inspired to emulate that concept. LEAF also stands for "leading everyone to an abundant future," a nod to solar energy as the key to sustainability. Decathletes from the 2002 and 2005 SD teams supported the 2007 project, as did many faculty advisors, community members, and industry professionals.

The UMD team aimed to erase boundaries between architecture and ecology. They designed interior living areas to spill over onto exterior decks. A living green wall connected the house with the landscape, and huge sliding glass doors on the south wall welcomed views of the outside into the house. An indoor waterfall was a striking feature of *LEAFHouse*. Crafted to represent the Chesapeake Bay region, a liquid desiccant wall system was encased in clear glass to create a delightful, relaxing ambience. This unique innovation was the "first of its kind" for a residential dwelling. With a nod to water conservation, rainwater was collected and recirculated in rain gardens, and grey water was filtered for irrigation use. Roof-mounted PV panels positioned at an angle enabled optimal solar exposure.

A Smart House Adaptive Control (SHAC) network monitored indoor light, temperature, and humidity, along with weather forecasts, as part of a computer package to guide energy use and maximize efficiency. This was a leading-edge innovation in 2007, and students who engineered the system were rightfully proud. The attractive building envelope included green vegetation and elements made of glass, wood, and corrugated metal. A sleek, contemporary interior echoed charming simplicity. UMD's second-place awards for the Architecture, Lighting, and Market Viability Contests were well deserved. UMD Decathletes excelled at sharing information and leading educational house tours, so their first-place finish in the Communications Contest was not surprising. *LEAFHouse* performed well during competition week, which helped the team secure second place overall in the competition. They were ecstatic!

Long line of enthusiastic visitors line up beside the local favorite, *LEAFHouse*, from the University of Maryland.

Aerial view of UMD's *LEAFHouse* with green wall and operable external louvers to provide shade and control excess heat from sunlight on hot days. The louvered doors in this photo can also open fully to allow fresh air and daylight.

Clerestory windows and translucent polycarbonate skylights containing nanogel (with high thermal value) created a brightly lit interior space for *LEAFHouse*. The Murphy bed shown here folded up into the wall to create more floorspace.

Reflections from Brittany L. Williams, AIA LEED AP BD+C, a practicing architect at the award-winning firm, Gardner Architects, LLC. She focuses on the synthesis of sustainable active and passive design strategies and a multidisciplinary approach at the residential scale. Brittany is also a Clinical Assistant Professor in the School of Architecture, Planning, and Preservation at the University of Maryland.

LEAFHouse Decathlete Experience

Serving as architecture team leader for the University of Maryland's Solar Decathlon 2007 entry was a challenging yet rewarding pursuit. Learning collaboratively with students from a host of other disciplines from across the university, designing and constructing an innovative house from scratch, and exhibiting our efforts on a worldwide stage proved to be a life-altering experience. While celebrating our second place win overall on the steps of *LEAFHouse* in the Solar Village was certainly memorable, but some of the most transformative experiences came on hot summer nights back on campus when we were building the house.

Watching our drawings come to life right before our eyes as we actually constructed the house revealed areas that needed further collaboration among disciplines. The rich, although sometimes tense, conversations to negotiate between performance, constructability, functionality, building code, competition rules, systems components, and aesthetics illustrated the necessity of this type of collaboration across areas of expertise. This enabled the design process to evolve and created space for innovation in our project. Due to this experience, I now advocate for collaborative multi-disciplinary design processes in my professional practice and in the classroom.

My participation as a Decathlete while working on my Master of Architecture degree was the single, most impactful experience in my academic and professional life. Nearly all my pursuits teaching in academia and in the architecture profession can be traced back to lessons I learned and connections I made during that experience. Solar Decathlon led me toward a career path where I view environmental stewardship as a top priority in the design process. I embrace the power of the built environment to guide our everyday actions in support of a more sustainable future.

Technische Universitat Darmstadt

Known as the "city of science," Darmstadt is a major center of scientific institutions, engineering centers, and high-tech companies.

Decathletes from the Technical University of Darmstadt (TUD) in southwest Germany traveled farther than any other team in the competition. TUD students assembled a unique, compact dwelling based on the onion principle, with several layers: external for shading, privacy, and power-generation with PV panels; middle layer with vacuum-insulated walls (east and west) plus large windows (north and south); and an inner core that contained mechanical and technical systems. This simple, flexible house relied on low-tech passive systems as top priority, and then incorporated active systems into the architectural design. Their main strategy was to make use of photovoltaics on the roof and all sides of the house to generate as much power as possible. Turned out that this was spot on—the TUD house functioned like a mini-power plant.

A well-insulated building envelope included phase-change materials (Micronal by BASF) that could store thermal energy in the ceiling and walls. Distinctive oak louvers on all exterior sides were quite sophisticated. The louvers integrated solar cells and tilted automatically to follow the path of sunlight during the day. This system ensured comfortable shading and maximized the production of solar power. To maximize flexibility and reduce energy loads, the interior of the house was set up in zones, instead of rooms. An unusual feature of the TUD house was a bed embedded in the wooden flooring that opened up for sleeping. What a clever way to make efficient use of space!

The distinctive exterior doors on the TUD house with operable louvers that contained solar cells could be opened to let in daylight or closed for privacy and thermal control.

The air space between exterior doors and interior glass walls moderated indoor temperature and light. Solar cells on the glass roof above, another example of BIPV, allowed daylighting while producing power for the house.

This entry "wowed" jury members and the public by pushing the envelope in just about every aspect. Small but powerful and elegant, the house from Darmstadt took first place in Architecture, Engineering, and Lighting Contests and third place for Appliances. It was not surprising that this "power plant house" tied for first place with seven other teams in the Energy Balance Contest. TUD Decathletes said they wanted to make an impact, and their first-place finish in the 2007 Solar Decathlon was a clear testimony that they had delivered.

Inside the TUD house, a retractable bed in the floor could be closed up during daytime, to free up more floor space for dining or other household functions.

There is another fascinating story about the developing relationships among Decathletes from TUD and Carnegie Mellon University (CMU). In the winter of 2007, TUD students took a trip to Pittsburgh to attend a workshop with Carnegie Mellon students, followed by a wonderful three-week tour of the U.S. Faculty advisors and students benefitted from those shared experiences, and that visit strengthened friendships between those teams that continued to grow over time. Who would have guessed that partnerships, marriages, and babies would later emerge from this group of Solar Decathlon participants?

Reflections from Barbara Gehrung, who holds a Master of Architecture Degree from Technical University Darmstadt (TUD), Germany. She is a Certified Passive House Consultant, AIA International Associate, and partner at Gehrung+Graham, LLC, an architecture firm specializing in energy-positive architecture and passive house

design. In 2006, she joined Professor Manfred Hegger's team of educators and researchers at TUD as Project Manager for Team Germany in SD 2007.

Talk about the connection between CMU and TUD and how it evolved over time.

In 2003, I was a summer intern at the Center for Building Performance at Carnegie Mellon University. There, I learned about the Solar Decathlon and initiated a collaboration and student exchange program between TUD and CMU for Carnegie Mellon's entry to the 2005 competition. Leading up to 2005, students and faculty from both universities held joint design workshops in Darmstadt and Pittsburgh. Three students from Darmstadt, Isabell Passig (Schäfer), Andreas Pilot, and I, went back to Pittsburgh and Washington, DC in the summer of 2005 to help build "Pittsburgh Synergy's" house and partake in the competition. For Solar Decathlon 2007, while entering as a separate team, we again had design workshops and field trips with participating students from Pittsburgh and Darmstadt. Seeing the "Pittsburgh Synergy Home" on CMU's campus and the experience from the 2005 competition gave TUD's team a sense of reality for the magnitude of the challenge. In addition to many lessons learned, that gave CMU students insights into European approaches and technology for sustainable and energy-efficient architecture. This collaboration formed friendships and changed lives beyond the competition: Two SD2007 student team members from Pittsburgh and Darmstadt are now married, and three Darmstadt alumni, including myself, are now living and working in the United States.

How did you and Cass Kawecki meet? How did SD participation influence your relationship?

Cass and I met in the summer of 2005, while I was helping to build CMU's Solar Decathlon house. Cass was an adjunct professor at CMU teaching an interdisciplinary course in real estate development. He saw me in a tool belt swinging the hammer, and we connected over our mutual interest in green, energy-efficient architecture. He took me on a first date during the competition in Washington, DC. One good thing led to another, and we got married in 2007. We now have two children, a dog, and a happy life.

How did you become project manager for TUD?

In 2001 TU Darmstadt identified energy-efficient building as a vital part of architecture education and created the *Unit of Design and Energy-Efficient Building* at the School of Architecture. Professor Manfred

Hegger, a pioneer in architectural practice in this field, became its chair and was asked to create a comprehensive curriculum for learning and research. I was fortunate to be among his first students. Hegger and his team led with openness for student impulse, and provided opportunity, space, and networking support to bring these ideas to fruition. Prof. Stephen Lee, of the *Center for Building Performance* at Carnegie Mellon University, where I interned during the summer of 2003, shared this spirit. When I first approached the two professors about a student exchange and collaboration for Solar Decathlon 2005, they gave my idea a chance. When Isabell and I returned from the competition in Washington, DC, Hegger was excited by our initiative to apply to Solar Decathlon 2007 as a team. He and Andrea Georgi-Tomas, senior assistant professor in his team, garnered the necessary support from the university. Just three weeks before graduation, we learned that our proposal had been successful. Fortunately, Isabell and I were hired to lead the project.

Describe the design process within the TUD team.

Professor Hegger and our project management team developed and organized a three-semester interdisciplinary curriculum, winning support from other faculty at the School of Architecture by providing related classes and credits. That made it possible for participating students to focus on designing and building our "Year 2015 Prototype Home." We recruited 25 students, mostly from architecture, but some from the School of Electrical Engineering, who committed to design and build TU Darmstadt's house. The overall design process was iterative, testing different strategies with accompanying energy modeling. It also included collaboration with other research and higher learning institutions and industry partners. Together with the students we determined that the Passive House standard would be the basis for a net-zero energy performing building, approaching energy from conservation first, with active systems second. A multi-stage architectural and energetic design competition, going in steps from designs by each student, then to group projects, determined the concept to be realized.

Once the design concept and guiding principles were determined, individuals or groups of students took on specific responsibilities, such as construction detailing, modeling performance scenarios under various weather conditions, modeling and specifying the mechanical systems, interior design, lighting design and controls, and managing the construction process. Weekly team meetings throughout and regular delivery deadlines provided structure and opportunity for feedback and correction. In parallel, Professor Hegger worked to win major industry sponsors, form an advisory board, and secure accompanying research grants from the German Federal Ministry of Housing and Urban Development, and Federal Ministry of Economy and Industry. Sponsorships, partnerships, and grants paid for construction materials and provided opportunities for modeling, testing and development, paid staff salaries, and covered the immense cost of shipping and overseas travel.

What was most challenging about preparing for SD 2007?

The most challenging aspect was the tight timeline to get funding and manage the logistics of getting the house built and shipped across the Atlantic Ocean (and back).

Who designed those amazing PV louvers on the outside of the 2007 house?

Throughout, the team's goal was to find synergies and truly integrate energy performance and function with the architecture. As such, the design had always featured moveable shutters for exterior shading and privacy. The idea to integrate PV with the shutters came during a team meeting, late in the project, in the spring of 2007. With only eight weeks left until shipping, the energy modeling determined the need for additional installed kWh. Thomas Wach, one of our students, took on the detailing of the shutters, while the project management team facilitated the material sourcing and collaboration with IT students from TU Munich for the programming of the motors. We negotiated with Schindler Façade Solutions to use their manufacturing line for a weekend, where a group of students assembled the shutters. The completed shutters arrived just in time to be loaded into the shipping container. The first time they were installed on the house, was on the National Mall at the competition.

What was your reaction to your team's first-place finish in the Washington, DC competition?

We entered SD 2007 hoping to make an impact on the future of integrated, net-zero architectural design and were excited about being part of an international competition. In light of the meticulous planning, immense team effort, and resources that went into our participation, we were hoping to do well and succeed on our mission. Winning, however, was beyond a dream come true.

How did SD impact your professional career and personal life?

Participating in the Solar Decathlon was a jump-start for my professional career: first, by generating a fascinating job with a lot of responsibility immediately after graduation; second, by being part of an international cutting-edge project in a field I aspired to work in. The Solar Decathlon and its Solar Village make this beautiful, comfortable, and earth-friendly way of living quite visible and real for the general public. It's now a worldwide effort. To this day, 15 years later, I am approached by clients or colleagues about our 2007 entry. It has been amazing to be part of a movement that went from 14 prototype homes in 2002 to becoming a desired product and achievable way of building in both residential and commercial sectors today. In addition, I met my husband through participation in the exchange with Carnegie Mellon for Solar Decathlon. The familiarity with American architecture and culture from this process facilitated my decision to move to the U.S.

Special Tribute to Manfred Hegger

As practicing architect, educator, and scientist, Professor Manfred Hegger had the courage to take paths unknown, and the openness and humility to allow others, students, researchers, colleagues, to join him. Thanks to his collaborative spirit and leadership, we, a team of young people, were able to make an international mark on the future of sustainable architecture and building -integrated renewable energy. Sadly, Manfred Hegger passed away too soon, in 2016. He foresaw many of the challenges (climate change, resource conservation, etc.) we face today, and with his work he influenced policy, research, and architectural practice with the goal of helping to address these challenges progressively and intelligently. While he is sorely missed, his ideas live on in the work of the many colleagues, researchers, and students he informed and continues to inspire.

Source: "SONNIGE ZEITEN – SUNNY TIMES, Solar Decathlon Haus Team Deutschland 2007 - Solar Decathlon House Team Deutschland 2007," Manfred Hegger (Ed.) ISBN 13: 978-3-328766-88-3

You might be wondering how a team excels in this design-and-build competition. By scoring the most points, of course. But how does a team accomplish that? Solar Decathlon includes 10 contests that are scored independently. The primary strategy is to score the most points *in every contest*, but that is easier said than done.

Typically, five contests are juried. Selected panels of accomplished professionals in architecture, engineering, building science, communications, and technology tour the houses at the onsite event. Each panel then discusses how well each team has met specific judging criteria. When the teams begin the planning process, they are well aware of these criteria, so they make decisions about which aspects of their finished house might impress jury members, in order to earn high marks in those contests.

The other five contests are performance based. Those contests measure climate control inside a house, hot water supply, and appliances, as well as energy production, efficiency, and usage. Reliability is paramount. If equipment or systems do not work well, teams lose points. The competition requires that teams use plenty of energy to prove to the public that solar houses can indeed provide all the energy needed for everyday activities in a typical American household, such as to wash and dry laundry; power lights, appliances, and electronic devices; cook and prepare meals; maintain a comfortable temperature indoors; and operate an electric vehicle. Those activities are worth a certain number of points (2 pts, 20, pts, etc.). If a team fails to perform any of the tasks, they lose points.

Weather is an added complication for performance-based contests. For example, on cloudy or rainy days, PV panels produce less power, so teams must conserve energy. If a task is worth only a few points, teams might decide to skip that task on a rainy day, in order to save power for tasks worth more points. That is an example of strategic planning. The term "net-zero energy" means that a building can produce at least as much energy as it needs. For the Energy Balance Contest, teams must show that their houses produce ample power during the competition, completing all required tasks using only solar energy. Results indicate that when teams apply intentional, calculated strategies to ensure appropriate balance of energy production and usage, they come out ahead in that contest.

Santa Clara University took third place overall at SD 2007.

After three consecutive entries in SD competitions, the UMD team was thrilled to win second place overall.

TUD Decathletes from Darmstadt celebrated with their first-place win overall at SD 2007.

Competition outcomes suggest that it is rather difficult for these one-of-a-kind houses to excel in all 10 contests. When innovative designs and new technologies are used, ample time is needed to test these ideas and systems. However, conceptualizing, planning, and constructing unique, state-of-the-art houses within a 2-year period is a formidable undertaking. SD Organizers have observed that exemplary team leadership and excellent project management make a big difference in contest outcomes. "Who's on first?" is often related to "Who's got the most adept leaders?"

SD Organizers remind the teams of an important design principle called KISS: "keep it simple, stupid." That phrase (and related variations) originated in the United States during the 1960s. This concept suggests that most systems work best if they are not overly complicated. But that is a tall order in a competition where innovation is key. SD teams had to learn about critical trade-offs. For example, to allow for clever new elements in their houses, teams might choose to keep other systems at a more basic level. Logical, robust decision-making became an essential cornerstone of Solar Decathlon success.

Decathletes from Georgia Tech exemplified the power of diversity.

Depending on the school, some teams leaned heavily on architectural design, some relied on high-quality engineering, and others demonstrated multidisciplinary strength. When the third Solar Decathlon came along, most teams realized that the real secret sauce is team diversity—the more diverse the better. To master and do well in all 10 contests, teams need strong representation from architecture and engineering, but also from communications, business administration, finance, marketing, public relations, graphic design, information technology, and computer engineering. Decathletes began to appreciate the wisdom of this quote from Stephen Covey, educator, accomplished speaker, and author of the best-selling book titled, *The 7 Habits of Highly Effective People* (1989): "Strength lies in differences, not in similarities."

Remarks from U.S. DOE Secretary Samuel Bodman at Solar Decathlon 2007 captured the essence of mastering such as ambitious challenge: "This is a difficult competition. It requires team effort – a steady one. Victory is not earned through a single endeavor. It comes only after a consistent display of excellence throughout the competition in a variety of disciplines."

Solar Decathlon, with its distinct emphasis on inventive thinking, demonstrated the insightful truth of these words from Michael Eisner, former chairman and CEO of the Walt Disney Company from 1984–2005, "Diversity is a great force towards creativity."

A summary of media outreach from SD 2007 indicated increasing excitement about the competition: 425 television airings, 47 radio spots, 311 online articles, 269 magazine articles, and 5 documentaries, with a total of 649,781,240 impressions (views or audience interaction with content). It was irrefutably evident that Solar Decathlon had surpassed expectations.

Reflections from Thomas Meyers, an architect who has been engaged with building safety for three decades on four continents. He served as U.S. DOE Solar Decathlon Building Official for nine competitions, and also at SD China in 2012 and SD Europe in 2019. He is past Chairman of the ICC International Residential Code committee and serves on the International Building Code Fire Safety Committee.

In 2004, I was given the opportunity to work on a program billed as being ideally suited for my building safety skills. I had graduated 10 years earlier from the University of Colorado Boulder with an architectural degree and spent nearly a decade working for city building departments. I was also a member of the ICC Evaluation Services Committee, which was responsible for reviewing the criteria used to evaluate new and innovative building materials and systems for acceptance by code officials. I had just joined a small consulting firm and enjoyed the new challenges of working on projects from the other side of the building department counter. The Solar Decathlon competition, planned for fall 2005 in Washington, DC, appeared to be very much within my capabilities. Little did I know what I was walking into. I was completely unaware of the eventual impact this unique experience would have on my life and professional career.

My formal role for Solar Decathlon is to evaluate the building code used in a particular SD event. Then, I review the construction drawing packages from participating teams for conformance. During an on-site competition, when houses are displayed in a Solar Village, I perform building code compliance inspections with an emphasis on public safety. My first experience was in 2005, and I continued to work at Solar Decathlon events in the U.S. in 2007, 2009, 2011, 2013, 2015, 2017, and 2020. I also supported SD China in 2012 and SD Europe 2019 in Hungary. As my role evolved somewhat, I found myself serving as a consultant to participating teams. Frequently, that involved assisting in negotiations with local authorities, giving guidance on successful execution of complex design strategies, helping with construction skills or techniques, and offering emergency improvisation as deadlines loomed, so SD houses would be ready for the big public event.

Prior to my involvement in Solar Decathlon competitions, I had only a passing interest in energy efficiency. This changed markedly, and nearly immediately—as I began working with Decathlete teams. As the competition evolved over time, it became apparent that these young students had a greater mission in mind. They knew their work was critical to the survival of future generations as climate change became a more urgent problem. This passed on to me, and I realized that Solar Decathlon engagement had transformed my own perspectives. Some SD performance features, such as advanced air sealing and passive solar strategies, were embedded into a renovated outbuilding that I used as my office in Northern Colorado. Upon selling that property, I decided that our next house needed to outperform anything available on the market. Looking back on multiple SD competitions, I gleaned the best features to incorporate into new construction for my personal use. This new residence, finished in 2020, performs extremely well and should be grid-positive after the PV installation is complete on our soon-to-be constructed "central plant" building that also serves as workshop and support building for our organic farming operations.

Reflecting on the innovations incorporated into Solar Decathlon houses, I'd say that some are 10–20 years in advance of current practice. A few examples are: the inclusion of phase change materials, advanced air sealing techniques, and HVAC capability and versatility, particularly systems that include desiccant moisture reduction strategies. Over time, we have seen some technologies used in SD houses move from bespoke experimental to readily available, off-the-shelf equipment. For me, the most notable example was the heat pump water heater for the 2007 University of Illinois team that was built right on the National Mall for that competition.

A Decathlete of the University of Illinois at Urbana-Champaign showing their home-made heat pump hot water heater at SD 2007; Photo credit: Tom Meyers.

Regarding the impact of Solar Decathlon on the construction industry, I believe the influence is apparent, although not directly obvious. By 2015, it was not unusual to see building science incorporated into designs and to find remarkably energy efficient houses on the market. Energy (heat) recovery ventilators are not uncommon, and thermal envelope efficiency is high. Part of that is driven by the energy code, but advanced codes are only adopted when communities support the concepts and restrictions contained therein. Solar Decathlon helped make that possible, bringing palatable efficiency to the general public through educational competitions on the national and international stage.

An impressive legacy of Solar Decathlon is the influence of this competition on the professional careers of Decathletes. I remain friends with a number of SD alumni, and for many, the ability to successfully execute a challenging project gave them the confidence to excel in their chosen careers. Many Decathletes have formed companies that specialize in construction that is technologically advanced, high-caliber, and energy efficient.

For me, Solar Decathlon has been transformational. Since my involvement, I personally subscribe to the SD mission of energy efficiency and resilience. I teach these concepts in my seminars, and I apply these principles in my own life. Valuable interaction with Decathletes has enriched and added purpose to my professional experiences. Without a doubt, SD engagement has been the most rewarding work in my entire career. It has also added breadth to the safety element of my work. I am no longer only concerned about fires, floods, and structural performance, but I am now a strong proponent of protecting and preserving earth's climate for future generations.

A CU-Boulder Decathlete sharing information about their house to visitors, including author Melissa King at SD 2007.

Solar Decathlon 2009

With the third SD competition behind him, Richard King turned his attention to proposals for the 2009 event. Interest was at an all-time high, with close to 40 universities on the list of wannabe participants. Solar Decathlon had become a well-recognized entity on college campuses across the United States, and students and faculty advisors were eager to get involved. That led to a new challenge: how to ensure fair evaluation of all team proposals and how to determine which teams were most likely to succeed with their SD project. Having a larger playing field to choose from was a "good problem" to have, but one that required multiple perspectives and careful decision-making. To determine which schools should be included in SD 2009, Richard and a select group of Solar Decathlon Managers from NREL reviewed and assessed all the proposals individually, followed by a full day of discussion. Many factors, such as prior SD participation, leadership, geographic location, financial resources, university support, team composition, and proposed design, influenced the selection process. For the 2009 competition, only about half the teams that wanted "in" were chosen to participate.

The final lineup for 2009 included nine returning teams (from SD 2007 and 2005) and 11 new teams (first-time experience). Participating Decathletes represented 15 different states, as well as Puerto Rico, Canada, Spain, and Germany. Here are the 20 collegiate teams selected for Solar Decathlon 2009.

- Cornell University
- Iowa State University
- Pennsylvania State University
- Rice University
- Team Alberta (University of Calgary, SAIT Polytechnic, Alberta College of Art & Design, Mount Royal College)
- Team Boston (Boston Architectural College, Tufts University)
- Team California (Santa Clara University, California College of the Arts)
- Team Germany (Technische Universität Darmstadt)
- Team Missouri (Missouri University of Science & Technology, University of Missouri)

- Team Ontario/BC (University of Waterloo, Ryerson University, Simon Fraser University)
- Team Spain (Universidad Polytecnica de Madrid)
- The Ohio State University
- Universidad de Puerto Rico
- University of Arizona
- University of Illinois at Urbana-Champaign
- University of Kentucky
- University of Louisiana at Lafayette
- University of Minnesota
- University of Wisconsin-Milwaukee
- Virginia Tech

What was going on in the United States at that time? The nation had experienced the Great Recession of 2008, but there was renewed hope for a brighter future. President Barack Obama was in the White House, with his outspoken appetite for a clean energy economy. Nobel Laureate Steven Chu, Director of the Lawrence Berkeley National Laboratory and professor of physics and biology at the University of California, Berkeley, had been named the U.S. Secretary of Energy. Dr. Chu, a fervent believer in renewable energy solutions, was strongly committed to moving away from fossil fuel dependency. In Richard King's mind, this seemed like prime time for the 2009 event. He was more than ready to pull out all the stops to facilitate a Solar Decathlon surge.

Based on valuable feedback from previous competitions, changes were needed to improve the quality of the experience for all involved. Solar Decathlon attracted huge crowds, and teams had to find better ways of managing long lines of visitors waiting for house tours. Industry representatives and consumers expected houses in the Solar Village to be grid-tied, since that was typical for solar-powered residences at the time. In addition, the balance of subjective versus objective contests begged for adjustment, in order to make them approximately equal in point value. After all, these houses had to be attractive and aesthetically pleasing, but they also had to work and work well.

Thanks to the generosity of Pepco, a regulated public utility that operated in the District of Columbia and parts of Maryland, houses in the 2009 Solar Village were net metered for the very first time. Each house in the competition was connected to a microgrid, and Pepco provided the electric meters and electric utility interconnection. Another loyal and generous sponsor, Schneider Electric, global specialist in energy management systems and supporter of sustainable development, played a major role in this effort. Schneider Electric supplied the Solar Village microgrid, plus design and site engineering services and equipment needed for safe and reliable electrical distribution to connect SD houses with Pepco. With the support of these sponsors, the new Net Metering Contest was launched. This contest demonstrated that solar homes can supply all the energy that homeowners need, plus deliver excess electric power back to a utility company. Net metering also eliminated the need for batteries in every house to store energy for use on cloudy days and in the evenings. At the time, batteries were large and bulky because lighter, lithium-ion batteries were not yet commercially available.

In October 2009, the National Mall came alive once again as the fourth Solar Village rose up under sunny skies.

Both of those sustaining sponsors, Pepco and Schneider Electric, delivered terrific public outreach and plenty of enthusiastic volunteers (human resources) to support activities in the Solar Village. In addition, many SD houses featured energy management systems and products from Schneider Electric and the Square D Foundation, such as lighting controls, metering solutions, inverters, building automation and control, and electrical distribution equipment. Schneider Electric also sponsored several of the collegiate teams.

Seen from a Smithsonian Tower in Washington, DC, 20 distinct solar rooftops made a visual announcement of the 2009 Solar Decathlon competition.

Sponsors are definitely the lifeblood of this competition, and their generous support makes a huge difference. Over the years, impressive growth and greater diversity have characterized SD sponsorship. SD sponsors offer sizable monetary donations, valuable resources, and significant in-kind contributions. There were 14 sponsors for SD 2002, 16 sponsors in 2005, and 21 sponsors in 2007. SD 2009 ushered in a remarkable increase, with a whopping set of 35 charitable sponsors. This terrific group included Applied Materials, BP Solar, Alliance for Sustainable Energy, American Institute of Architects (AIA), American Society of Heating, Refrigerating, and Air Conditioning Engineers (ASHRAE), Dell, Dow

Corning, National Education Association (NEA), Meteor Solar LED Lighting, Ox Blue Inc., Perkins & Will, Popular Mechanics, the U.S. Green Building Council, and others. Some sponsors have covered the cost of receptions, meals, water in the Solar Village, and supplies for teams and SD organizers. Some have offered expertise and service as Solar Decathlon Jury Members and Observers.

Support from many dedicated sponsors enriched the quality of the educational experience, boosted public awareness, and elevated the messaging about renewable energy and sustainable development. Being located in Washington, DC, the nation's capital and home base of many associations and non-profit organizations, was a tremendous asset. In short, Solar Decathlon took on the mantle of a "home-grown event" for Washingtonians, who were justifiably proud to host this unique collegiate competition on a biennial basis.

Striving to improve the competition, SD Organizers considered how the 10 contests might be modified. As noted, Net Metering was added as a new contest, asking teams to prove they could produce enough power for daily activities that were utilitarian. They could also earn bonus points for producing surplus energy. Another new performance-based contest was Home Entertainment. This contest was created to show that houses in the Solar Village made enough solar electricity to power activities of a more social nature, such as dining with neighbors or watching movies with friends. Each team hosted Decathletes from other teams for two required social events in their house. One evening included a dinner that was prepared and served to neighboring Decathletes. A second evening included shared conversation and watching a movie together.

For SD 2009, teams could earn a total of 1000 points in the following 10 contests:

Juried

- Architecture—100 points
- Engineering—100 points
- Market Viability—100 points
- Communications—75 points
- Lighting Design—75 points

Monitored

- Comfort Zone—100 points
- Appliances—100 points
- Hot Water—100 points
- Net Metering (new)—150 points
    - Does the house produce enough energy to power all its needs?
    - Does the house produce more energy than it consumes? How much more?
- Home Entertainment (new)—100 points
    - Does the house work well for preparing, serving, and sharing meals?
    - Does the house work well for shared entertainment, such as home movies?

As in every Solar Decathlon, distinguished jury members are an essential part of the assessment package for each participating team. These esteemed leaders played a critical role in the 2009 competition.

- **Architecture:** Kevin Burke, Jonathan Knowles, Sarah Susanka
- **Engineering:** Richard Bourne, David Click, Ted Prythero
- **Market viability:** James Ketter, Joyce Mason, Paul Waszink
- **Communications:** Maureen McNulty, Jaime Van Mourik, Alan Wickstrom
- **Lighting design:** Nancy Clanton, Ron Kurtz, Naomi Miller

Find out more about these leaders at:
https://www.solardecathlon.gov/past/2009/juries.html

After visiting every house and watching team presentations about their projects, jurors met to discuss observations and talk about their impressions. Deliberation was not an easy task, and jury teams clustered together for hours to engage with careful reviews and evaluation. With each successive competition, the houses and the technologies they employed were more sophisticated. Fortunately, jury members brought deep knowledge and extensive experience to the daunting task of evaluating SD teams to choose winners. Special award ceremonies announced final outcomes from each jury panel. Decathletes waited with great anticipation. After each announcement, the ranking of all teams might change, and the "leaderboard" display on the National Mall would show the rearranged order. Often, points awarded to teams were remarkably close. Ten contests that unfolded over a period of time created a fascinating competition!

Nestled between the Smithsonian Museums, U.S. Capitol, and Washington Monument, the innovative Solar Village was a noticeable nod to the future.

An important goal of this collegiate competition was to advance the public's general knowledge about clean energy, "smart" eco-conscious solutions, and energy efficiency. Visitors at the onsite event were eager to tour houses in the Solar Village, but additional learning opportunities were available. The "Anatomy of a House" exhibit showcased innovative building technologies and schematic representation of a grid-tied PV system. The "Get Smart: Take Charge of Your Energy" exhibit aimed to help people understand choices they could make as smart energy consumers. Informative, free workshops were also offered inside the main tent. U.S. DOE staff and SD sponsors presented a wide range of topics geared to educating the public about high-tech consumer products, green energy jobs, how solar works, green building standards, sustainable home design, and more.

The communications team at NREL focused on maximizing public outreach, and Carol Laurie (NREL staff) was a major contributor to this compelling content. Webmaster Amy Vaughn deserves huge credit for the user-friendly SD website and timely social media presence. Anyone who attended SD 2009 will tell you that just being there expanded your awareness of the latest and greatest energy-related technology and design solutions. As an added bonus, the exuberance of Decathletes in sharing their projects was infectious with everyone around them.

Brimming with happiness, Melissa and Richard King were delighted with the fourth successful Solar Decathlon.

The dynamic nature of the Solar Decathlon enabled change over time: some planned, and some serendipitous. Each team in the

2009 competition relished the opportunity to highlight their own distinctive corner of the world. House design strongly embraced regional roots, cultural traditions, and demands of local climate. Although certain design parameters, such as passive solar, may apply universally, others function best when adapted to a particular region, culture, set of meteorological conditions, or geographic site. With a nod to a keen "sense of place," students tuned into their own locale when planning for the 2009 competition. Here is a brief look at a few of the amazing entries.

University of Louisiana at Lafayette

Newcomers to Solar Decathlon, *BeauSoleil* (which means *sunshine* in Cajun French) from the University of Louisiana at Lafayette honored the culture of French-speaking Acadians who originally settled in southern Louisiana. Most team members hailed from that region, and they had survived Hurricane Katrina in 2005. As a result, they decided to build a structure that combined passive solar design and energy-efficient innovations with hurricane-resistant elements such as sliding louvered panels and structural insulated panels (SIPS). Planned in "dogtrot" style, the house had a center breezeway flanked by a kitchen and living room in north–south orientation to capture refreshing breezes in a warm climate. A unique transparent NanaWall® system on a track that rotated 360 degrees could enclose or open this delightful space.

Perhaps the most memorable aspect of the house was the unabashed friendliness of its designers. Every team member personified southern hospitality, and a visit to their house guaranteed a warm welcome. Decathletes also hosted fun evening get-togethers with music and dancing to liven up the Solar Village. To add to the enjoyment, the BeauSoleil Band from Louisiana gave a free concert on the porch of their house during competition week. The public must have felt their congenial spirit, because *BeauSoleil* received the most votes in the People's Choice Contest. The team was ecstatic to earn that coveted mark of distinction: the favorite house of visitors. *BeauSoleil* was also honored with the first-place award in the Market Viability Contest.

BeauSoleil from the University of Louisiana at Lafayette used solar concentrators to focus sunlight on a cone-shaped vessel that heated up to make popcorn.

As kernels popped, the popcorn came shooting down a tube to fill a bowl so that visitors could enjoy a great snack after touring *BeauSoleil*.

BeauSoleil honored the Southern "dogtrot" style with an open courtyard in the center to welcome cooling breezes.

Team Ontario/British Columbia

Moving on to a completely different climate in more northerly latitudes, Team Ontario/BC constructed *North House* to withstand the Canadian climate. Its sleek, contemporary style boasted floor-to-ceiling windows on the south, east, and west sides, plus vertical solar panels to frame the window glass. After all, snowy winters in that region could spell trouble for rooftop PV panels. The north side functioned as a utilitarian "Swiss Army wall" with hooks, cubbies, and other utilitarian elements for storage purposes. Since it was a small house, students maximized every square inch and designed ways to "hide" furniture and other interior elements when not in use. For example, to make more living space available, a retractable bed was stored in the ceiling, but it would be lowered at night for sleeping. Visitors loved that clever feature!

Decathletes assembling *North House* from Team British Colombia/Ontario, one of the two Canadian teams in the 2009 competition.

North House boasted quadruple-glazed, floor-to-ceiling glass windows on three sides to enable passive solar gain, plus retractable shades hidden above and behind the PV panels to block sunlight during the day or retain heat at night.

North House had a flat roof for PV panels and hot water collectors.

Visitors were intrigued with how Decathletes maximized use of space in *North House,* including a bed that was hidden in the ceiling during the daytime and lowered at night for sleeping.

Team member Lauren Barhydt explained that they intended to display "a celebration of the small home." The team was proud of the "living interface" computer system they developed to give homeowners visual feedback about energy use and house performance. They felt that people who understand what is happening with the mechanical systems of a house will make more eco-friendly decisions about its operation. The *North House* team was thrilled to receive several awards: second place in the Communications Contest, third place for both Comfort Zone and Net Metering, fourth place for Hot Water, and fifth place for Lighting Design.

Rice University

ZEROW House from Rice University in Houston, Texas, was an intentional representation of an affordable, practical, energy-efficient row house. Often referred to as a "shotgun" house, the team leaned on this traditional concept to design a dwelling appropriate for low- and middle-income families. They succeeded in meeting the U.S. Housing and Urban Development Guidelines for affordability. The Rice Decathletes built a house that was affordable for 80% of those with median incomes in that region.

ZEROW House had two main sections: wet core and light core. The wet core contained plumbing and mechanical systems, with connections to the kitchen and bathroom. The light core relied on natural lighting and boasted a bright "light cove" outfitted with LED light strips on the walls and ceiling. This ingenious design made the relatively small interior space feel much larger. Committed to doing more with less, their *ZEROW* House minimized energy needs with energy-efficient appliances and lighting fixtures. Plus, these Decathletes planned to give back to the community with an arrangement to move their house to a permanent location in Houston's Third Ward. The Rice Decathletes literally jumped for joy with the announcement of their second-place awards for both Architecture and Market Viability.

Kudos to Rice University Decathletes from Houston for their *ZEROW House*, which was technologically advanced yet affordable for middle-income families. Their clever use of light made the physical space seem much larger than it actually was.

Cornell University

The unusual *Silo House* hailed from Cornell University in upstate New York. This team was determined to craft a unique dwelling. So after being selected for SD 2009, they set up a design contest for their dwelling. They then chose the most unconventional of the 15 submissions. *Silo House* was inspired by the numerous grain silos that appear in the rolling hills landscape of the Finger Lakes Region in New York State. Three individual plug-and-play modules shaped like cylinders (or silos) surrounded a courtyard area. One silo was the bedroom, one was the kitchen, and one was the living room. Module walls facing the shared courtyard comprised a clear NanaWall® system that offered both daylighting and insulation.

Silo House from Cornell University included three cylindrical modules (bedroom, kitchen, and living area) that faced onto a center courtyard.

Cornell has cold, snowy winters, so the exterior silo walls were filled with closed-cell spray foam insulation with an R-30 rating, while the roof and flooring were made of SIPS (structural insulated panels) to ensure that the house stayed warm and cozy. Corrugated steel cladding with reddish hues covered the exterior walls, which gave the house an appropriately rustic look. To reflect its intentional post-agrarian style, students added native grasses all around the *Silo House*. The Cornell team succeeded in its quest for originality, and long lines of curious visitors waited patiently for a narrated tour of the *Silo House*. These Decathletes went above and beyond in that role, and their second-place award in the Communications Contest was well deserved.

The unusual silo design from Cornell University beckoned throngs of visitors eager to get a glimpse of this innovative prototype.

University of Minnesota

Decathletes from the University of Minnesota designed their *Icon Solar House* to function well in a cold climate. They relied on passive solar concepts, a tight building envelope, and spacious areas outside the exterior walls wrapped in insulated, triple-paned glass to capture bright light and the sun's heat. The team set up a bifacial PV system that allowed some light to pass through translucent roof-top panels, which could then be reflected onto the backside of the solar cells to make more power. Radiant floor heating provided warmth, and desiccant material (for dying purposes) in the solar hot water system helped control humidity inside the house.

With a focus on widespread consumer acceptance, *Icon Solar House* aimed to combine a "familiar look" with eco-conscious systems to generate strong market appeal. With a large team of more than 150 students from diverse fields of study, their broad knowledge base gave them a leading edge in the competition. They captured first place in both Lighting and the coveted Engineering Contest, as well as third place in Appliances and Home Entertainment. The Goldy Gophers (mascot) from the University of Minnesota were thrilled with their fifth-place win overall in the competition.

The walls and roof of *Icon Solar House* from the University of Minnesota were designed to combat frigid winter temperatures in that northern climate. Exterior walls were rated at R-50, and the roof at R-70.

The east side of the prototype from the University of Minnesota featured large windows and a covered porch for outdoor enjoyment.

Team Alberta, which represented the University of Calgary, SAIT Polytechnic, Alberta College of Art & Design, and Mount Royal College in Canada, featured structural insulated panels (SIPs) with R-44 insulation and a ground-source heat pump to keep the house comfortable on cold days. The dwelling was assembled with sustainable post-and-beam timber framing.

The *Black and White House* from Spain had a rooftop covered with solar panels that could be tilted and pivoted to track the sun throughout the course of a day.

LumenHAUS from Virginia Tech, which won third place in Architecture, sparkled with exceptional beauty after dark. Photo credit: Virginia Tech.

As luck would have it, Mother Nature was uncooperative for the last couple of days of the competition. Surprisingly, wet weather did not dampen Decathlete spirits or enthusiasm, nor did it deter the public from touring the Solar Village in droves. Hundreds of people showed up with umbrellas and rain boots and stood in long lines to learn about these unique net-zero-energy dwellings.

A rainy day in the Solar Village.

Who Took Top Honors in 2009?

Team California represented Santa Clara University and California College of the Arts. Their gorgeous *Refract House* won third place overall, showing impressive expertise and ranking in the top three positions for almost every Solar Decathlon contest. Their unique bent tube design with three modules (living room, bedroom, kitchen) surrounded a courtyard area and controlled light very effectively. This team impressed SD jury panels and won first place for Architecture and Communications, second place for Engineering, and third place (with a tie) for Market Viability.

Decathlete Allison Kopf on the steps outside the *Refract House* from Team California at SD 2009.

Allison Kopf, entrepreneur and investor, is Chief Growth Officer at IUNU, an Artificial Intelligence company serving the Controlled Environment Agriculture (CEA) industry. Named one of Forbes "30 Under 30" (2019) and "Woman of Impact" for Global Women Fresh (2021), Allison is a sought-after speaker on the future of agriculture and at conferences such as TEDx, Forbes AgTech Summit, and Alltech Ideas, and investment

partner at XFactor Ventures. She was Project Manager for 3rd Place Team California at Solar Decathlon 2009.

It's difficult to describe my feelings when I heard Richard say, "for a team that broke out of the box..." in announcing the third-place winner of the 2009 Solar Decathlon. Instantly, everyone on Team California knew we had placed third in the competition. Our *Refract House* broke the mold of the traditional modular square house; we wrapped the home's angular components around an external courtyard creating a unique living space. "Breaking out of the box" was our unofficial slogan.

I remember falling to my knees and shedding a few tears. I wasn't the only one. The whole tent was filled with Decathletes, parents, sponsors, and supporters who had spent the last two years working harder than we've ever worked before. Exhaustion, excitement, pride, and countless other emotions bubbled up in that moment as Team California took the stage to accept our trophy.

The Solar Decathlon is one of the most impactful experiences of my life. As a 19-year-old student, I had to learn how to manage people, raise money, get building permits from a major city, complete project deliverables, and so much more. After two years of designing and building our fully functional homes, we shipped our houses to the National Mall in Washington, DC to create the Solar Village. Before the on-site excitement could begin, many moving pieces had to come together in just the right order.

Santa Clara University also participated in Solar Decathlon 2007 and placed third in the competition. For that team, getting the house from California to the nation's capital proved to be one of the biggest challenges, and one that didn't come with points. The truck carrying that house broke axels on the trip, so it arrived late. We learned from that experience and made the call to ship our house early. That meant we needed a staging area, and we were lucky to find a suitable place with our Jesuit friends at Gonzaga College High School in Washington, DC.

For us, the anxiety wasn't so much the truck axels; it was the glass. Our house had multiple single-panel glass features that were installed in California and had to be shipped as-is to Washington, DC. When we arrived at Gonzaga, we all crossed our fingers and prayed as we unscrewed the wooden panels affixed to our house. Luckily, everything had stayed in place, and we were set to move the house to the National Mall.

What I remember most about the first night when the houses were arriving was the camaraderie. As we donned our safety equipment in the dark, waiting for our houses to get to the National Mall, we got our first opportunity to meet the teams we'd soon be competing against. Across from our house in the Solar Village was Team Illinois, who later edged us out for second place. To our left was Team Puerto Rico. A little further down the street was Team Germany, who became first-place winners of the 2009 competition. We spent that evening wandering the Solar Village to meet our new neighbors as we prepped to unload our houses.

To this day, some of those competitors are good friends and colleagues. Joe Simon, who led the Illinois team is now at NREL and recently launched the EnergyTech University Prize. I had the pleasure of serving as a judge for that competition and seeing all the incredible work Joe's been doing in this industry. A few years ago, I visited a few of the Decathletes from Team Germany in Germany. Just the other day, one of the growers at a customer site mentioned that he too had participated in the Solar Decathlon when he was in college, a few years after I participated.

Team California beaming with pride after showing *Refract House* to U.S. Secretary of Energy Steven Chu.

Many of my Team California colleagues went on to become founders of successful companies, earn PhDs in fields like artificial intelligence, join top architecture firms, and more. It's been absolutely inspiring to see how far-reaching the Solar Decathlon is. It's a training ground for students interested in sustainability and innovation. Solar Decathlon created a network of students who worked toward a similar goal for years. That experience established a community of thousands of young professionals capable of solving our energy crisis and passionate about doing so. Any time someone reaches out who also participated in the Solar Decathlon, I immediately want to connect.

Solar Decathlon isn't just about the students. It is also an incredible opportunity to work together with sponsors, meet leaders in the energy industry, and speak directly to the public. On opening day of the 2009 event, I had the honor of delivering a welcome address at the opening ceremony alongside Lucas Dixon from Ohio State University. U.S. Secretary of Energy Dr. Steven Chu spoke next. Later, he spent time touring our house and speaking with the team to learn about our incorporated technology and innovation. It was a surreal experience to show our house to the honorable U.S. Secretary of Energy.

During the competition, Congressman Mike Honda invited the Santa Clara team to do a presentation in the U.S. Capitol. Imagine a group of young students coming right off a construction site, ecstatic to present to Members of Congress about our solar house. It was a sight to behold! I was so inspired by the experience that years later I completed an internship in Representative Honda's office.

One of my best memories about the Solar Village was the public tours. It rained during our competition - often. I remember thousands of people coming through our house with muddy shoes and wet umbrellas. But it didn't matter. Every day, we were out on the front lawn and inside the home telling our story to anyone who would listen. And every day, lines would form filled with folks who wanted to hear it. At one point, I lost my voice.

To this day, folks still contact me to ask for details about our house because they want to incorporate something into their own. Just the other day a couple reached out asking about the house plans because they were in love with the *Refract House* and wanted to build a solar-powered tiny home on their land.

Competition sponsors were incredible, too. In the Solar Village, sponsors aren't just sponsors; they're part of the whole experience. During the on-site event, we worked alongside engineers from Schneider Electric, Applied Materials, Pepco, and more as "peers." Many of us went on to work for these companies. I launched a partnership with Schneider Electric with my own company last year. During our first call to discuss how we could think differently about energy management in commercial greenhouses, I pulled out a book written about our

Refract House with a page dedicated to the Square D products we used, which are Schneider Electric products.

This aerial view of *Refract House* emphasizes the unusual design, which proved to be an effective layout for comfortable and convenient daily life.

The public demonstrated keen interest in learning more about *Refract House* with an inside tour from Team California.

By the end of the 2009 competition, I had been in Washington, DC for four weeks. During that month and the preceding two years, I learned more than I can fathom and built some of the strongest and longest-lasting connections I have. Some of my best friends are Decathletes, and some of my mentors and mentees are Decathletes. Father Engh, who was President of Santa Clara University at the time and the first person I ever pitched to for capital for the Solar Decathlon house, was the officiant at my wedding. It's hard not to see how the Solar Decathlon influenced my life and career.

I chose to study Physics at Santa Clara University and wanted to explore a career in sustainability after the Solar Decathlon. I went on to become an early employee at a startup working to revolutionize the way we grow our food. After that, I founded a software company called Artemis. We built the underlying infrastructure needed to scale the indoor agriculture industry. We won the prestigious TechCrunch Disrupt Cup in 2015, raised venture capital, scaled the business and team while growing to become the largest software company in the industry. Our food system and our energy system are deeply entwined, and I've focused my career on building solutions to some of these massive sustainability challenges. In 2021, Artemis was acquired by IUNU, where we're using AI and computer vision to go even deeper into these challenges.

I try to stay involved in the Solar Decathlon and with Santa Clara University to give back to the experience that has shaped me. I had the pleasure of serving the 2017 Solar Decathlon as an Innovation Juror and the opportunity to see the impact of the Solar Decathlon on so many other students. It's an honor to be a part of this community.

Gable House from the University of Illinois at Urbana-Champaign won second place overall. This team mastered the performance-based contests, and their house was an excellent example of energy efficiency and reliability. The design echoed a traditional look to honor its regional midwestern roots, while the systems demonstrated exceptional ingenuity. *Gable House* won first place in Hot Water, Appliances, and Home Entertainment, second place in Lighting, Net Metering, and Comfort Zone, and fifth place for Engineering.

Gable House from the University of Illinois at Urbana-Champaign, made of 100-year-old barn wood, took second place at SD 2009.

surPLUShome from the Technische Universität Darmstadt successfully defended the title of the 2007 TUD champions by winning first place at Solar Decathlon 2009. With an intentional "focus on the façade," this two-story cube dwelling was completely covered with solar cells, producing more than double the amount of power needed. *surPLUShome* took first place in Comfort Zone and Net Metering, second place in Hot Water, third place in Architecture and Lighting, fourth place in Engineering and Home Entertainment, and fifth place in Market Viability and Appliances. The team embraced energy efficiency and sustainability with a distinctive solar aesthetic that proved to be a winning combination.

If numbers tell a true story, SD 2009 was a smashing success. There were more than 307,000 house visits during the 10-day competition and over 923 million media impressions for the fourth Solar Decathlon. Fifty-seven VIP tours were offered to 417 special visitors, and 34 legislators from Capitol Hill made the trek down to the National Mall to see these unique student-built houses. Nationwide, the competition generated more than 730 television stories and 74 radio interviews. An energetic set of 770 enthusiastic volunteers supported the SD organizers to make it all possible. To sum it up in one word, Richard King said, "Wow!"

First-place winner *surPLUShome* from Team Germany attracted huge crowds.

Special Award: Civic Innovators

In the spring of 2010, a few months after the disassembly of the 2009 Solar Village, Richard King got a phone call from Chase Rynd's office at the National Building Museum (NBM) in Washington, DC. He was delightfully surprised to find out that the U.S. DOE Solar Decathlon had been selected as the recipient of a special award for "Civic Innovators." On an annual basis since 1986, the NBM has chosen individuals and projects for distinguished recognition with an Honor Award for exemplary service and accomplishments. In 2010, the NBM recognized the contributions of three groups *"that have enhanced our nation's civic life through the commitment to vibrant, healthy communities"* (2010 Honor Award Program, National Building Museum). The Solar Decathlon, along with Founders of the New Orleans Habitat Musicians' Village (Harry Connick, Jr, Branford Marsalis, Ann Marie Wilkins, Jim Pate) and Perkins+Will (design firm), would be honored at a black-tie gala event in the historic NBM Great Hall on May 11, 2010.

The Great Hall of the National Building Museum was elegantly staged with exquisite floral bouquets, gorgeous table settings, and resplendent lighting. Following a delicious dinner for the honorees and other invited guests, Dr. Kristina Johnson, Under Secretary of Energy, accepted the Honor Award for the U.S. DOE Solar Decathlon. Chase Rynd shared these words: "The Solar Decathlon continues to prove its astounding contribution to improving the Nation's built environment and civic experience. This innovative model to educate the next generation of architects, engineers, and builders with renewable energy, energy-efficient, and environmentally-responsible systems, and the resulting research and development and lasting impacts are extraordinary and clearly deserving of our Civic Innovators Award." After more tributes and speeches, everyone was treated to a fabulous musical performance with Harry Connick, Jr. and members of the New Orleans Habitat Musicians' Village. This was an event to remember! As an added bonus, small-scale model houses from teams selected for upcoming Solar Decathlon 2011 were to be on display at the NBM through the summer of 2010.

The classic sophistication of the Great Hall at the National Building Museum in Washington, DC was a dignified setting for the annual Civic Innovator Awards ceremony in May 2010.

Richard King with Solar Decathlon Organizers Byron Stafford and Wendy Burt at the National Building Museum.

Melissa and Richard King are all smiles at the Civic Innovators Award ceremony.

Reflections from Susan Piedmont-Palladino, an architect, writer, curator, and professor based in Washington, DC and currently the Director of Virginia Tech's Washington-Alexandria Architecture Center (WAAC).

Expanding the Audience

A seasoned curator shared his theory of museum audiences with me when I began my work as a Consulting Curator at the National Building Museum. My first task was to polish and complete the text for an ambitious exhibition, *Big and Green: Toward Sustainable Architecture in the 21st Century*, which opened in January 2003. Museum audiences, he told me, come in three flavors--streakers, strollers, and studiers—and we have to keep them all in mind when we plan exhibitions and write the interpretive texts. I have since added a fourth, the serendipitous, or if we let go of the alliteration obligation, the unintentional visitor. Each poses its own challenges for curators: the streaker speeds through to check the visit off their to-do list; the stroller ambles along looking for what might catch their eye; the studier reads every word, examines each artifact, and may leave a comment questioning or even correcting a source. The serendipitous, unintentional visitor is a bit different. They may have come to see a different exhibition, tagged along with someone else, or stepped inside just to get out of the rain. Catching them is a prize because they didn't expect to learn anything yet may walk away changed.

I was thinking of all of them as I planned the exhibition *U.S. Department of Energy Solar Decathlon 2011 Finalists: A Special Presentation* during the summer of 2010, but I was really hoping to intercept the latter group. That summer, a visitor might have come to see *A Century of Design: The U.S. Commission of Fine Arts, 1910-2010*; or a very different type of visitor might have come in early July to see the wildly popular *LEGO® Architecture: Towering Ambition.* If they paused to step into the brightly painted ground floor gallery, they would have experienced a clever "you are here" moment, along with a reminder that in 2011 "you could be there." The design for this exhibition was the work of the relatively young architecture firm E/L Studio. Elizabeth Emerson and Mark Lawrence—the E and the L respectively—laid out the models of the houses in the same order that they would be on the National Mall and, more importantly, because the gallery had the same orientation as the Mall, the houses were also in their solar positions.

As curator, I had the enjoyable task of interpreting the exhibition's contents and explaining to our many flavors of audiences why this work is so important and exciting. I introduced the 2010 exhibition to the public with these words:

"Ten years ago this spring, the U.S. Department of Energy invited colleges and universities to compete in the inaugural Solar Decathlon, a competition to design, build, and demonstrate a solar-powered house. Since the first Solar Decathlon in the fall of 2002, the program has unleashed the creative power of design and engineering students to rethink the role of energy efficiency—and solar power in particular—in home design and raised public awareness on the topic. The Solar Decathlon challenges student teams to integrate reliable and efficient solar power with excellent design, resourceful engineering, and affordable systems. With Solar Decathlon 2011 still more than a year away, these teams have already competed against their peers for one of the twenty places in the Solar Village on the National Mall—and are hard at work designing their houses. The teams of this fifth Solar Decathlon are the most diverse yet, representing five countries on four continents, and they illustrate the promise of solar energy in a variety of geographic locations and climates. With their different design approaches and target markets, these twenty competitors show that solar-powered living can be for everyone."[1]

The exhibition, however, tucked into two bays on the ground floor of the cavernous Building Museum, was missing the sun. Emerson and Lawrence cleverly used the shallow domes of the gallery to mark the path of the sun as it would be when the Decathlon took place in the fall of 2011. In addition to interpreting the work exhibited, I also had to interpret the exhibition itself. I hoped to turn the visitors' attention to the sky, to the portrayal of the sun, marked above them, and to the arrangement of the model houses:

"What are the dots? The pattern of orange dots on the walls and ceiling represents the path of the sun throughout the year, while the set of gold dots illustrates where the sun will be in fall 2011, when the U.S. Department of Energy Solar Decathlon begins. The pattern on the floor is a clue to the houses' future locations on the National Mall. The gallery space itself represents in miniature the setting of the houses on the Mall and their relationship to the sun."

The take-away message, that "solar-powered living can be for everyone," was finally part of the competition itself that year: for the

---

[1] *U.S. Department of Energy Solar Decathlon 2011 Finalists: A Special Presentation*, National Building Museum, 2010

first time the ten contests included an Affordability Contest. Design innovation needs to be shared to larger and larger audiences, if we expect the status quo to change. The exhibition and the shift in focus to affordable, achievable solar living were twin steps forward in reaching the public at large, especially to people, the serendipitous unintentional visitors who may not already be insiders to the renewable energy space. "Communications" has been a long-standing contest in Solar Decathlons, and I felt like I had entered that particular contest with this exhibition.

That was Then – This is Now, as we begin again in 2021

I think of the 20 years of Decathletes who have taken up these challenges, as the scale of the problem has increased from single 800-square-foot houses to office buildings, mixed-use, and multi-family housing. Those Decathlon alumni are now in professional practice, of one kind or another—teaching, researching, designing, building, inventing—and taking Decathlon lessons into those activities. In the 2002 Rules and Regulations we explained that this competition differed from athletic decathlons in that it would be a team effort, not an individual effort. A more nuanced difference, however, and perhaps a more important one, might be that an athlete inevitably grows weaker in the years after competing, but Solar Decathletes grow stronger, wiser, more skilled, and more powerful.

The first of the ten contests in 2002 was called "Design," and that was not an accident, although nowhere in the original text had I offered what I consider to be the best, most inclusive definition of design. Nobel Laureate Herbert Simon states, "Everyone designs who devises courses of action aimed at changing existing situations into preferred ones." He goes on to explain that "The natural sciences are concerned with *how things are*. Design, on the other hand, is concerned with *how things ought to be*."[2] All Solar Decathlons here in the U.S. and across the globe focus directly on how things ought to be. Everyone who participates is engaged in design action. From where I sit in the dark autumn of 2021 receiving updates on COP26 meetings in Glasgow on a misleadingly weightless communications device that I carry in my pocket (misleadingly weightless because it depends on the massive energy draws of data centers), it is clearer now more than ever how things ought to be. Maybe it's time to add a new contest to welcome additional Decathletes who can bring expertise to a new SD Contest called "Policy-making." Their challenge would be drafting new rules and regulations for the real world.

---

[2]Simon, Herbert, *The Sciences of the Artificial,* Third Edition, 1996, p. 111–167

In the meantime, I have my advance team for U.S. DOE Solar Decathlon Design Challenge 2022: a few highly motivated graduate students, a volunteer consultant or two, and a seasoned teaching assistant with one Design Challenge already behind her. We gathered for a kick-off meeting and are already dreaming of the possible and the necessary, of "how things ought to be." We batted ideas around for an hour. What was the first one? Questioning the question.

Solar Decathlon 2011

After four phenomenal competitions and a noteworthy track record, the Department of Energy hailed Solar Decathlon as a spectacular educational program. Decathletes proved they had what it takes to lead the nation—and the world—in demonstrating how to tread more lightly on the earth's resources. They were primed and ready to ensure a brighter future. On September13, 2010, Richard King was flying high when he received official approval from the National Park Service to hold the fifth Solar Decathlon on the National Mall in October 2011. He shared the exciting news with teams selected for that competition, and they were ecstatic.

Six of the 20 teams chosen for SD 2011 were collaborative ventures among two or more schools, demonstrating that applicants understood the value of joining forces with institutions known for different strengths. The following teams representing 14 states and five nations on four continents were thrilled to be chosen for the next Solar Decathlon:

- Appalachian State University
- Florida International University
- Middlebury College
- Victoria University of Wellington, New Zealand
- The Ohio State University
- Parsons New School of Design and Stevens Institute of Technology
- Purdue University
- The Southern California Institute of Architecture and California Institute of Technology
- Ghent University: Team Belgium

- University of Calgary: Team Canada
- Tongji University: Team China
- University of South Florida, Florida State University, University of Central Florida, University of Florida: Team Florida
- Massachusetts College of Art and Design and the University of Massachusetts at Lowell
- Rutgers, the State University of New Jersey, and the New Jersey Institute of Technology: Team New Jersey
- The City College of New York: Team New York
- Old Dominion University and Hampton University: Tidewater Virginia
- University of Hawaii
- University of Illinois at Urbana-Champaign
- University of Maryland
- University of Tennessee

During the fall of 2010 Decathletes dug in with gusto, brainstorming design ideas, engineering possibilities, and advanced technologies. Teams were determined to make the fifth Solar Decathlon the best yet. Richard King and his group of leaders continued monthly all-team conference calls to be sure everyone had all the information they needed. Eager sponsors signed on to support the 2011 event, and local news outlets featured complimentary stories about hometown teams from their particular neck of the woods. This was a great example of the familiar maxim: "All systems go, full speed ahead."

No one would have predicted the high-stakes drama that would unfold in the coming months.

An Abrupt Change

Suddenly, without warning, in December 2010, permission to set up a Solar Village on the National Mall was revoked. Richard King received notice from DOE supervisors that the Department of the Interior had revoked his permit for Solar Decathlon 2011 to take place on the National Mall. The main reason? Grass! Apparently, the National Park Service had big plans to revitalize 684 acres on that central site in Washington, DC, to the tune of $600 million. Officials were not pleased with the after-effects of this popular event, which aimed to advance public education of green-energy solutions.

Yes, trucks and cranes that delivered equipment and supplies required to build these solar houses disturbed the lawn of "America's Front Yard." Yes, temporary walkways installed for thousands of visitors to make house tours over a 2-week period created bare spots in the grass. However, designated funds were spent for lawn cleanup and re-seeding as soon as the Solar Village was disassembled. In good faith, the SD organizers did their best to restore the site after each competition. Unfortunately, some government officials had different perspectives.

Decathletes across the country were justifiably upset. When they heard the shocking news, many of them were at the International Builders Show in Orlando, FL to display small-scale models of their green homes. Here they were, just nine months from the start of the competition, when the rug was pulled out from under them. Plans were in full swing, and houses were under construction. Any move to a different site would necessitate a backward shift. Houses would require redesign to meet the demands of a new location. Here they were, answering President Obama's call to the nation to kick the fossil fuel habit. Here they were, using alternative energy sources for a spectacular showcase of solar-powered homes. And suddenly, grass was prioritized over their innovative projects.

Students were unwilling to sit back and do nothing. They shared their disappointment with each other and mobilized to take action. They contacted representatives on Capitol Hill, reached out to newspapers, and stressed the significance of keeping the Solar Village on America's main stage in Washington, DC.

Elisabeth Neigert, team leader for the Southern California Institute of Architecture and California Institute of Technology, offered these words in an interview with the *Washington Post* (article by Darryl Fears, February 12, 2011): "It's very frustrating. The Mall is an international platform that calls the world's attention." Neigert noted that if the National Park Service would keep the Solar Decathlon in its original location, she and her fellow Decathletes would help re-sod the area to repair any damage from the event.

In speaking with John McCardle of *The New York Times* (article on January 14, 2011), Addison Godine from the Middlebury College team said, "Throughout, we've all been dreaming about building these solar-powered houses on the Mall. Any other location seems like it will be less."

This unanticipated announcement got plenty of attention on Capitol Hill, where funding for the competition originated. Twelve U.S. Senators, including John Kerry (MA), Barbara Boxer (CA), Patrick Leahy (VT), Kirsten Gillibrand (NY), Robert Menendez (NJ), Dan Inouye (HI), and Bernie Sanders (VT), sent a letter to Energy Secretary Steven Chu and Interior Secretary Ken Salazar on February 2, 2011. They asked him to support Decathletes by allowing Solar Decathlon 2011 to take place on the National Mall. On February 4, 10 U.S. Congressman, led by Edward Markey (MA), Ranking Member on the Committee for Natural Resources, also sent a letter to DOE Secretary Chu and Interior Secretary Salazar. They requested a reversal of the Interior Department's January decision to revoke the permit. These U.S. Senators and members of the House of Representatives believed that this collegiate competition and showcase of solar homes should indeed take place on "America's most popular Front Yard."

Event sponsors stepped up to the plate to express their concern as well. Chip Dence of the National Association of Home Builders sent a letter to President Obama, asking him to be an advocate for the Decathletes. He said that this unexpected turn of events was like "changing the rules in the middle of the game." Dence believed the competition should take place on the National Mall, as originally planned.

In the meantime, Richard King set out to investigate alternative locations for the competition. Not an easy task. The event required a large open area (about 11 acres), mostly flat, with unobstructed views of the sky (to capture sunlight for solar power). The site had to be navigable for vehicles and cranes, so that teams could deliver house parts to their assigned lots. This area would need ample parking and access to electricity and internet infrastructure for a 3-week period. It also had to be in or near a city with easy accessibility for the public.

Knowing that SD houses were designed specifically for Washington's latitude and fall climate, Richard hoped to identify a spot in the nation's capital. He contacted RFK Stadium, but there were planning a 50th anniversary celebration, so they were out. He checked out National Harbor, Maryland, next door to Washington, DC, but that site lacked sufficient space. Richard and his team from NREL investigated 11 possible locations in the DC metro area, but none offered everything the Solar Village needed. He also considered seven sites beyond the city, including the parking lot for Dick's Sporting Center (stadium) in Denver, Colorado, and Sheep's Meadow in New York City's Central Park. None of them were the right fit, and they all had limitations based on physical space, timing, etc. What could be done?

Apparently, the efforts that the Decathletes set in motion to keep the venue in the nation's capital made an impact. With support from news releases and media stories unleashed in the public domain about the Solar Village being kicked off the National Mall, decision-makers recanted their earlier position. On February 23, the U.S. DOE announced that Solar Decathlon 2011 would be allowed to stay in Washington, DC. However, the site chosen for the event was different—it would be West Potomac Park, where baseball fields were sandwiched between the Potomac River and the Franklin Delano Roosevelt Memorial. It was not perfect, but it was an acceptable solution to the problem. Teams let out a collective sigh of relief, realizing that their cooperative endeavor had paid off.

Meanwhile, Richard King and SD Organizers switched gears to begin intense planning for the new location. They had to map out a completely different setup for the Solar Village. They had to notify sponsors about the change, including those charged with assembling the required infrastructure. They had to figure out how to get thousands of visitors over to the riverside site, since Solar Decathlon would no longer straddle the Smithsonian Metro Station. The organizers had some concerns, but they were confident they could make this work for another successful Solar Decathlon.

Newcomer to the 10 Contests

The 10 contests for 2011 were similar to the contests for the 2009 competition, except for two changes, one minor and one major. Lighting was integrated into the Architecture Contest, which was not a big deal. A new Affordability Contest was added to the mix, which altered the ground rules considerably. DOE Secretary Steven Chu, a scientist grounded in reality, wanted Solar Decathlon teams to show the public that building a net-zero-energy home did not have to break the bank. Due to special features and materials, a couple of houses in 2009 had very high price tags, which led some people to believe that sustainable, solar-powered homes were not affordable for the general public.

A professional estimator was hired to develop the estimated cost of construction for each house in the 2011 competition. If a house cost $250,000 or less to build, the team got full points (100) for the Affordability Contest. If a house cost more than $600,000 to build, the team received zero points. For houses with construction costs of $250,000 to $600,000, a sliding scale applied for awarding points. This contest forced teams to "do more with less." In addition, it demonstrated to the public that innovative homes powered with sunlight could have a reasonable price tag.

These are the 10 contests for Solar Decathlon 2011:

Juried

- Architecture—100 points
- Engineering—100 points
- Market Appeal—100 points
- Communications—100 points

Monitored/Calculated

- Affordability (new)—100 points
- Comfort Zone—100 points
- Hot Water—100 points
- Appliances—100 points
- Home Entertainment—100 points
- Energy Balance—100 points
    - o A bidirectional meter on every house determines the net energy produced and consumed during the competition.

A new set of enthusiastic jury members and panelists served this competition to determine how well teams met the criteria for each of the four juried contests and the Affordability Contest. As in earlier events, these renowned professionals came to observe and evaluate, but every one of them left the Solar Village rather amazed at how their own deep knowledge and understanding had been stretched in new directions. The following experts participated in Solar Decathlon 2011.

- **Architecture:** Paul Hutton, Michelle Kaufmann, Robert Schubert
- **Engineering:** A. Hunter Fanney, William Rittelmann, David Springer

- **Market appeal:** Susan Aiello, Brad Beeson, Joyce Mason
- **Communications:** Ryan Dings, Mark Walhimer, Stacy Wilson
- **Affordability:** Matt Hansen, Ric Lacata

Learn more about these experts here:
https://www.solardecathlon.gov/past/2011/juries.html

Another seismic shift had occurred between the 2009 and 2011 competitions. Smartphone sales had skyrocketed, with more than 488 million smartphones sold worldwide in 2011. That year, market penetration of smartphones in the United States was 31%; it rose to 44% the next year. In some areas, mobile devices were on the cusp of becoming standard fare, altering information access with astonishing speed. This development made learning about the Solar Village a whole lot easier—and faster. In a nod to resource conservation, SD Organizers wanted to cut down on the volume of paper used for the competition program distributed at each event. Thus, it was down-sized from 8½ × 11 inches to 4 × 9 inches and included QR codes for all participating teams (brand-new feature). With this technology, visitors could scan those QR codes on the spot to discover more details about Solar Decathlon houses.

The 2011 Solar Village under construction at West Potomac Park in Washington, DC, with the Potomac River in the background.

Aerial view of the 2011 Solar Village as teams assemble their houses to prepare for opening day.

Solar Decathlon 2011 opened to the public in mid-September. The layout at West Potomac Park was quite different, with two main pathways and four streets that crossed between them, giving the public good views from all sides.

As a result of the move to West Potomac Park, the layout for the Solar Village was reconfigured. Instead of one central street lined with solar houses on each side, the 2011 event had four main streets with houses on both sides. This meant that many of the lots had neighboring houses next door, across from them, and behind them. The Decathletes and SD Organizers appreciated this change because it facilitated communications and movement through and around the space. In essence, this proximity gave the Solar Village the look and feel of a tightly-knit community.

September is typically one of the driest months in Washington, DC, so SD Organizers expected plenty of sunshine during the construction period in mid-September 2011. To their dismay, Mother Nature defied those expectations and delivered torrential rainfall the week before trucks and cranes rolled in. Due to the low elevation of West Potomac Park, the soil acted like a drain field for water from higher surrounding areas. The result? Thick mud. Although precipitation had stopped by the time construction began, it took a while for the soggy ground to dry out. Of course, that did not deter the enthusiasm of Decathletes as they built net-zero-energy homes at the new site. After all, they had succeeded in gaining access to a highly visible location in the nation's capital, and that gave them reason to smile!

Once again, Solar Decathlon sponsors stepped up to the plate to donate money, in-kind contributions, publicity, and human resources (volunteers) at the Solar Village. Sustaining sponsors included Dow Corning, Lowe's, M.C. Dean, Pepco, and Schneider Electric. Supporting sponsors and resource partners included ASHRAE, Applied Materials, BP Solar, Bosch, DC-Net, Dow Solar, Meteor, MGN Solar, National Association of Home Builders, Perkins+Will, Popular Mechanics, Powerhouse Solar, Alliance for Sustainable Energy, National Education Association, *Solar Today*, Wells Fargo, and more.

As in every SD event, an impressive number of enthusiastic volunteers turned out to share vibrant smiles, a warm welcome, informative tours of the Solar Village, and behind the scenes help. For example, Schneider Electric encouraged their employees to "show up" for onsite support, and many Federal employees working in the nation's capital were happy to be part of this innovative showcase. In all, 730 volunteers contributed more than 4000 hours to SD 2011. Easily identified by sunny yellow T-shirts, their cheerful presence reminded all that educating the public about clean energy solutions is not only exciting, but also "everyone's business."

Lowe's, a sustaining sponsor of SD 2011, provided a mobile store equipped with basic tools and other hardware where Decathletes could purchase or borrow needed items.

This Lowe's store was a huge benefit for all the teams because it helped conserve precious time and energy. Instead of driving around town to locate materials or equipment, Decathletes had easy access to this mobile unit right in the Solar Village.

Volunteers in yellow T-shirts were an integral part of every Solar Decathlon event, providing friendly greetings and valuable information to visitors. Over 400 volunteers helped make SD 2011 run smoothly.

2011 Team Highlights

The prototype houses at Solar Decathlon 2011 certainly came in many varied shapes and sizes, showing that Decathletes could literally "think outside the box" when designing net-zero-energy dwellings. That particular competition embodied boundless creativity in terms of the physical layout of solar homes. No longer content to squeeze every element into a cube or rectangular shape, some teams arranged the necessary components of their houses into clever configurations that reflected a specific theme, tradition, or cultural heritage. In so doing, the Decathletes focused intentionally on the significance of a dwelling as more than a place to live. A home could be a sanctuary, a place where someone can relax, sleep, rest, fulfill nutritional needs, learn, socialize, grow, and be in harmony with the natural surroundings. Solar Decathlon teams had discovered numerous ways to honor the richness of a home while remembering to tread lightly on the earth's resources.

Team Tidewater

Old Dominion University and Hampton University joined forces to build *Unit 6 Unplugged*. This clever dwelling was designed as part of a multi-family building designed to include six modular units. The team wanted to show that each energy-efficient unit would be more affordable for urban residents if there could be cost-sharing for the infrastructure. With door and window sensors, remote transmitters for light switches, and motorized windows on the large porch that allowed it to be an open or closed space, *Unplugged* was an attractive demonstration of practical, user-friendly smart house features offered at a reasonable cost. This uber-friendly team from Virginia was absolutely thrilled when U.S. Senator Mark Warner stopped by to tour their home that oozed with southern charm.

Old Dominion University and Hampton University teamed up for SD 2011. Here, one modular unit for their house is lowered onto the assigned lot in the Solar Village.

Team Tidewater's *Unit 6 Unplugged* is designed to blend seamlessly into an urban neighborhood. This modular unit would be part of a larger, multi-family building.

Virginia Senator Mark Warner talking to Decathletes during his visit to their *Tidewater House*.

Appalachian State University

The Appalachian State University team embraced concepts of independence and ingenuity with their beautiful *Solar Homestead*. Six outbuildings connected to an outdoor area covered with a trellis of bifacial solar cells for filtered daylighting. To honor their heritage, the team wanted this "Great Porch" to be a reflection of the traditional lifestyle of early settlers in the rural mountain areas of North Carolina.

The two-bedroom *Solar Homestead* offered comfortable living and dining spaces, plus a fully conditioned "Flex Space" for use as a studio, office, or guest room. This house featured an on-demand solar thermal hot water system and phase-change material in a Trombe wall to help condition the indoor temperature. These enterprising "Mountaineer" Decathletes created unique reflective hats that visitors could unfold and wear to display the principle of bifacial PV. This whimsical marketing tactic was highly effective, and it did not take long for everyone to identify *Solar Homesteaders* roaming around the Village. This team showed palpable joy with their second-place award in the Communications Contest and third-place award for Architecture.

This photo of the *Solar Homestead* from Appalachian State during the assembly phase showing the "Great Porch" with bifacial solar panels on the roof that will serve to connect several outbuildings, which reflect the rural mountain lifestyle typical of early settlers in North Carolina.

Handmade wood siding provided distinctive rustic appeal for the *Solar Homestead* from App State.

The *Solar Homestead* included six outbuildings that formed a self-sufficient ensemble connected to the outdoor living space protected by a trellis of bifacial PV cells.

The team from Appalachian State University created cool foldable hats with a message as part of their communications plan to educate children and adults about solar energy. Each hat was labeled with *The Solar Homestead,* plus this statement: "I am part of the solution."

Richard and Melissa King were invited to an evening dinner party at the App State house, along with their neighbors in the Solar Village. These gatherings were part of the Home Entertainment Contest and featured delicious food typical of the host team's region.

Florida International University

The ultra-sleek entry called *perFORM[D]ance House* from Florida International University was planned for empty-nesters, such as a middle-aged Hispanic–American couple who enjoy living in a subtropical climate. Responsive to environmental conditions, the contemporary open-pavilion design featured operable louvers for privacy, as well as shading and protection from high winds, to make it responsive to environmental conditions. A retractable glazing system on windows, natural cross-ventilation, and a reflective white exterior reduced energy demands to keep the dwelling cool. Exterior decks, gardens, and a biofiltration pond surrounding the house created an ideal setting for casual outdoor living.

Expectations Surpassed | 237

The white exterior of the dwelling from FIT was reflective, which helped keep the house cooler on hot, sunny days. The flat roof with expansive overhangs created light airy open spaces on the patio that surrounded the house.

The *perFORM[D]ance* house from Florida International University included wide overhangs with adjustable louvers that could be folded down to cover the large glass windows and doors in case of hurricanes or inclement weather.

Middlebury College

The team from Middlebury College overflowed with spirited enthusiasm, and their steady upward movement during the course of the competition was quite stunning. They went from 15th place on Day 1 to 13th place on Day 2; from 8th place on Day 8 to 5th place on Day 6, and then finished in 4th place overall. Their house, called *Self-Reliance*, was intended as a family friendly dwelling that could endure the chilly New England climate. Triple-paned windows with cork insulation in the frames, a stacked ventilation system, air-to-air heat exchange, and an indoor green wall for growing fresh produce helped make the house "self-reliant." The team selected building materials with consideration for lifecycle costs and minimal environmental impact. No matter when visitors showed up at their house, this team offered a warm welcome, informative tour, and joyful exuberance. They were thrilled to capture first place in several contests: Market Appeal, Communications, and Home Entertainment.

Self-Reliance from Middlebury College was a simple, durable structure with a gable roof to allow snow and rain to slide off easily.

In addition to gardens on their patio, the Middlebury team boasted a green wall in the kitchen for growing vegetables and herbs.

Decathletes from Middlebury College in Vermont with their never-ending smiles quickly earned a reputation as potential winners of a congeniality competition.

Team Canada

The house from the University of Calgary, called *TRTL* for *Technological Residence, Traditional Living*, was a unique structure that drew inspiration from native Americans. Its elongated dome-shaped appearance was a nod to the Treaty 7 Tribes from Alberta, Canada. The south-facing windows and PV system on the roof honored the sun as the source of energy and life. The dwelling's structural insulated panels (SIPs) have high resistance to fire and mold, and other durable materials used for construction were projected to have a life cycle of 75–100 years.

The Native American color palette, artwork, and interior design motifs were prominent features inside the house. After it was assembled in West Potomac Park, *TRTL* received special blessings from Reg Crowshoe, a former Indian Chief, during a memorable traditional ceremony at the Solar Village.

This side view of *TRTL* from the University of Calgary shows the SIP layers that encase the dwelling.

The elongated dome shape was so distinctive that everyone could easily identify *TRTL*.

In a prominent nod to Mother Nature, *Technological Residence, Traditional Living* incorporated colors and materials from the natural world, as well as traditional Native American motifs.

The Canadian population includes more than 600 native groups, and many of those people lack quality housing that is affordable. *TRTL* was a proposed solution to this ongoing challenge.

Reg Crowshoe performed a special native American dance to commemorate the completion of *TRTL* at SD 2011.

The Southern California Institute of Architecture and California Institute of Technology (Cal Tech)

Perhaps the most unusual design at the 2011 Solar Village was the house from Cal Tech called *CHIP*. Its distinct shape and puffy vinyl-coated fabric mesh covering that functioned as "outsulation" attracted many curious visitors. Everyone wanted to know more about this prototype dwelling. Inside the house, several levels created space for various daily tasks, such as sleeping, food preparation, eating, and relaxing. Faced with the high cost of land and housing in Southern California, this team aimed to showcase a cost-effective solution that incorporated innovative technology, such as real-time energy usage and 3D cameras to activate automated light control. They were elated with second-place awards for Engineering and Home Entertainment, a third-place award for Affordability, and a sixth-place finish overall in the competition.

The unusual shape of the CHIP house from Cal Tech made it possible to park a car under the shade of the north side.

The solar system for the *CHIP* house was secured to the water-tight membrane on top of the roof.

This photo shows the custom-made "outsulation" on CHIP that protected the house with a weather-proof barrier on the outside.

Expectations Surpassed | 245

The Sci Arc Team posing on the southern side of CHIP, which could be expanded in that direction to create more living space for the dwelling.

Team China from Tongji built the *Y Container* that recycled six shipping containers for an easy-to-transport-and-assemble solar house. This dwelling addressed a thorny problem in many Chinese cities where independent living is often unaffordable.

Top Three Teams for SD 2011

The top three teams for Solar Decathlon 2011 aced the competition by showing prowess in many of the 10 contests. They invested time and talent wisely to prove that careful, collaborative planning and savvy use of resources (including human brainpower) can yield awesome results.

Victoria University of Wellington

The beautiful house from New Zealand, named *First Light*, represented a "kiwi bach" holiday home that integrated indoor and outdoor living in a gorgeous, light-filled dwelling. This team won first place in Engineering, Hot Water, and Energy Balance (tie); second place in Architecture; third place for Market Appeal; and third place overall for SD 2011.

Fun-loving Decathletes from Victoria University of Wellington enjoyed playing outdoor games in front of their solar house.

First Light from New Zealand captured third place overall at SD 2011. Their attractive wooden structure with sheep's wool insulation offered refreshing open spaces and well-crafted interior rooms.

Purdue University

The *INhome* from Purdue, short for *Indiana Home*, was designed for the typical Midwesterner with a traditional look. Interior finishes were made of recycled material, an air purification system ensured healthy indoor air, and a smartphone could control the centralized system for heat, lighting, energy usage, and door locks. The team took first place for Energy Balance (tie), second place for Affordability and Comfort Zone, third place for the Hot Water Contest, and second place overall for SD 2011.

Purdue Decathletes installing the PV system on the rooftop of their structure.

INhome from Purdue University was designed as an ultra-efficient dwelling with market appeal for a typical midwestern family. This house, which cost $257,854 to build, won second place in the Affordability Contest.

Many visitors appreciated the more traditional exterior of the Purdue house, including its wide porches with railings and a side garage. *INhome* won second place overall at SD 2011.

University of Maryland

No stranger to Solar Decathlon, the University of Maryland (UMD) had competed in previous events on the National Mall. However, the 2011 entry exceeded expectations to shine brightly in every way. UMD's *WaterShed House* drew inspiration from the Chesapeake Bay ecosystem, showing by example how a dwelling can manage stormwater, filter pollutants from greywater, and reduce water consumption. This stunning architectural beauty featured a green roof to slow rainwater runoff, liquid desiccant waterfall to control indoor humidity, and automation system to monitor house functions and indoor conditions, such as temperature and lighting.

WaterShed House focused squarely on sustainability to demonstrate both maximum efficiency with minimal environmental impact. Final standings were testimony to the team's superb achievement: first place in Architecture, Hot Water, and Energy Balance (tie); second place in Market Appeal and Appliances; third place in Communications, Comfort Zone, and Home Entertainment; and fourth place in Engineering. It was no surprise when the final score for the Maryland team was 20 points ahead of the nearest competitor, winning first place overall in SD 2011.

Prefabricated modules for *WaterShed* from the University of Maryland arrive at West Potomac Park.

The green roof and split butterfly roofline of *WaterShed* helped collect storm water runoff.

The breath-taking beauty of *WaterShed* was planned holistically, with an emphasis on water conservation, storm water management, and water recycling. The design of the dwelling aimed to showcase the path of a water drop.

Even at night, the dramatic visual appeal of *WaterShed* was evident to all.

The University of Maryland took first place in the 2011 competition. Ironically, this rainbow appeared overhead the day after that major announcement.

Amy Gardner, Professor at the University of Maryland School of Architecture, Planning and Preservation, Fellow of the American Institute of Architects, and founder of Gardner Architects LLC, has twice led the University's winning entries in the Solar Decathlon. She shares these words about the WaterShed team experience:

Succeeding as a team in the U.S. DOE Solar Decathlon 2011 meant the world to us! We were able to share our vision for a more sustainable future with millions around the world! Throughout the course of the experience of designing and building *WaterShed*, we all learned many things as a team. Here are some I would like to share: 1) Universities can be perfect laboratories for developing innovations in sustainability; 2) There is a real desire for houses like *WaterShed*; and 3) Houses like *WaterShed* have the power to teach. And yes, there's a bright future for houses like *WaterShed*!"

Media outreach and the public's response to SD 2011 were remarkable, with more than 47 million Solar Decathlon website views, a 21% increase over SD 2009. During the 11-day competition, there were 357,000 house visits, and 1500 participants in 51 free workshops offered in the Solar Village. Digital, print, radio, and television impressions totaled over 16 million. Solar Decathlon 2011 made an impact, and people were paying attention.

During the disassembly period at West Potomac Park, Richard King watched with a bit of nostalgia as teams took their houses apart for shipment back to home turf. He knew that the next competition would not return to Washington, DC. The U.S. DOE had issued a call for cities, and Richard was anxiously waiting to see which location might be a suitable spot for the next collection of superstars with their ingenious solar-powered houses. From his earlier search in the winter of 2011, he was well aware that an appropriate place might be rather hard to find, but he was determined to find one!

Reflections from Robert Schubert, Professor Emeritus of Architecture and former Associate Dean for Research in the College of Architecture and Urban Studies at Virginia Tech and creator of the Student-Initiated Research Grants Program at VT to encourage experiential learning. He initiated and was a faculty advisor for Virginia Tech teams that participated in the Solar Decathlon, including their LumenHAUS for SDE 2010 in Madrid and FutureHAUS for SDME 2019 in Dubai. Both houses won first place in those competitions. Schubert and the Virginia Tech LumenHAUS team were chosen as one of the nine recipients of the 2012 American Institute of Architects Honor Award, the only time that a university has received this prestigious award.

Virginia Tech's involvement with the Solar Decathlon Competition started in the year 2000 during a conversation with Jann Holt, Director of the Virginia Tech's Washington Alexandria Architecture Center. A new program sponsored by the U.S. Department of Energy called the Solar Decathlon was in development. Jann suggested that the College of Architecture and Urban studies consider entering the competition. Jann Holt and Henry Hollander were consultants to DOE to help with the organization and planning of the very first Solar Decathlon. Holt, knowing of my interest in renewable energy and building design, thought this was something Virginia Tech might benefit from. At the time, I was Associate Dean for Research in the College of Architecture and Urban Studies and thought this might be an excellent hands-on, experiential learning opportunity for our students. I needed more information, so I attended several workshops conducted by DOE to introduce this innovative collegiate competition.

The idea was simple: a solar-powered house would be fabricated at respective collegiate institutions and brought to the National Mall in Washington, DC, where the houses would compete with one another. Sounds simple enough, but little did we know where this would take us. Now almost 22 years later and with a legacy of four houses and five competitions under our belt, Va Tech's latest iteration, FutureHAUS, is currently on display at the 2020 World Expo in Dubai, UAE under the supervision of faculty member Joseph Wheeler. FutureHAUS, the first-place winner in the 2018 Solar Decathlon Middle East exhibition, was invited by the Dubai Electricity and Water Authority (DEWA), the Sustainability Energy Partner of Expo 2020, to have the house on display at this prestigious international event.

Over the course of 20 years and five competitions, there have been many lessons learned, and I'd like to share some of those observations. I would also like to emphasize that this is clearly a team effort of numerous disciplines, faculty, students, and industry, in which opinions and perspectives may differ. My observations are a product of being a professor of architecture for 43 years, an associate dean for research for 23 years, a judge in the 2011 Solar Decathlon architecture competition, and part of the Virginia Tech Solar Decathlon interdisciplinary team for 22 years.

It's important to note Virginia Tech is a Land Grant University that encourages and supports the transfer of knowledge to the Commonwealth of Virginia and an international stage. This acknowledges the supportive environment in which the Virginia Tech Solar Decathlon grew up. In addition, there is a strong interdisciplinary emphasis to work across disciplines at our university. Since the expertise to pull-off a Solar Decathlon entry did not reside in just one discipline, it made working across disciplines a necessity. This provided a rich educational proving ground to expand the horizons of our students. The nature of competition seems to bring out the best in a team effort. Add the context of an international stage, and this demonstrates what is possible to a broader audience. The national and international setting led students to work harder and push the boundaries of their comfort zone to best represent Virginia Tech and the USA.

With the emphasis on the integration of systems in Solar Decathlon projects, this encouraged students to grow and broaden their perspectives, especially when it comes to understanding the interrelationship between architecture, building science, and engineering systems. It was interesting to observe the influence of interdisciplinary crossover when architecture students and engineering students would enter into each other's worlds to participate in decision-making and fabrication. Architects started to think more like Engineers, and Engineers started to think more like Architects. That's a positive development!

The nature of a Solar Decathlon Competition involves objective and subjective evaluation. For many students, this was the first time they came in contact with the consequences of their design decisions. There exist multiple metrics to reflect success and failure of a design, but for many, this was the first time that design intention reflected on paper was experienced in reality. This unique hands-on learning opportunity opened up a new level of understanding for students. Natural student leaders emerged from this melting pot of activity. If you were to consider the number of students involved in Solar Decathlon-related curricula across the university, you would be looking at close to a hundred. Out of those hundred, a core team of a dozen student leaders would emerge naturally. It was extremely rewarding to see this self-identified student leadership team evolve over time, from the early vestiges of design and fabrication, to the actual competition that took place on the National Mall and abroad.

Lessons Learned: Legacy of Four Solar Decathlon Projects from Virginia Tech

On reflection of Virginia Tech's 22-year engagement with the Solar Decathlon, there were numerous "lessons learned." I think these six are worth sharing.

Simple is Better, but Not Always Easy

According to jazz musician Wynton Marsalis, "Genius always manifests itself through attention to fine detail. Works of great genius sound so natural they appear simple, but this is the simplicity of elimination, not the simplicity of ignorance." If one were to look at the four Solar Decathlon houses from Virginia Tech, you would see an evolution in the progression and refinement of the solution. At the risk of a broad generalization, it could best be characterized as "refined simplicity." If you were to compare the 2002 solution with the 2009 LumenHAUS solution, you would see a significant reduction in the "parts count."

You would also see a transportation strategy much simpler in approach, allowing the house to be transported as a single unit, instead of multiple modules to be assembled and disassembled before transportation could take place, as in the 2002 house. This simplicity or refinement came about from a team of faculty and students focused on the eloquence of the solution through simplicity of elimination. The opportunity to do four houses and have continuity of faculty and student input has allowed this refinement to take place.

Don't Underestimate the Complexity of the Transportation Strategy

I would say by far the most complex part of a Solar Decathlon project involves the transportation strategy. Transportation touches on many aspects and influences such tangibles as: the size and weight of the delivered unit(s), the overall finished ceiling height, the amount of time required for reassembly/disassembly, the ability to display the unit in different site conditions, and whether a crane or other heavy equipment is required for final assembly. Our 2009 Solar Decathlon entry, LumenHAUS, used a transport system inspired by the double-drop lowboy trailer. Made up of removable and reusable components (axle/wheel assembly on the back and hitch hookup on the front), this allows for simple truck transport, flexible suspension, and higher height clearances than conventional modular transport methods. The transport strategy allowed us to place LumenHAUS in different public display venues, such as the National Mall in Washington, DC, Times Square in New York City, the European Solar Decathlon in Madrid, Spain, the Millennium Park in Chicago, Illinois, the Farnsworth House in Plano, Illinois, and several key locations on the Virginia Tech campus and in the Blacksburg, Virginia community.

Standardization is Key

Standardization had a pronounced influence on the design of the environmental conditioning strategy employed, starting with our 2005 house and refined in the 2009 LumenHAUS. Convention may suggest discrete solar systems to be used to condition interior space and heat domestic hot water. The most common application is to use solar thermal systems for hot water heating and space conditioning, with photovoltaics used to generate solar electricity.

Our strategy was to first use passive conditioning strategies, and when thermal comfort conditions could no longer be maintained, solar generated electricity would be used to power a geothermal heat pump system to provide both domestic hot water and space heating/cooling.

Standardization allowed us to simplify the system and use one type of controller, which also provided the ability to recover waste heat from the process of cooling interior spaces and reutilizing waste heat to provide domestic hot water. We did not need to provide solar access for a solar thermal array; instead, we were able to use prime solar exposure area and occupy it with solar photovoltaic panels to provide us with additional power generating capacity. This could be fed back to the utility grid, where it contributed to our overall power generation and energy credit.

Bridging the Cultural Divide

While the School of Architecture and Design played the major role of coordinating the three Solar Decathlon projects, there is no single discipline responsible. All the SD projects were generated by an interdisciplinary team of architects and engineers. In those respective disciplines, there was the School of Architecture and Design, including programs in Industrial Design, Interior Design, and Landscape Architecture; the Department of Building Construction; the College of Engineering, including the Departments of Mechanical Engineering, Civil Engineering, Computer Science, Electrical and Computer Engineering; and the College of Business. Those disciplines share the common goal of creating a Solar Decathlon project, but each has their own approach and value system. Therefore, the project is richer because of that diversity. For a management system to be successful, it must have flexibility and acknowledge that diversity.

It's important to recognize the level of commitment afforded by the academic reward system of each respective college and work within those parameters. An academic culture based on a studio model, such as Architecture, has a different way of working and required time commitment, when compared to Engineering, which is more classroom-oriented. It would be easier if each shared a common pedagogical structure, but working within the structural differences allowed the cultural diversity to flourish, which provides the true strength of the team.

Strategies to Create Milestones

One of the most difficult elements to master in a Solar Decathlon project is time management. A great investment is choosing a person dedicated to time management and responsible for time management software. In addition to this rather obvious resource, are built-in milestones that focus the team's energy and provide the opportunity to accelerate progress. One such strategy used for the 2005 and 2009 houses was to place the project into a public context when it was three-quarters of the way through completion. As soon as the building was far enough along to allow transportation, it was moved to the parking lot of a Lowe's Home Improvement Center near Christiansburg, Virginia for the 2005 project and to a shopping mall in Blacksburg, Virginia for the 2009 project.

This allowed us to test the transportation strategy well in advance of moving the houses to the competition site in Washington, DC. It also accelerated progress, to make it possible to move these houses. They became public exhibitions where the community could observe and visit, providing the students with opportunities to cultivate their skills providing explanations and public tours. That capacity is one of the cornerstones of the Solar Decathlon competition on the National Mall, the ability of the student team to deliver an effective explanation, whether it is to the public or to the various assessment juries connected to the competition.

Cultivate Your Resources

None of our Solar Decathlon projects could have been done without the help and support of the industries that participated. This is probably the most important resource, offering both in-kind material support and a knowledge base that can't be found elsewhere. After 12 years of working with industries connected to Solar Decathlon projects, it finally dawned on me what one of their primary motivations for industry participation was. I originally thought it had solely to do with product placement, and while that is important, I later realized it was getting access to some of the best and brightest students the university has to offer. Students involved with Solar Decathlon projects are typically creative, smart, self-motivated, and ambitious; all qualities which you would like to have in a future employee.

At Virginia Tech, relationships forged with industry through Solar Decathlon projects continue to evolve, providing a rich and varied resource that helps shape and influence a research agenda in the School of Architecture and Design.

Consider the End Use of the Project, Well Beyond the Competition

One aspect that can benefit the project is having long-range vision of the end-use of the solar house. This requires thinking about the project well beyond the Solar Decathlon competition. Investments can be made early on in development which may not be competition specific. For example, sensors and instrumentation that support research plans after the competition are better addressed and installed when building assemblies are accessible during construction. Investment in the transportation strategy allowing rapid deployment and set-up can facilitate taking the project to the public in different venues.

As a land-grant university, Virginia Tech has a commitment to support outreach activities for its primary audience, the constituents of the Commonwealth of Virginia, as well as national and international audiences. Transportation methods and module dimensions that can accommodate transporting over the interstate highway system and secondary roads can make the difference for which audiences the project can be shared with. Investment in the chassis system and the flexibility it has provided (see "Don't Underestimate the Complexity of the Transportation Strategy") has been paid-back many times over and will continue to support future outreach activities.

I cannot over-emphasize the importance of adequate documentation. The deliverables required by the Solar Decathlon competition provided the foundation for essential documentation. The as-built documentation and the operation and maintenance (O&M) manual become primary tools to accompany the project into the future. With a constantly changing team of faculty and students, proper documentation is paramount for the continued successful operation and maintenance of the project.

Solar Decathlon Virginia Tech: In Conclusion

These lessons learned reflect but a small sampling to highlight the more important influences I have observed over a 20-year time period. These strategies are mainly focused on project development. If focused more on the actual competition, a whole new list could be generated. And, if one were to poll any faculty members who have participated in Solar Decathlon projects, both at Virginia Tech and other schools, you might get a completely different list.

I am indebted to the U.S. Department of Energy and Organizers of the Solar Decathlon, in particular, Richard King, for providing this tremendous opportunity for an intercollegiate competition highlighting the progressive use of the sun's energy to mold and shape our future houses. I speculate that no one would ever have estimated the profound influence this intercollegiate competition would have in molding curriculum and influencing two decades of young professionals, while also sharing with the public such fresh and vital approaches in residential construction that embrace the use of solar power.

Chapter 6

Europeans Take Up the Charge

During the first decade of the 21st century, Solar Decathlon overcame many obstacles to attain a strong reputation as an outstanding educational program in the United States. For schools of architecture and engineering, participating in a Solar Decathlon was a coveted opportunity for college students to gain hands-on experience with the challenges of innovative residential design and construction. Motivation was heightened with the prospect of winning a national competition that rewarded ingenuity and creativity.

Solar Decathlon Europe 2010

While all this was happening in North America, government leaders, university faculty, and building professionals from across the Atlantic Ocean became increasingly interested in this unique program. International participation in the Solar Decathlon gained plenty of media attention, and the idea of expanding the competition in other parts of the world took hold. Thanks to the enthusiasm and unwavering commitment of several key individuals, Spain took the lead in spearheading efforts to create a similar event in their own country. Tremendous credit goes to Javier Serra Maria-Tomé and

---

Solar Decathlon: Building a Renewable Future
Melissa DiGennaro King and Richard James King
Copyright © 2024 Jenny Stanford Publishing Pte. Ltd.
ISBN 978-981-5129-47-2 (Paperback), 978-981-5129-13-7 (Hardcover), 978-1-003-47759-4 (eBook)
www.jennystanford.com

Sergio Vega Sanchez for initiating this expansion, which became a real tipping point for the Solar Decathlon. Find out more about their involvement in the interviews that follow.

Javier Serra is a Spanish architect and retired official of the Spanish Government with more than 40 years of experience in the field of building regulations, solar energy, and building quality in different departments of the Spanish government. Javier led the organization of Solar Decathlon Europe (SDE) and served as director of the competition for its first two editions in Madrid, Spain: 2010 and 2012.

Tell us about your career and how it led to your engagement with solar energy.

I had been working for the Government since 1972 as an architect and was approaching the end of my career in 2005. I had developed experience drafting building regulations, managing building quality, incorporating sustainability, etc. and was enthusiastic about solar energy. My active participation in many national and international events (European Union, United Nations...) and programs regarding passive solar building, energy efficiency, and solar energy increased my fondness for solar energy. With this background, I got involved in the promotion of solar and passive energy building in Spain.

At the beginning of my career in 1979, I took responsibility for drafting the first thermal regulations on new buildings. Years later, at the beginning of 2000, I started to prepare the new Building Code. I became Deputy Director for Innovation and Quality in Building at the brand-new Ministry of Housing, created in 2004 by the new government after general elections in Spain. By then, we were busy in the preparation of the future building Code, aligned with a Performance-based approach, which was developed in many advanced countries.

When did you first hear about the Solar Decathlon?

I first heard about the Solar Decathlon in Fall 2005 from friends who were teaching at the School of Architecture at Madrid Polytechnic University (UPM). They were part of the student team that participated in the 2005 competition in Washington, DC with the *Magic Box* house. They were the only team from Europe, and their participation had media echoes in the press, television, and radio.

In Spring 2005 the Building Code, which contained compulsory use of solar energy in buildings for the first-time, was approved by the

Council of Ministers. Flat solar panels to heat domestic hot water were required in all buildings, and I was a personally committed to that requirement. Then I had my first contact with a new UPM Team that had been selected to participate in Solar Decathlon 2007. I fully endorsed this university endeavor, and the Housing Ministry granted funding for *Casa Solar*, their two-year project. I believed this was a great way to promote innovation and energy efficiency. The plan was for the team to build two prototypes: 1) one house was designed for easy transport, so it could be exhibited at different places in Spain; it was intended to be a future laboratory for experimentation at the university, and 2) another house was designed for the Solar Decathlon competition in Washington, DC.

How did Spain become interested in hosting a Solar Decathlon?

The *Casa Solar* project was progressing, and the transportable prototype was shown in various exhibitions in Madrid. The most important one, promoted by the new Spanish government, was ESPAÑA SOLAR. This event aimed to show the potential of renewable energies in our country. The Spanish president, ministers, and many other authorities saw the *Casa Solar* display, which stirred up great interest in the endless potential of using solar energy in buildings. By July 2007 a new government was in place, and a new Minister of Housing was nominated. I had to inform them of our current projects.

In August 2007 Professor Sergio Vega, faculty advisor for the UPM team, contacted me. Professor Vega and the UPM Team were at the NAHB Research Center in Maryland to assemble *Casa Solar*, their solar house for the Solar Decathlon on the National Mall. Richard King and his staff visited them, and UPM told U.S. DOE officials they were interested in bringing the Solar Decathlon to Europe. Professor Vega discussed his preliminary communications with staff at the Spanish Embassy and UPM authorities to explore the possibility of a future SD competition in Europe. All of them were in favor of this idea.

Consequently, in the middle of our summer holidays, I had quick conversations and written correspondence with my contacts in the Ministries of Energy and Environment and staff at the Presidency of Government to get their opinions. Happily, all of them gave me the green light for "Solar Decathlon Europe" on one condition: We, in the Ministry of Housing, should take the lead in organizing this challenge. So, the UPM team had the initial idea, previously 'cooked' with our Embassy and DOE, and my role was to make it true. All this happened in barely six weeks, but that was just the beginning.

U.S. DOE Assistant Secretary Alexander Karsner and Spain's Under Secretary of Housing Fernando Magro signing the Memorandum of Understanding for Solar Decathlon Europe, with the first competition scheduled for 2010 in Madrid.

The next step was drafting a Memorandum of Understanding (MOU) between Spain and the United States. We had to overcome the legal and administrative burdens required. Finally, and happily, the official signing took place on September 18, 2007 on the National Mall in Washington, DC, right next to UPM's *Casa Solar*. I was over the top with happiness watching this ceremony. Solar Decathlon Europe would become a reality.

What were your thoughts when you visited the Solar Decathlon in Washington, DC?

Before taking the lead role of this SD export to Spain and Europe, I studied the information available on former SD competitions. The SD website helped me a lot and gave me a good idea of its scope. Nevertheless, that 'virtual visit' to SD was nothing like my first real visit to the National Mall. It was fantastic and exciting to see the Solar Village just before its opening. There were plenty of cranes, trucks, and busy Decathletes delivering materials and getting ready for the event, and I was able to have close contact with SD Organizers. During my stay, I observed the competition and SD award ceremonies, which was very exciting. I enjoyed the spirit and good atmosphere among Decathletes who were working hard on their imaginative ideas and giving enthusiastic explanations to thousands of visitors. No doubt, that made me an eager Solar Decathlon fanatic!

How did you and your team plan for Solar Decathlon Europe in 2010?

By April 2008 we had a new government, and Ms. Beatriz Corredor was appointed as Minister of Housing. She immediately fell in love with the concept of SDE and became our first major supporter. That circumstance helped me a lot to secure sufficient funding for the competition in Madrid. To go forward, we prepared a multi-annual contract with UPM for the following fiscal years; the Ministry of Housing would provide the governance, steering, and funding, and UPM would provide technical expertise. They had helpful experience from participating in the Solar Decathlon in the U.S. I think this is similar to the way that U.S. DOE worked with the National Laboratory of Renewable Energy (NREL) for U.S. SD competitions.

In 2009 the UPM Organizing Team went to Washington to work closely with U.S. SD Organizers to prepare for the 2010 event in Madrid. Minister Corredor participated in the opening ceremony for the 2009 competition on the National Mall and then enjoyed visiting the Solar Decathlon houses. U.S. DOE Secretary Dr. Stephen Chu honored the Spanish *Black and White House* with a visit and tour. There was excellent media coverage about this in Spain.

Javier Serra and Richard King became good friends during their collaboration on Solar Decathlon events.

Later that year we found a suitable place to hold SDE 2010 in Madrid. An agreement was made with Madrid Municipality, and the "Villa Solar" would be located in 'Madrid Río,' a new linear urban pedestrian and green area gained when the city buried busy highways in tunnels underground. That site was wonderful! It was beside the Casa de Campo, Campo del Moro, and Royal Palace, with lovely views and easy access by public transportation, buses, metro, and train.

What was the most challenging aspect of hosting the first SDE?

The task of hosting SDE was not easy. I had to overcome many difficulties, but the most notable were getting funding, engaging with public and private stakeholders who could contribute financially, and persuading Madrid officials to host the competition twice (2010 and 2012). Also, an unfortunate and unexpected international situation occurred in 2008: the subprime affair that provoked a global economic crisis. This lasted until 2014 in Spain, and that made it difficult to secure funding for SDE 2012.

What worked well for the first SDE in 2010?

The Spanish Organizing Team did a wonderful job planning the inaugural Solar Decathlon Europe in June 2010, and the event was a complete success. Every problem was resolved quickly, and cooperation with the private sector worked well. The Villa Solar was privileged to have visits from majors, ambassadors from participating countries, our heir to the throne Prince Felipe of Asturias, and thousands of spectators. The impact was enormous: 270,000 visitors to the SDE website, more than 100,000 on-site visitors that represented 157 countries during the competition (June 7–27). In addition, UPM organized other related EU projects that aimed to create social awareness in many European countries. The firm support from Minister Corredor and her staff for the first SDE event was essential for its success.

By October 2010 we had good and bad news from our government. The bad news: the Ministry of Housing as such disappeared; the good news: Ms. Corredor would continue in the government as Secretary of State for Housing and Urban Planning. Therefore, she would continue supporting the next SDE edition in 2012.

Javier Serra on location during the assembly phase of Villa Solar 2010 in Madrid.

Who and what was helpful for the inaugural Solar Decathlon Europe competition?

The strong relationship between SDE staff and UPM team members was paramount. Despite their youth, UPM students performed as real professionals, providing skills and solutions for every problem that arose. They had learned a lot as Decathletes in SD 2007 and as organizers-in-training in SD 2009.

The leadership of Sergio Vega, Professor in the School of Architecture at UPM, was fundamental because he had advanced understanding of construction and project management. Finally, the unwavering support and expertise of the U.S. Solar Decathlon staff and Director Richard King was crucial.

What are your best memories from the two SDE competitions in Madrid?

My memories from both competitions are endless. The close contact with team members was rewarding. I was so happy seeing the wonderful response of multitudes of visitors coming not only from Madrid, but from all over Spain and other countries. I loved the award ceremonies. The media coverage was also fabulous.

Prince Felipe VI, currently the King of Spain (in center), visits Villa Solar 2010 in Madrid with Beatriz Corredor (next to the Prince), Richard King (beside Ms. Corredor), Javier Serra (far right), Pascal Rollet (front row, right end), Sergio Vega (back row with sunglasses), Joe Wheeler (left of Pascal), and others.

One memory I hold dear is very special: the SDE visit of our King Felipe VI, heir to the throne at that time. I felt that he was close and comfortable chatting with Decathletes when he toured the houses. I was also very happy to see the U.S. Solar Decathlon Director Richard King and his wife Melissa among us all the time, enjoying a SD competition in Europe for the first time.

The special prize SDE received as the Best European Project for Communications from the European Commission during EU Sustainable Week in 2011 was fantastic. Mr. Ottinger, the Energy Commissioner, presented this award in a ceremony in Brussels. Without question, the success of SDE 2010 compensated for all the difficulties and disappointments we had, particularly the ones we faced in trying to keep and celebrate the second edition, SDE 2012.

Share your thoughts about the second event in Madrid, SDE 2012.

The SDE 2012 call for participation was launched early in 2011. Unfortunately, the 'Madrid Rio' site could not be the location for the second Villa Solar, but we secured another suitable place nearby in the 'Puerta del Angel' site at the Casa de Campo Park. It was very convenient, well connected, and had beautiful views. By the end of November that year we had new general elections in Spain.

The election meant a new government from a different party with new ministers and new staff in all ministries. That circumstance, together with the economic crisis, meant budget cuts. Therefore, SDE 2012, which was in an advanced stage of development, was in jeopardy. We managed to save it by reducing by 50% the subsidy granted to participating teams. Thankfully, this compromise solution allowed us to hold the competition in September. Like its predecessor, SDE 2012 was a huge success by every measure. Personally, it was a big relief for me.

In your opinion, how has SDE been beneficial for Spain and Europe?

Almost everyone knows that Spain is a blessed country regarding solar radiation, the best by far in Western Europe. Up to now, many may have thought that photovoltaic energy could be captured only in solar farms on the ground in the countryside (very common in 2010). Through public education via SDE, many people realized the endless potential of this energy, which is ready to power homes and buildings "freely." This was achieved by showcasing the net-zero-energy solar houses at the Villa Solar.

Obviously, other European countries who participated in SDE realized the powerful educational awareness experience of this event. At the same time, it was a good opportunity to experiment and show the endless talent, ingenuity, and skills of students of architecture and engineering from all over Europe and the world. This applies to industry as well. SDE was beneficial for all.

What other thoughts would you like to share about your involvement with Solar Decathlon?

This is a personal feeling. I was a 61-year-old veteran when SDE came into my professional life in 2007. I had the right to enjoy a decent retirement, earned after more than 35 years serving as a government official. So, I no longer needed to add another 'medal' to my career. Nevertheless, the idea of organizing SDE after having visited it captivated me so much that I decided without hesitation to take this step, ignoring all the risks and uncertainties it might involve.

This decision was consistent with my background as an official dedicated for years to the improvement of building quality and the promotion of energy efficiency and sustainability in construction. It was such a new and exciting challenge that I felt I simply could not miss it.

On the other hand, the decision to lead SDE in August 2008 helped me later to overcome an unexpected personal circumstance, the sudden loss of my wife at the beginning of that year, just when we started to organize SDE 2010. The contact with young people on the UPM organizing team, the frantic work involved, the various trips to Washington with my minister, with SDE Team, the many conferences we organized to disseminate news of the event, all this helped me overcome this personal situation. SDE helped me to finally feel finally happy and with a sense of duty fulfilled.

One of the best fruits I have received as a reward for my effort as SDE Institutional Director has been gaining many friends. I am especially proud to enjoy the friendship of the 'true kings' of the Solar Decathlon, Richard and Melissa King, whose infinite support and close friendship have helped me so much along these years. Without them I doubt that I would have agreed to undertake and succeed in this wonderful adventure.

Javier Serra with Melissa and Richard King, who treasure a close friendship.

SDE 2010, the first Solar Decathlon held outside the United States, was a significant milestone. Spain's initiative to host this inaugural event in Madrid made it possible for Richard King's vision to take flight on an international stage that would eventually make an impact on five continents and the Middle East. SDE 2010 demonstrated the transcendent power of this university competition to inspire future architects, engineers, and builders across the globe to question "what is" and pursue "what can be" for sustainable living solutions.

In Richard King's words, "The beauty of expanding the Solar Decathlon on an international level is that it can motivate more students to dream outside the box in addressing today's challenges. We inhabit one planet, and we all face the same threats posed by climate change. By working together around the world, we multiply our potential to make a difference and optimize our capacity to succeed. By encouraging both innovation and public education, we magnify the probability that people everywhere will begin to embrace essential, widespread change, step-by-step. The greater your understanding, the more willing you may be to accept new ways of doing and living."

A view of SDE 2010 Villa Solar along the Manzanares River in downtown Madrid, Spain (looking north at the Puenta del Rey), in the shadow of the Almudena Cathedral and the Palacio Real. This gateway to the city of Madrid was a well-connected location with easy access to transportation systems (subway, buses, and trains).

Villa Solar 2010 took place in a beautiful, historic, pedestrian-friendly setting in central Madrid by the river (looking south at the Puenta Segovia).

Thanks to the due diligence of those who were passionate about the Solar Decathlon, SDE 2010 became a reality. The competition included 17 teams that represented seven countries on four continents.

China
- Tianjin University
- Tongji University

Finland
- Helsinki University of Technology

France
- Arts et Metiers Paris Tech
- Ecole Nationale Superieure d'Architecture de Grenoble

Germany
- Bergische Universitat Wuppertal
- Stuttgart University of Applied Sciences
- University of Applied Sciences Rosenheim
- Fachhochschule fur Technik and Wirtschaft Berlin

Spain
- Universidad de Valladolid

- Universidad CEU Cardenal Herrera
- Universidad de Sevilla
- Universidad Politecnica de Cataluna
- Instituto de Arquitectura Avanzada de Cataluna

United Kingdom
- University of Nottingham

United States
- University of Florida
- Virginia Polytechnic Institute and State University (Virginia Tech)

The inaugural SDE competition took place on June 18–27, 2010, beneath bright, unwavering, sunny skies in Madrid, Spain. During this 10-day period of delightful weather, over 100,000 visitors of all ages came to experience Villa Solar. The high-level of enthusiasm, conviction, and passion for this endeavor among the Decathletes in Madrid matched what had transpired at SD events in the United States. In addition, that same eager curiosity observed at the U.S. competitions was evident among the public who came to tour and learn about these innovative dwellings focused on eco-conscious living at Villa Solar.

Over 100,000 visitors of all ages attended Villa Solar 2010 in Madrid, Spain.

Refocus from the University of Florida, one of the two teams from the United States at SDE 2010, featured a pleasant breezeway that separated the private and public areas of the dwelling and also provided excellent cross-ventilation and natural light.

Visitors lining up to tour *Refocus*, from the University of Florida. The house had operable shading devices that doubled as protection for glass windows and doors during hurricanes.

Fab Lab House, from the Institute of Architecture Avanzada de Cataluña with shaded patio under the dwelling.

An interior view of *Fab Lab House*, where the supporting structure, flooring, shelving, and rooms were computer designed and machine cut to fit together perfectly.

The solar system for *Fab Lab House* was made of flexible PV panels to follow the roof curve.

Napevomo, which means "Do you feel well?" came from Decathletes studying at Arts and Métiers Paris Tech, France. This house had a parabolic concentrating collector on the roof to focus sunlight onto the photovoltaic cells.

The sun's radiation was concentrated so intensely on the solar cells that they looked white-hot. This design generated enough electricity and hot water (from tubes behind the cells) for the entire *Napevomo* house.

SML House from the Universidad CEU Cardenal Herrera, Spain, had an amazing roof system of solar collectors. In the back section of the roof, they could switch from solar hot water to electricity by sliding PV panels over the hot water collectors, or by sliding PV panels back to expose the solar thermal collectors to the sun to produce hot water. Depending on the need, they could adjust the amount of electricity and hot water their roof system generated.

Luukku from Helsinki University of Technology, Finland, reflected the design of a Finnish summerhouse built primarily of wood, which serves as both a thermal and moisture barrier for the dwelling. This house featured thick insulation to seal the house from extremely cold outdoor temperatures in that northern climate.

Armadillo Box from l'Ecole Nationale Supérieure d'Architecture in Grenoble, France, included a central core (kitchen, bathroom, utilities), skin (frame with tight insulation), and outer shell (to protect the house from the elements). Construction materials were renewable, recyclable, or from nature. Armadillo means "little one in armor" in Spanish.

In 2010, Tongji University was the first university team from China to enter a Solar Decathlon competition.

The first SDE showcase of sustainable houses turned out to be a surprisingly close competition. As teams jockeyed for position, it looked like a tight race, and when the final scores were announced, only four points separated the top three teams. Virginia Tech's *LumenHAUS* was honored to take first place with 812 points. *Ikaros Bavaria*, the house from the University of Applied Sciences Rosenheim, won second place with 811 points, and *Home+* from Stuttgart University of Applied Sciences won third place with 808 points. Of course, the winning teams were thrilled, but spirits soared among all Decathletes at the final awards ceremony. Once again, the Solar Decathlon design-and-build program substantiated the exceptional value of multidisciplinary teamwork combined with hands-on construction experience, along with a full-scale public exhibition of unparalleled ingenuity and craftsmanship.

Home+ from Stuttgart University of Applied Sciences, Germany, had solar cells mounted on the roof, plus more on the east and west exterior walls. This team placed third overall in the 2010 competition.

Team Stuttgart used polycrystalline silicon solar cells, which entered the market in 1981. Polycrystalline solar panels are made by melting fragments of raw silicon together. Melted silicon is then poured into square molds, cooled, and cut into square wafers for solar panels. Panels made from polycrystalline silicon are about 13–16% efficient. Single-crystal solar cells are made by forming higher grade silicon ingots into bars and cutting them into wafers. These cylindrical ingots are further cut on four sides to make silicon wafers. Panels made from single-crystalline silicon are 15–21% efficient.

Team Stuttgart had a skylight filled with suspended solar cells to complement the exterior PV cells.

Visitors gathering to tour the *Ikaros Bavaria* house from the University of Applied Sciences in Rosenheim, Germany, which won second place overall in the 2010 competition. This modular design is insulated with vacuum insulation panels, which make it highly energy efficient. The stunning zig-zag façade has a distinct purpose: to shade the dwelling and allow for optimization of the changing direction and angle of sunlight during the day and throughout a year.

The Rosenheim team uniforms honored their heritage: men in lederhosen (short leather trousers) and women in dirndls (wide below-the-knee skirts). This team won the second-place award at SDE 2010.

LumenHAUS from Virginia Polytechnic Institute and State University in Blacksburg, VA, took first place overall at SDE 2010. This glistening architectural beauty was designed as an open pavilion with large sliding stainless steel screens covered with metal disks that functioned as part of a four-layer shading system. Those special sliders on the south wall, a modern version of outdoor shutters, could be opened for daylighting and closed to block sunlight and maintain privacy.

The team from Virginia Tech accepting the first-place award for *LumenHAUS* at SDE 2010.

Reflections from Sergio Vega, an architect, professor, and researcher in the School of Architecture at the Technical University of Madrid (UPM). With broad professional experience, Sergio has led multiple national and international research projects. He was Director of Solar Decathlon Europe for the first two competitions held in Madrid, Spain: SDE 2010 and SDE 2012.

When and how did you first find out about the Solar Decathlon?

I first heard about Solar Decathlon 2002 through a UPM solar panel start-up company. The UPM team competed for the first time in SD 2005 with the "Magic Box" house. Spain was the first non-American university to enter the competition. The UPM team included professors and researchers from the Solar Institute, the School of Architecture, and the Department of Construction and Architectural Technologies. Although I was not a member of that team, I witnessed first-hand the development of the project, its construction, the competition, and the educational potential it represented. At the end of 2005, with the new call for teams for the SD 2007, UPM decided to participate again. I was appointed as Faculty Advisor for this competition with the house called *Casa Solar*.

The *Magic Box* house from UPM in Madrid, Spain, was the first international entry in Solar Decathlon 2005 on the National Mall in Washington, DC.

As a faculty advisor for the UPM Team in SD 2007, what were your impressions of the competition?

Participating in Solar Decathlon 2007 and leading a team of students and professors representing UPM was a professional challenge of the first magnitude, due to the technical, logistical, cultural, and linguistic difficulties involved, especially with a high budget that had to be self-financed with the resources that we generated. It was a great honor and responsibility to represent my university and my country. This experience had a big impact on my professional career. My impressions of the competition exceeded my expectations. I was surprised by the intensity of the work and personal commitment required. Thanks to the active commitment of our fantastic team, all went well. What captivated me the most was the potential for innovation, education, and social awareness.

Why were you enthusiastic about bringing Solar Decathlon to Spain and Europe?

From the beginning of my participation in Solar Decathlon, I loved the competition and the media interest it generated. Before I went to the U.S., the Spanish government organized a social awareness event on solar energy called "España Solar," which featured *Casa Solar* (house for SD 2007). Our team approached the Spanish President's cabinet about the possibility of bringing this competition to Europe. They showed interest, and we agreed to advance the idea after the 2007 Competition.

During assembly of *Casa Solar* in Maryland at the National Association of Home Builders (NAHB) Research Center, I seriously reflected on a possible Solar Decathlon Europe event. After analyzing all the pros and cons, I proposed to Mr. Richard King the possibility of bringing the competition to Europe and adapting it to European design sensibility. He welcomed the proposal enthusiastically, so we held meetings to explore this further. With the active participation of the Spanish Embassy in the U.S., a Memorandum of Understanding (MOU) was developed between the Spanish and American governments. Spain would host the first two editions of Solar Decathlon Europe. This MOU was signed by both governments at *Casa Solar* on October 18 during the SD 2007 competition. Richard King's enthusiasm and our eagerness to take this step were key drivers for this agreement.

Signing the MOU for SDE 2010 in Madrid on the National Mall in Washington, DC. Photo credit: Sergio Vega.

What was the most challenging aspect of organizing SDE 2010?

After SD 2007 and my return to Spain, the Spanish Ministry of Housing asked UPM to organize the 2010 and 2012 Solar Decathlon Europe competitions in Madrid, and the Rector of the university appointed me as Director/Project Manager. Strategic planning occurred, objectives were identified, methods to achieve them were defined, a time line was created, and a budget was proposed. The agreement between UPM and the Spanish Government was signed, and work began.

A big challenge was to define the best strategies to achieve the proposed scope and objectives, including an understanding that the Solar Decathlon is not just the Competition; it must make public visitation attractive enough to raise awareness among all ages, from children to mature professionals. We needed to select an adequate and balanced management team, mostly university students (including many who had competed in SD 2005, 2007, and 2009), some UPM professors and researchers, and professionals who specialized in specific areas, such as communication, logistics, etc.

Another huge challenge was financing the project. The Spanish Government only ensured the explicit cost of the competition. The on-site event and all communication activities depended on sponsorships, both in cash and in kind. For SDE 2012, when the budget was more restricted, we developed a multi-scenario system in which dozens of activities were planned, and depending on the activation deadline of each activity, were offered or not according to the money available at that time. As a result, some activities did not see the light of day. The creativity, freshness, and commitment of our young team ensured that the competition went smoothly and successfully, despite those difficulties.

The ten contests for SDE were different from SD competitions in the U.S. What prompted the organizers to make these changes?

The main reason for adjusting the contests in the jump to Europe was to adapt them to European perspectives and priorities. We wanted to emphasize sustainability, energy efficiency, innovation, and social awareness. The European organization considered some changes in the rules and regulations to promote innovation and sustainability. Differences in the ten contests of Solar Decathlon 2009 and Solar

Decathlon Europe 2010 are shown in this table. I believe that the adjustment of the ten contests brought our competition much closer to the social and technical challenges of today's European society. These ten contests have remained fairly stable throughout the five editions of Solar Decathlon Europe held so far, with slight adjustments from one edition to another.

Ten Contests - 1000 Points			
U.S. Solar Decathlon 2009	Point Value	Solar Decathlon Europe 2010 and 2012	Point Value
Architecture	100	Architecture	120
Engineering	100	Engineering and Construction	80
Net Metering	150	Electrical Energy Balance	120
Communications	75	Communications and Social Awareness	80
Market Viability	100	Industrialization and Market Viability	80
Comfort Zone	100	Comfort Conditions	120
Lighting Design	75	House Functioning	120
Hot Water	100	Energy Efficiency	100
Appliances	100	Innovation	80
Home Entertainment	100	Sustainability	100

What made the SDE 2010 and 2012 events such a big success?

Well over 100,000 people visited the 2010 competition, and more than 2,000 national and 5,000 international media impacts were recorded. The success of the SDE 2010 competition held in Madrid had its origin in multiple causes, but if we had to summarize it in two words beyond Richard King's vision, we would talk about methodology and team. One of the key drivers was the team that competed in SD 2007; they later became the "core" organizers of SDE. These Decathletes worked side-by-side as "shadows" of the American organizing team for U.S. SD 2009 in Washington, DC. Many lessons learned from this "shadow experience" were integrated into the European competition. Another factor was the decisive commitment of the Spanish government, as represented by Javier Serra. His unconditional and unwavering support led to the success of SDE 2010 and SDE 2012, despite the problems arising from the change of government party in the 2011 elections.

For SDE 2012, there were 220,000 visitors during the competition, with about 5,000 children and teenagers, 2,000 university students, 6,000 professionals, and more than 45 educational activities repeated daily (more than 200 activities in total). Records show 1,886 national media impacts and many unmeasured international impacts, with ~150,000 website visits during the competition. For SDE 2012, suggestions for improvement were derived from the *Critical Analysis and Lessons Learned Report* written about SDE 2010. The second event was better in many respects, despite budget cuts.

Publication of two books described the SDE 2010 and 2012 competitions. In addition, many articles in scientific journals, a special issue of *Energy & Buildings*, and four PhD dissertations with a focus on Solar Decathlon houses were published. Multiple visits from ambassadors, politicians, housing ministers from the European Union, and a special award from the EU during *Sustainable Energy Week* in 2011 created additional impact.

Tell us about how the EU became involved with SDE. How did that affect the organization and location of the competition?

The Spanish Government and UPM, as organizer of the competition on its behalf, always had genuine interest in the competition and wanted continuity beyond the two editions organized in Spain. They wanted to promote the best legacy of Solar Decathlon Europe, in order to take advantage of this body of knowledge for the future. The Spanish government awakened the interest in SDE, with the objective of the EU becoming the leader of future competitions in Europe. Support of the European Union and the United States for the joint commitment to SDE was demonstrated by the joint press release from the E.U.-U.S. Energy Council in Washington, DC on November 28, 2011. It states that "The E.U. and the U.S. intend to cooperate on continuing Solar Decathlon Europe Competitions, transforming them into an initiative to foster sustainable economic development by creating markets on both sides of the Atlantic, and for integrating innovative technologies and renewable energy sources into new and refurbished low impact buildings."

An MOU signed by the U.S. and French Governments in the spring of 2012 authorized a 2014 edition of SDE in Versailles, France. UPM was one of the co-organizers, transferring its experience to the new organizing team. After 2014, no country offered to host another edition in Europe. Four professors from France, Germany, the Netherlands, and Spain then formed an SDE Secretariat to provide leadership for the European Commission. Finally, the Energy Endeavour Foundation (EEF), a Netherlands-based non-profit business entity endorsed by the U.S. Department of Energy (DOE) to steward SDE, became custodian of the SDE Rules and SDE brand. EEF offers the structure and framework for future SDE competitions.

Solar Decathlon Europe 2012

As stipulated in the agreement between the United States and Spain, the second international event was held 2 years later in Madrid. SDE 2012 took place on September 14–30 in a new location, Casa de Campo Park in Puerta del Angel, not far from the original Madrid Rio site. Eighteen teams from 11 countries displayed bold new house designs with a wide variety of shapes, sizes, and configurations, each one an exemplary model of energy efficiency and sustainability.

A special feature of this edition was connecting all the houses with a smart microgrid provided by Schneider Electric, a major sponsor of the competition. This "intelligent" distribution network with close proximity between units of production and consumption allowed for greater efficiency, while balancing supply with demand in real time. Due to the clever interface of this system, visitors could view graphs of actual energy use in the solar village at any given time. Visitors could also see how much excess power was being produced and directed back to the city's power network, leading to better public understanding of renewable energy management.

The elevated site at Casa de Campo for SDE 2012 had excellent views of the Royal Palace and ample space for visitors eager to learn about these innovative net-zero-energy homes.

Competition houses at Villa Solar 2012 created a distinctive display of diverse architectural designs.

True to its reputation as a magnet for the public, SDE 2012 attracted thousands of curious visitors.

SDE 2012 also aimed to increase social awareness about renewable energy, with a concerted effort to expand messaging and educational activities for European audiences. The competition was an excellent springboard to draw attention to new materials, technologies, and sustainable building practices. There was something for everyone onsite, from workshops and presentations for adults to hands-on engagement, games, and age-appropriate contests for children. The organizing team did a fabulous job of preparing and managing a diverse menu of excellent learning opportunities for all visitors. Special recognition goes to two outstanding professionals who accepted leadership roles at the Solar Decathlon competitions in Madrid, Spain: Edwin Rodriguez-Ubinas and Claudio Montero. Edwin's impressive work ethic and managerial strengths, plus Claudio's organizational skills were a winning combination. Furthermore, their capacity to get along with anyone and everyone contributed greatly to the remarkable success of the first Solar Decathlon events outside the United States.

SDE 2012 featured hands-on children's design-and-build activities made of recycled materials, as well as a display of children's creations from a "make your own city" workshop.

Families loved the "maker space" at SDE 2012 that demonstrated children's creativity.

The competitor lineup for SDE 2012 included the following universities, with some teams representing more than one university or country.

Brazil
- Universidade Federal de Santa Catarina, Universidade de São Paulo

Denmark
- Technical University of Denmark

Egypt
- American University in Cairo

France
- Arts et Metiers Paristech Bordeaux
- École Nationale Supérieure d'Architecture de Grenoble
- Ecole Nationale Supérieure d'Architecture Paris, Malaquais, Universitá di Ferrara, Ecole des Ponts Paristech, and Politecnico di Bari (France and Italy)

Germany
- RWTH Aachen University
- University of Applied Sciences Konstanz

Hungary
- Budapest University of Technology and Economics

Italy
- Universitá Degli Studi di Roma TRE, Sapienza Unviersitá di Roma, Free University of Bozen, Fraunhofer Italy

Japan
- Chiba University

Portugal
- Universidade do Porto

Romania
- Ion Mincu University of Architecture and Urbanism, Technical University of Civil Engineering of Bucharest, Politehnica, and University of Bucharest

Spain
- Universidad CEU Cardenal Herrera de Valencia
- Universidad del País Vasco; Euskal Herriko Unibertsitatea
- Universitat Politécnica de Catalunya
- Universidades de Sevilla, Jaén, Granada, and Málaga
- Universidad de Zaragoza

Prispa house from the Bucharest team was a simple yet elegant structure that honored traditional Romanian lifestyle at an affordable cost.

Ekihouse from the team representing a research institute in the Basque region of northern Spain had an over-hanging roof on the south side to allow shading in the summer and passive heating in the winter. The movable, double-layer façade had glazing on the interior and perforated steel panels on the exterior to help control sunlight and adjust to various weather conditions.

Ekihouse featured exterior planters with sunflowers to demonstrate the importance of living greenery to add natural beauty and provide space that could be used for edible plants.

FOLD house from the Technical University of Denmark featured walls and roof made of two layers of durable Kerto wood from Finland with Rockwool Aerowolle as insulation sandwiched in between the slabs for thermal efficiency. To honor Denmark's world-famous building blocks, the team created a LEGO® wall panel study area next to the kitchen.

Sumbiosi house from Bordeaux University in France included Fresnel lenses that concentrate sunlight on PV cells to decrease the size of rooftop panels and a tracking system to increase efficiency, which allows for cogeneration of electricity, hot water, and heat transfer fluid. The house also recycled greywater for use in non-potable functions.

Omotenashi house from Chiba, Japan, had a green wall for cultivating crops, an intermediary space connecting the exterior with the interior of a home.

Europeans Take Up the Charge | **297**

The rooftop of the Chiba house covered with tile-shaped solar panels resembled the shape of a traditional Japanese roof. This house was fabricated with units made by robots.

Casa π Unizar from the Universidad de Zaragoza, Spain, stood out with its sliced cylinder design to reduce heat loss between the interior and exterior.

An aerial view of *Med in Italy* (bottom left), which took third place at SDE 2012, with its enclosed patio for fresh air and privacy. The masonry walls around that space provided thermal mass for winter warmth, plus absorptive power to keep that area cooler in summer.

The final scoring outcomes of this competition were more widely dispersed than they were at SDE 2010, with greater differences between participating teams. For example, 11 points separated the first and second place winners, with a 34-point difference between the second and third place winners, and a 28-point difference between the third and fourth place teams. *Canopea*® from the Rhones Alpes team in Grenoble, France, captured first place overall; *Patio 2.12* from Andalucia, Spain, won second place; *Med in Italy* from Rome, Italy, took third place; and *Ecolar* from Konstanz, Germany, came in fourth place. After such dedicated hard work to design, build, and showcase these unique houses, there was plenty of well-deserved celebration in the Villa Solar following the awards ceremony!

The team that built *Med in Italy* created clever decorative sunflowers that contained solar cells on display in their patio.

Patio 2.12 from the Andalucia Team (in the foreground) took second place in the 2012 competition. The house consisted of four prefabricated modules surrounding a central patio. On winter days, the covered patio captures sunlight to function like a greenhouse, and on summer days, folding glass panels open that area to allow natural ventilation created by the vertical walls.

Canopea from l'Ecole Nationale Superieure d'Architecture de Grenoble, France, was a stunning, beautiful nanotower that mastered every aspect of the 2012 competition to win first place overall. This ingenious model suggested a new solution for urban density in alpine regions where available land for buildings is limited.

Faculty advisor Pascal Rollet talking about the design of Canopea with Richard King at SDE 2012.

Canopea, the winning house from Grenoble, France, had comfortable community gathering space on the top floor. The main floor was composed of three boxes that included a technical core with HVAC, plus primary kitchen, living, and bedroom areas in a flexible configuration for work, entertainment, and sleeping.

Canopea Decathletes celebrating their first-place victory at SDE 2012.

Reflections from Pascal Rollet, Full Professor of Architecture at École Nationale Supérieure d'Architecture de Grenoble. His award-winning architecture firm in Paris, Lipsky + Rollet, received the 2005 Équerre d'Argent (Silver Square) for best building of the year, as well as nominations in international competitions, such as the Benedictus Award 2005 and World Architectural Festival 2008. Professor Rollet is a member of the Architecture, Environnement & Cultures Constructives research team and Chair of the Habitat du Futur at Les Grands Ateliers Innovation Architecture, a French research center dedicated to "learning by doing" and "learning by building" in architecture, art, and engineering. He was Director of Solar Decathlon 2014 in Versailles, France, and is a committee member of the European Energy Endeavour Foundation, which promotes Solar Decathlon in Europe. Professor Rollet was Faculty Advisor of Team Rhône-Alpes at SDE 2010 and again at SDE 2012, where his team won first place in the competition.

Why did you want to get involved with the Solar Decathlon?

The School of Architecture in Grenoble has been involved with experiential learning since the 1980s. Professors and researchers of CRATerre[1] conducting the R&D program about raw earth construction were strongly committed to the "learning by doing" principle that they had experienced themselves during their own studies in the 1970s in Paris. I was personally trained by them as an architect (1979–1985) through experimental building operations in Africa, where earth construction was specifically adapted to the local conditions (existing natural resources, poor financial means, high know-how). I also studied at Berkeley, where I discovered the concept of *Environmental Design* created by William Wurster and Catherine Bauer. Even though I did not study with Christopher Alexander (my initial plan), who was on a sabbatical while I was there (1988–1989), but mostly with Lars Lerup and Stanley Saitowitz, the spirit of Wurster Hall probably echoed what I had learned before with CRATerre and infused silently in my mind.

When I was back from the States, I was hired as a young professor in Grenoble (after a national selection process). I joined the CRATerre

---

1CRATerre: Center of Research and Application Terre: Patrice DOAT (architect), Hugo HOUBEN (engineer in nuclear physics), founding members of CRATerre—soon to be joined by Hubert GUILLAUD (architect), Nathalie SABATIER (ethnologist), Anne-Monique BARDAGOT (socio-ethnologist)—were teaching and researching in Grenoble. They have been my professors and mentors.

team for my academic career. I was then asked to take responsibility of the master program. We decided to call it *"Architecture, Environment & Building Cultures,"* based on ecological architecture and a hands-on approach of materials and construction. The program was organized in two years.

- The first year was based on a polytechnic approach of materials called *"Propédeutique Matériaux"* to develop a basic pedagogical agenda: learn the constructive and symbolic logic of stone, earth, wood, steel, composites, textiles, and glass before designing anything. Learning this logic through practice and manipulation, along with a two-week design-build studio for small pavilions realized at Grands Ateliers.
- The second year was based on a longer design-build studio called *"Habiter-Léger-Pas Cher,"* during which students were put in a real situation to design and potentially build Affordable and Light Housing projects after a holistic analysis of the geophysical (natural resources, population, industrial development) historical, ethnological, and economic analysis of the context.

This studio operated from 2002 to 2008. It produced a series of concepts that were to be used later in experimentations, such as the nanotower concept which was developed within the Canopea® project (2010–2012). So, between 1988 and 2008 we had been practicing design-build studios coaching, finding sponsors to help us fund the prototypes or real-size experiments, and putting together multidisciplinary teams composed of architects, engineers, builders, and communication people. We were ready for the Solar Decathlon without knowing that it had existed since 2001!

When the French Ministry told us that a representative of UPM (Sergio Vega) had come to Paris to advertise the European version, we had a shock: this was meant for us, but the application deadline was just two weeks away. We scrambled and put together an application booklet within a week. We also found a major partner, with Vincent Jacques Le Seigneur, General Secretary of the National Institute for Solar Energy (Institut National de l'Energie Solaire et aux Energies Alternatives or INES) who had known about Solar Decathlon for a long time and had dreamed about putting together a French Team but never succeeded with his usual partners in the university. The match was perfect, and we were able to deliver the application just in time, after threatening the director of our school of a collective resignation if he did not sign the application form for an adventure that he considered as pure madness (which was true... but we did not expect what was to come at the time).

In conclusion, we decided to get involved with Solar Decathlon because it was the right competition format at just the right time for us.

In your opinion, what made this competition a worthwhile endeavor?

It gathers all the constraints that future architects, engineers, builders, developers, facility managers, maintenance crews and politicians will encounter during their professional life. It forces students, teachers, researchers, and professionals to work together from the beginning to find the best optimized solution to the issues they have decided to address. It promotes a way of designing houses, neighborhoods, cities, and territories in a holistic-systemic approach that is necessary today, if we want to rise to the challenge of climate change. It got started through the prism of energy, but it went far beyond this issue (which remains central) to question the basic concept of what human habitat is all about. The cleverness of the competition lies in the formula of ten contests (decathlon concept), which must all be considered together when designing a place to live. These are the contests for SDE 2014:

1. **Architecture:** How are space and matter organized to provide hospitality and safety to human bodies, as well as emotions and a sense of purpose to the human soul, through geometry, proportions, materiality, forms under the light casting shadows... with respect to history of the arts and building know-how?

2. **Engineering and construction:** How are technical problems coherently solved for the benefit of the architectural and social project, while considering the constraints of local context (natural hazards, climate, resources, technologies, performances, economy, sufficiency)?

3. **Solar energy systems:** How, like all vegetal species, does humanity turn to the main source of free energy that is naturally given to her on this planet: the one coming from the sun! How to value it for what it is: "the giver of light and all presences" as Louis Kahn once said.

4. **Electrical energy balance:** How does humanity achieve an optimized and balanced use of solar energy, without wasting it? How our species will understand that balance is the most important issue in life.

5. **Comfort conditions:** How can architects, engineers, and builders get together to guarantee good living conditions to any individual wherever s/he lives on planet Earth? How do we take care of every one of our kind?

6. **House functioning:** How can architects, interior designers, engineers, and industrial partners get together to provide the modern comfort that free people (and mainly women) from the burden of household daily maintenance? How do we do that without polluting our environment?
7. **Communication and social awareness:** Potentially the most important part of the competition: spread the word that it is possible to design and build more sustainable architecture then what the Modernists' mirage has tried to sell to all mankind. It can't be that only one solution can apply to such a diverse species as ours!
8. **Industrialization and market viability:** How do we produce, fabricate, and assemble things to improve our daily life because our two hands and our brains are meant to do so? Before being *sapiens*, we are first *homo faber*. In the process: how do we understand the ancient Greek word used to designate this issue: *oikonomia* which literally translates into, "How do we wisely take care of the common house?" Meaning not "How do we make profit out of all these ideas," but "How do we provide to every member of the species what is necessary to live a decent life?" (see United Nations definition of a "decent life").
9. **Innovation:** How can the human brain push its own limits and create new solutions inspired by experimentations, imitations, and improvements through renewed variations? Just like life!
10. **Sustainability:** How does humanity define its place in the environment and control its survival with respect to the planet and other species without depriving them of their rights?

As a faculty advisor working with multidisciplinary teams of students for SDE 2010 and 2012, what is the most powerful aspect of the Solar Decathlon as a learning experience?

It forces people focused on a discipline to realize that other disciplines exist and are as important as the one you are fascinated by. For architects who tend to consider that they are the only discipline that matters, since it has always been considered the first of the Major Arts, and since they oversee worlds that do not exist yet, it has been a surprise to participate in a competition where architecture is only one criterion out of ten. For engineers, who tend to consider that they are the most serious people since they scientifically solve problems, it has been a surprise to discover that an artistic architectural concept can make the difference in the end. For both architects and engineers who tend to consider that communications people are most of the time "talking in the wind," it has been a shock to realize that their ideas or concepts needed to be conveyed in a proper way so that the public can understand them, and that the spread of knowledge is, right now, more important than its permanent extension.

What important lessons did your student team learn from SDE 2010 that helped the next team create a winning house for SDE 2012?

The rules are very important because they define the boundaries within which the game must be fairly played. But sometimes the rules impose a limitation because the ruler has not seen what was really at stake at that moment. This was the case in 2010. As required by the SDE2010 Rules, all teams had brought buildings designed as objects to Madrid. But award-winning architect Glenn Murcutt, who was a jury member, saw the flaw and commented on that during his SDE juror's speech: "Here in Europe, the urban fabric is the question, not the building." I heard that loud and clearly! It was suddenly obvious to me that the next project had to speak about how sustainable buildings would fit into cities. So, the next step was *Canopea®*, a nanotower designed for the limited urban space in Grenoble, France.

Priztker Prize award winner Glenn Murcutt, an Australian architect known for innovative, environmentally friendly designs, announcing the winners of the Architecture Contest at SDE 2010.

Why do you think *Canopea®* was a significant development for sustainable architecture?

Canopea® was a major step because we addressed the issue of a sustainable building participating in the urban metabolism of a district. The nanotower was not designed as an isolated object interacting with other independent objects, each of which needed energy supply from a higher source. The project was designed as a single piece of a living puzzle in which each component had a specific energy profile allowing it to participate in multiple exchange processes between different pieces. For example, energy could be produced by PV panels sitting on top of buildings and stored in the batteries of cars parked in a vertical parking lot. Conversely, the refrigerator of an apartment could suck electricity out of the battery of a car to operate when the cost of electricity would be too high, due to consumption peaks. The junction of electric network engineering—which has been at the core of science and technologies developed in Grenoble since the nineteenth century[2]—and computer sciences had given birth to the *"smart grid"* concept.

At the time, we were working with researchers of the Nuclear Energy Research Center (Commissariat a l'Energie Atomique aux Energies Alternatives or CEA), INES, and industrial partners, such as Schneider Electric, who were just developing these technologies. In 2010, the *Presqu'île* new eco district project developed by the City of Grenoble had just been issued. It was based on the construction of a low temperature water loop allowing energy exchanges in both ways (feedings and withdrawals) between public buildings, heavy infrastructures such as the CEA nuclear reactor, the Isère and Drac River regulating plant using huge pumps to control the water flows, industries, and housing programs according to their needs at different time of the day. This idea of a city fabric functioning like an ecosystem characterized by a specific metabolism was extended by the *Canopea®* engineering team[3]

---

[2] By the end of the 19th century, because of the use of the power of running water descending from the mountains, forced into ducts to activate turbines producing direct current, Grenoble soon became the scientific and industrial center of what was then called the "houille blanche" (white coal), namely electricity. An international Exhibition was held in Grenoble in 1910. The university was soon to develop specialties in electric technologies, electrotechnics and electrical engineering. Companies such as Merlin & Gerin were created after WWII to produce heavy electric equipment. In 1992, Schneider—a French society specialized in mechanics—buys Merlin & Gerin and becomes Schneider Electric in 1999.

[3] Thomas JUSSELME and Laurent TOCHON, with the active participation of Guillaume PRADELLE.

at several levels, such as the electrical balance, transportation system, and information systems. Today, these concepts developed ten years ago are still waiting for large-scale applications.

What features made *Canopea®* an outstanding example? How has that house made an impact? Where is it now?

The most outstanding feature was the concept of the nanotower. The students' brilliant idea was to stack eight individual apartments presenting several spatial qualities of an individual house to create a small tower (last floor lower than 28 meters high for fire code safety reasons) topped by a transparent shared space covered with silk-screened glass PV panels. The result is a bizarre combination of urban sprawl and urban density; somehow, the Frankenstein crossing of Wright's Broadacre City and Le Corbusier's *Plan Voisin*!

The second brilliant idea—in all modesty—was mine: if you cannot bring a whole tower to Madrid, nothing in the rules said that you cannot bring the top of it. Everybody assumed that we had to build "a house." But it was not because the American origin of the competition always spoke of "solar houses" that Europeans could not address the issue of urban density and vertical collective architecture. Remember what Murcutt said! So why not just bring a sample? It was a scientific competition after all. Scientists are used to making demonstrations based on tests conducted on samples.

So, we designed a complete cluster of three nanotowers with their shared elevator shaft, shared urban farm, and shared parking lot. We implemented the project in a virtual reality tool that was developed by a team of Arts & Crafts Engineers at the Arts & Crafts School in Cluny, and we were able to show different interactive panoramic simulations during the tour. But we built only the top apartment and the rooftop shared space that perfectly fit the solar envelope authorized by SDE rules for Madrid's 2012 Solar Village. The impact was tremendous, and it worked so well that every edition after SDE 2012 has used the idea. I remember other dazzled faculty advisors of competing teams scratching their heads and silently wondering why they did not think about that themselves.

Canopea® has been rebuilt in the *Presqu'île* district of Grenoble, the place it was initially designed for, and it has been used as the house of an urban project for ten years. When the SDE 2012 project was finished, we had to dismantle and relocate it. But Team AuRA students who engaged in SDE 2021–22 in Wuppertal suggested basing the new project on reuse of those materials, since it had become such an issue with the rise of consciousness of resource limitation. Maxime Bonnevie, director of Grands Ateliers (former 2010 Decathlete for the *Armadillo Box®* and former 2012 Project Manager of *Canopea®*) who was also in charge of finding funds, decided that we should reuse components from *Canopea®*, rather than rebuild it a third time. The prototype was dismantled during the summer of 2021, and every bit has been reused in the next SDE house, *Agathe and Sophie*. PV panels were reflashed at INES (they only lost 2% of their production capacity in ten years), the washing machine, refrigerator, and kitchen appliances have been reused on the upper deck of A&S, and so on. What we learned in the process is that reuse is possible, but needs to be thoroughly planned, in order to be efficient. It is not so easy to put this into practice after decades of the culture of consumption.

What motivated you to change your role from being a faculty advisor for two SDE competitions to serving as the director of SDE 2014 in France?

The answer is simple:

- The faculty advisor of a team speaks to a maximum of a hundred students. Maybe to another hundred more colleges and local politicians.
- The director of the competition speaks to a thousand students and faculty members, plus another thousand more politicians and tens of thousands of the general public.

If you want to spread the word to a wide audience, the second position is the place to be. I did my part.

What were the biggest challenges of SDE 2014? In what ways did that competition excel?

The biggest challenge was to create a coherent multidisciplinary team in a highly selective technostructure. People from CSTB are high-level engineers often coming from *Polytechnic* or *Mines* French *"Grandes Écoles."* They think that they know everything about energy, building, technology, and finance! They often don't care about communication and the cultural aspects of the competition. They barely tolerate architecture because of its heritage aspect that they cannot control; but that can't be denied, especially in a city such as Paris or a site such as Versailles that had been chosen. Mostly they thought that they knew better than the Spanish team who had organized the previous SDE editions. They wanted to show that to the world and make a demonstration of their prowess (before the Germans did).

I had a really hard time convincing them that we needed Sergio Vega, Edwin Rodriguez-Ubinas, and Claudio Montero (from Spain) to help us and transmit what they had learned in Madrid. It was even tougher to have them admit that we could have Peter Russell, a German expert from Aachen, to scout what we were doing in preparation for a possible next edition of SDE in Germany. My view was to build a strong European chain of knowledge to educate the young European generations in the spirit of the ancient *Peregrinatio Academica*. The opinion of CSTB and French government representatives was based more on their traditional industrial rivalry with Germany. It is only because the French team won the 2012 competition that these people accepted having architects on the team. A sad reality. In fact, the French technostructure is still suffering from a harsh Napoleonian hangover. Napoleon was trained at Polytechnique and was a French Revolution general before declaring himself Emperor. You can still see that today in the way French politics works.

Only one man was aware of the importance of architecture in the picture, and fortunately, he was at the head of CSTB at the time. Alain Maugard (*Polytechnique* and *Ponts & Chaussées* engineer himself) had followed the SDE 2012 competition because we invited him to be part of the Team AuRA Scientific Committee. He was very impressed by the way we were able to put a multidisciplinary team together and succeeded at having architects, engineers, and builders working together. He understood the unifying power of a good architectural project, so he pushed the CSTB administration in the right direction to create a coherent organization for SDE 2014.

What were the greatest accomplishments of SDE 2014?

The first great accomplishment is to have been able to make the competition an international event, not just a European one. Twenty teams representing 16 countries from three continents and 26 universities participated. We had four teams from Asia (Thaïland, India, Japan, Taiwan), five from North and South America (two USA teams associated with French and German universities, and one from Mexico, Chile, Costa Rica), and 11 teams from Europe: two from Germany, two from Spain, two from France, and one from Switzerland, Italy, Denmark, Netherlands, and Romania.

The second great accomplishment was to gather top-notch jury members coming from all over the world: China, Germany, USA, Spain, Italy, South Africa, and France. All jurors were from top-ranked universities and national institutions, and NGOs, such as the China Academy of Art, TU Munich, TU Vienna, UC Berkeley, University of Pretoria, U.S. DOE, EPFL, French *"Cité de l'Architecture,* and the president of CSTB. From the beginning, my belief was that the level of a competition is set by the level of the jurors. The jury panel is what architects first look at before getting involved in an international competition. So, I started to seek jury members as soon as 2012. That is how I was able to get Wang Shu as leader of the architecture jury.

The third great accomplishment was to get five demonstrators built for the event. I particularly admired the Darmstadt *Cubity* students' house, which was just a beauty. For this project, I had the great privilege to meet Professor Manfred Hegger just a year before he passed away. He was a prominent mentor for all European architects involved in sustainable architecture, especially with the *Mont Cenis Akademie* project in Herne-Sondingen, Germany that he designed with Françoise-Hélène Jourda and Gilles Perraudin. I was also very proud of the AuRA pavilion that a group of freshmen from ENSAG designed and built in a year to showcase *Canopea®*.

The last great accomplishment, but not least, was the creation of a huge Solar Village in the park of Château de Versailles, with a lot of space and quite a magnificent setting, even though it was a bit far away for public visitation. I regret that we were not able to build the special "Decathlete Village" that we had planned, which would have made the solar community experience complete. Following the UPM example in Madrid, I managed to get students from Grenoble and Versailles involved as members of the organization as safety managers and inspections teams.

As an architect with tremendous experience over many years, how do you view the impact of Solar Decathlon competitions in the U.S., Europe, and all over the world? How has SD influenced architecture and design?

The impact of SD is not yet visible on the international architectural scene because it is a cutting-edge phenomenon. The architectural world is usually very slow to evolve, and more so in a global context of denial of the emergency related to climate change (despite IPCC alarming reports). Architecture generally changes 30 years after the avant-garde appearance. So, we are about to see real influence from Solar Decathlon. It has been working under the radar among younger generations. I think that we shall see an acceleration of the effects in the next decade, mostly because of the trifold combination of rampant crises: Energy + Water + Migrations, which goes with Economic Slow Down + Food scarcity (if not famine) + Xenophobia.

Today's level of confusion makes it difficult to see what needs to be done to free us from a kind of indifference (standing on the sidelines). It seems that we need an electroshock to change. As a top official of the Ministry of Housing participating in the political management of SDE 2014 once told me: "Pascal, you have to understand that the Republic reacts only when on the verge of falling off a cliff!"

In this chaotic situation, schools of architecture have been implementing sustainable architecture studios for twenty years now. In Grenoble, we have experienced a lot of resistance to set up such a curriculum because most of the teachers were slow and remained stuck in the modernist culture devoted to Le Corbusier or in a strange B version of postmodern "Decon" parametric architecture. Now everybody has started to realize that it was the right direction because real pressure is coming from students. They want to build with earth, straw, wood, and ecological materials. They speak about reuse, sufficiency, cooperative housing, participation, and they are reluctant to follow the old ways. They plan alternative ways of working and living. The younger generation could sometimes sound paradoxical: they ask for a slower and more eco-responsible life, but they are hooked on their smartphones, and they contribute to climate change through a continuous exchange of gigabits activated by energy-guzzling data centers and digital farms.

The deep change that Solar Decathlon started in architecture and building science lies in the methodology: the competition showcases everything that needs to be done in architecture and design to create adapted living spaces. The brilliant idea is to put together ten contests into one competition to evaluate ideas and concepts through a wide-angled design-build process. Students who have experienced that are transformed. Some of them have testified that they learned more in two years than during all their student life. I can see the difference in my own office when I hire a Decathlete, compared to other regularly-trained young architects.

Regarding the AuRA Team, I am particularly proud of the fact that all former Decathletes have found jobs within a three-month period after the competition. But what I am most proud of is that the prominent figures of the initial team have become important cogs of the academic and research machine that we have been building for 30 years. Nicolas Dubus, who was my longtime teaching-assistant before becoming full professor in charge of AE&CC master's degree at ENSAG, is currently President of the Board of Directors of the School and Faculty Advisor of Team AuRA for SDE 2022. Maxime Bonnevie, who was a master's student Decathlete on *Armadillo Box* in 2010, project manager of *Canopea* in 2012, and project manager of *Terra Nostra* in 2016, has become Director of the Grands Ateliers in 2017. He's the Executive Director of the *Agathe & Sophie* Challenge in 2022.

Jean-Marie Le Tiec, Sébastien Freitas, and Hugo Gasnier, who were team members in 2010 and 2012, have established their own successful professional practice in the Grenoble area and are now teaching at ENSAG. Several students of Team AuRA 2022 in Wuppertal are currently teaching assistants with me in the first year at ENSAG. We are stretching the chain and weaving an architectural and constructive culture for the next decade. In this respect, we are following in the steps of Richard King, and I am thankful to Richard for having initiated this track.

Christophe De Tricaud, student team leader in 2012, has been Team AuRA project manager in SDE 2022 and is on track for a professorship track with a PhD in progress. Life is a chain. I always said that SDE was a cross between an International Fair, an edition of the America's Cup, and a rugby match! Showcase the most advanced ideas in architecture, engineering, building, solar technologies, and way of life, then run a cutting-edge prototype (even frugal), and pass the ball!

What do you anticipate for the "next generation" of architecture regarding innovative solutions that might help fight global climate change? What might be coming up in the future design world?

The million-dollar question... and the topic of my essay for my *Habilitation à Diriger des Recherches* (HDR publication planned for 2023, work in progress). I am expecting change on three different scales:

Territorial scale

- The issue is *relocation of human settlements* in order to decrease urban pressure. This has to be done without artificializing any more agricultural or natural land. It means building and densifying what is already here. We are currently working on the concept of distributed urbanism (as in SDE Wuppertal). We are examining and learning again from the English Garden Cities example. We research the renewal of rural areas and look at Peter Calthorpe's "Pedestrian Pockets" proposal.

- This relocation principle calls for a *new circular economy* providing most of the primary goods in a radius of less than 100 km around a given area. The question of decentralized industrial production is at stake. Factory 2.0 is already obsolete, so we must go for Factory 2.1.

- "Beyond sustainability, *sufficiency is the theme*. The issue is not to find a balance between nature and our modern way of life, but to reduce our needs and find a way to live without moving around all the time. The shift in transportation modes is THE issue that will affect our lives the most. I wonder if we will go back to the 19th century at a time when people of my family were feeling adventurous just by going to marry a girl from a village in the next valley. Or, if technology is going to provide new solutions using very little energy and no CO_2 emissions.

City scale

- We have to plan, design, and develop *metabolic cities* on existing cities. They are living milieu, and they must house every species. Major change for most of us. If we want to have birds singing in the trees, we have to again get used to bugs, spiders, mosquitoes, worms, and other sloppy animals. (Not to speak of these various viruses…)
- This calls for less density but more intensity. How do we intertwine existing urban fabric with vegetation, urban farms, water, and traffic?
- We have to create a greener urban fabric, so that every citizen can breathe, relate to trees and plants, and spend leisure time without having to travel far in the countryside. In a way, Frank Lloyd Wright's Broadacre City was a much more futuristic vision than Le Corbusier Plan Voisin.

Building scale

- We have to touch the Earth gently! We need to cherish *lightness*: less matter, less energy, dismantling, and reuse capacity are the new qualities of space. In a way, we must learn to live outdoors again. Native people can teach us how to use tent-like buildings. No more caves.
- We need to use local materials such as earth, straw, mud, wood, and stone. We must recycle and reuse what is still usable.
- All building envelopes must be 100% active: they must collect water and harvest energy (BIPV, windmills), while providing insulation and ventilation systems and good solar protection. In an unstable climate, most of these envelopes must constantly adapt to their environment and to fast-changing climate conditions. I call these envelopes IO$_2$V skins (*peau Isolante à Ouverture et Opacité Variable*). Works almost in English too: Skin with Insulation, and Variable Opacity Openings (IVO^2). You tell me.
- We think that the Core-Skin-Shell composition could help to design and produce such buildings in large numbers.

- The production chain can be organized in an industrial way to prefabricate *central cores* providing tech rooms (electricity + HVAC systems) and all rooms dealing with interior waterproofing (bathroom, WC, kitchen) to be delivered on the site by truck as 3D modules.
- Local companies (masons, carpenters, window makers, sheet rocker, plumbers, electricians, cabinet makers) can build on site the *IO$_2$V skins* with local materials (made of wood, textile earth, straw, etc.).
- Specialized companies could tour a large territory to install prefabricated elements that need to be assembled to create a *100% active shell* which protects the Skin and Core.

If we put all this together, I think we'll get a new architecture inspired both by architects such as Buckminster Fuller, Richard Neutra, Charles & Ray Eames, Pierre Koenig, Raphael Soriano, Ralph Rapson, William Wurster, Glenn Murcutt, Renzo Piano, and by ancient nomadic architecture, such as the Polar Circle Nenets's yurts or the Lake Inle fisherman's villages in Myanmar. Add on top BIPV and electric drones... and this might be our future. The real challenge is to be able to DO ALL THIS AT ONCE and synthesize all this knowledge in a common vision. To do so, we must enlarge our brains and elevate ourselves to a much higher level of consciousness and intelligence. Before relying only on AI, I would try to enhance ourselves by training in a holistic approach, practicing polymath analysis of situations, and collecting ideas born through collective creation. Solar Decathlon is a prototype of this kind of process, and this is why I love it. Long live the Kings!

Solar Decathlon Europe 2014

The Versailles, France, location for SDE 2014, just 14 miles from Paris, heightened everyone's anticipation for the next competition. Steeped in historic significance, the Palace of Versailles was the residence of French kings for 100 years, and the gorgeous palace gardens are among the most visited sites in all of Europe. Formerly the seat of the French Parliament, the palace is also where the Allied forces and Germany signed the 1919 Treaty of Versailles at the end of World War I. Securing a special place for SDE houses on the grounds at Versailles was a remarkable triumph, thanks to the tireless efforts of Madame Catherine Pegard. Named president of the Public Establishment of the Palace, Museum, and National Estate of Versailles in 2011, Ms. Pegard realized that Solar Decathlon Europe represented a great opportunity to link the past, present, and future.

Preparation of the site for SDE 2014 in Versailles.

With hundreds of visitors coming to view the beauty and majesty of Versailles every day, why not offer them a chance to experience and learn about *La Cité du Soleil*®? Ms. Pegard enlisted valuable support from Francois de Mazieres (Mayor of Versailles), Beniot Parayre (Ministry for Ecology, Sustainable Development, and Energy), and Bruno Mesureur (Director of Marketing and International Development), and together they pulled out all the stops to move the needle from conceptual dream to reality. Professor Pascal Rollet, faculty advisor for teams in SDE 2010 and 2012, stepped into a new role as Director of the 2014 competition. His impressive background as an esteemed architect, innovator, educator, and leader made him uniquely qualified to take the helm of this exciting new edition.

Europeans Take Up the Charge | 317

During the assembly phase, large equipment arrived to deliver construction materials and modular sections to La Cité du Soleil site at Versailles.

An aerial view of La Cité du Soleil at Versailles for SDE 2014.

SDE 2014 attracted universities from all over the world, with 20 participating teams representing 15 nations in Europe, North America, Central America, South America, and Asia. Many teams were collaborative projects with students from more than one university; three of the 20 teams established partnerships between universities from different countries. This international event boasted distinctively diverse houses that integrated highly innovative solutions for energy efficiency and sustainability.

- **Denmark:** Technical University of Denmark
- **Costa Rica:** Costa Rica Institute of Technology, Cartago
- **France:** ENSA Nantes, ESB, Audencia Group, Audencia Nantes, Ecole des Mines Nantes, ISSBA, IUT Nantes, Architectes Ingénieurs Associés, Atlansun, Institut des Matériaux Jean Rouxel, Medieco, Novabuild, SAMOA, and SCE
- **France:** Université Paris-Est, ENSA Paris Malaquais, ENSA Marne la Vallée, ESTP Paris, Ecole des Ponts Paris Tech, ESIEE Paris, ENSG, and IFSTTAR
- **France and Chili:** Université de la Rochelle—Espace Bois de l'IUT and Universidad Tecnica Federico Santa Maria—Valparaiso
- **France and the United States:** Université d'Angers and Appalachian State University

- **Germany:** University of Applied Sciences Frankfurt am Main
- **Germany:** University of the Arts Berlin and Technical University of Berlin
- **Germany and the United States:** Rhode Island School of Design, Brown University, and University of Applied Sciences, Erfurt
- **India:** Academy of Architecture and IIT Bombay
- **Italy:** Universitá Degli Studi di Roma TRE
- **Japan:** Chiba University
- **Mexico:** Universidad Nacional Autonoma de México, the Center of Research in Industrial Design, the School of Engineering, and the School of Arts
- **The Netherlands:** Delft University of Technology
- **Romania:** Technical University of Civil Engineering Bucharest, University Politehnica of Bucharest, and Ion Mincu University of Architecture and Urbanism
- **Spain:** Universisad de Castilla—La Mancha and Universidad de Alcala de Henares
- **Spain:** Universitat Politècnica de Catalunya—Barcelona
- **Switzerland:** Lucerne University of Applied Sciences and Arts—School of Engineering and Architecture
- **Taiwan:** National Yang Ming Chiao Tung University
- **Thailand:** King Mongkut's University of Technology Thonburi

SDE 2014 aimed to stimulate housing designs that addressed three main challenges: energy, environment, and society. In refining the rules for this competition, six major issues became the focus:

- **Density:** Multi-family housing (rather than single-family housing) to reduce the environmental impact of dwellings
- **Mobility:** Solutions that combined housing and transportation systems
- **Sobriety:** Limiting demand for energy, even for renewable energy resources
- **Innovation:** Applied to all aspects of human habitats
- **Affordability:** Sustainable urban housing solutions that are affordable
- **Local context:** Dwellings that reflect the climate and culture of their region

With those issues at the forefront, SDE 2014 included the following 10 contests:

Juried

- Architecture
- Engineering and Construction
- Energy Efficiency (also measured)
- Communications and Social Awareness
- Urban Design, Transportation, and Affordability
- Sustainability
- Innovation

Measured

- Comfort Conditions
- House Functioning
- Electrical Energy Balance

As in every edition of Solar Decathlon Europe, distinguished professionals were invited to serve on jury panels that represented their particular areas of expertise. After visiting the competition houses and reviewing required documentation, these jury members presented awards to the winning teams. SDE 2014 Decathletes rose to the occasion in creating extraordinary examples of design-and-build resourcefulness. This event was another stunning showcase of architectural elegance and engineering excellence.

Signs to direct visitors in front of the two-story townhouse model (*Reciprocité*) from a combined team of Decathletes representing Appalachian State University (the United States) and Université de Angers, France; note the signs in front of the entrance that provide information about the house as people wait in line to enter.

An interior view of the main living area from the second-floor bedroom in *Reciprocité*.

Reciprocité featured flexible, open areas with plenty of daylighting to accommodate the basic functions of daily living.

Techstyle Haus from the Rhode Island School of Design, Brown University (United States), and the University of Applied Sciences in Erfurt, Germany, had a metal frame covered with a textile sleeve made of water-resistant Sheerfill® fiberglass, a fabric roof similar to some stadium covers. The sleeve is coated with a photocatalytic material that contains an inner sandwich of insulation.

The Inside Out team posing for a photo in front of their *Techstyle Haus*.

The elegant interior of *Techstyle Haus* featured flexible, light-filled spaces with contemporary furnishings.

Adaptive House from King Mongkut's University of Technology in Thailand was created as a model of how architecture can adapt to natural disasters. Intended for a flood-prone district in Bangkok, the soaring two-story structure sat on top of above-ground steel piles, so residents can stay safe during flooding. The house is made of various water-resistant materials, such as bamboo planks for patios.

The outcomes of the 2014 competition were quite unpredictable, especially due to the unusual weather that July. A soggy low-pressure system was stuck over northwestern Europe for much of the 10-day event, with several days of heavy cloud cover and steady rain. Teams had to strategize how to outsmart Mother Nature's adverse conditions to keep their houses fully operational without expending too much precious energy. Some teams were well prepared, and their houses demonstrated consistently strong performance, despite less-than-optimal sunshine.

As in the 2010 edition, the final scores for SDE 2014 were remarkably close—a true nail-biter experience. The Italian team from Roma Tre University pulled off a dramatic first-place win with 840.63 total points, but they barely squeaked ahead of the French Atlantic Challenge Team that ended up with a total of 839.75 points. The Delft University team from the Netherlands challenged both the front runners with a close third-place finish of 837.87 points in all. Needless to say, the audience at the final awards ceremony was abuzz with anticipation, and the announcement of winners led to raucous, rousing shockwaves of animated cheering as jubilant teams clambered onstage. What a memorable celebration!

Team Pret-a-Loger from Delft University of Technology, Netherlands, in front of their *Home with a Skin*, the third-place winner at SDE 2014. This team showed how to apply a second skin to an aging rowhouse, to avoid demolition of the building, along with a solar greenhouse on one side for fruit and vegetable production.

Home with a Skin also featured a green roof.

Philéas from Team Atlantic Challenge (ENSA Nantes and other universities in France) demonstrated a clever renovation of an old, vacant industrial building to give it a new life as space for multiple purposes: housing, offices, restaurant, and ventilated greenhouse for growing vegetables. This team won second place in the 2014 competition.

Phileas featured a central atrium and four sloped, south-facing roof surfaces covered with photovoltaic cells to maximize natural light and solar energy.

Team Rhome in front of their house called *RhOME for denCity*, which won first place overall in the competition. The dwelling is part of an "urban regeneration project" that will provide sustainable homes in a community in Rome, Italy, near the site of ancient aqueducts.

Team Rhome celebrating their first-place victory at the final awards ceremony.

Organizers of SDE 2014 sharing smiles at the final award ceremony; left to right are Edwin Rodriguez, Pascal Rollet, Claudio Montero, and Louise Holloway.

Richard King giving a speech at the closing awards ceremony.

Reflections from Michel Orlhac, an engineer with expertise in marketing, mergers, and acquisitions and former Vice President for the Global Marketing Division of Schneider Electric, dedicated many years of his career to promoting Solar Decathlon competitions worldwide. His idea for global sponsorship helped develop enthusiastic support amongst Schneider employees for SD teams and events.

You have been a strong supporter of Solar Decathlon. How did you hear about this competition?

I heard about SD the first time through my colleague Isabelle Malinconi, who took me to SDE 2010 in Madrid. At that time, I was Strategic Marketing manager for the T&D (Transmission and Distribution) Business Unit for Schneider Electric.

In your opinion, what makes Solar Decathlon a powerful mechanism for communications, marketing, and public outreach?

The key mechanism is the physical presence of the Solar Village with innovative houses, where you can invite company customers and employees, but also reach the general public. Since everyone lives in a house, everyone is touched by this building project focused on housing.

For example, SDE 2014 in Versailles was a perfect mix of business and competition. Schneider Electric (SE) had more than 100 volunteers working at the on-site event, including 20 people who came from outside of France. Over the two-week period, the SE display tent received over 6000 visitors, and most were potential customers.

Across many years of support for ten SD competitions around the world, Schneider Electric has sponsored 32 teams, including four of the winning teams. This has been a successful endeavor that has benefitted the company in many ways. It also became standard practice for SE to seek out the top students from Solar Decathlon competitions as potential new employees, especially the electrical engineers, architects, and communications experts. The hands-on nature of the design-build-showcase process provides excellent work-force development that benefits and connects university students and the working world. Private industry engagement is key for further development of SD competitions, in order to bring government leaders, academe, business leaders, young talent, and the public together to rally around a common purpose and contribute to overcoming climate change.

Michel Orlhac speaking with Prince Felipe VI (now King) at Villa Solar 2010.

Employees of Schneider Electric have been highly engaged volunteers during SD competitions. What feedback did the company receive from those volunteers?

The feedback has been always extremely positive, including from those who managed the volunteers. I like quoting one volunteer from our IT department who volunteered at Versailles for SDE 2014, where the weather was very rainy over the two-week period. He said, "I have been very happy to wade through mud with SE customers for this special event."

Michel Orlhac with Richard King at the Schneider Electric display tent at SDE 2014 in Versailles.

Why do you think the Solar Decathlon is a worthwhile program?

SD is a wonderful initiative to fight climate change because buildings are a major sector that releases large amounts of CO_2 into the atmosphere. The Solar Village is a great way to share proven solutions with building professionals and the general public, so they can experience these first-hand to better understand those innovations.

A fundamental common goal is missing in today's society, since various political situations around the world show that individualism is growing stronger. Solar Decathlon is a perfect program to offer a significant common goal to students worldwide, as well as highlighting the value of sustainability in buildings, a key component of our footprint on the earth.

What are your most memorable moments from SD involvement?

No doubt, the most memorable moment for me was the beginning of the final awards ceremony at SDE 2014 in Versailles, where I organized and participated in singing the European anthem "Ode to Joy" with everyone in the audience. We began singing in German (the original script), then continued in French, then in Italian, and finally in English. There were about 2000 people singing together, mostly Decathletes from all over the world. One French man came up to me later and said, "I've never before felt so proud to be a European." This, by the way, was one of the most memorable moments of my entire 37-year career, along with my first Paris Marathon in 2010 that included greetings from the Schneider Electric CEO Jean Pascal Tricoire after running 42 km through the city!

Michel Orlhac joining in a rousing rendition of "Ode to Joy" with Decathletes onstage at SDE 2014.

Another memorable experience was during the design process in the U.S., many months before the competition, where I interviewed all the participating teams. My job was to identify which ones I would recommend for the local Schneider Electric organization to sponsor and to select some key leaders that I would recommend as new hires for the company. I'd like to mention that after six years of retirement, I am still connected with Solar Decathlon colleagues and stay updated on the competitions around the world. I'd like to see the competition continue in the future.

Another crucial leader in the triumph of SDE is Louise Holloway, Director of the Energy Endeavor Foundation. Her tenacious efforts to ensure the continuation of this competition in Europe led to the creation of a new governing body responsible for securing and planning new editions in different European locations.

Louise Holloway, Director of the Energy Endeavour Foundation (governing body of SDE) and recent winner of Eurosolar's European Solar Prize, is a Paris-based entrepreneur, creative director, and award-winning university educator in the fields of communication strategy and project management. Her emphasis is on how communication, technology, architecture, and entrepreneurship reach the smart citizen in our regenerative urban landscape.

How did you get involved with Solar Decathlon Europe? What motivated you to engage with the competition?

Pascal Rollet, SDE14 Competition Director, brought me onto the team of Organizers after we met through colleagues at l'Ecole Nationale Supérieure d'Architecture de Versailles in France. Those academic colleagues were the link, as they were proposing concepts for Decathlete residences. I was motivated to participate in SDE by the combination of youthful vitality, healthy competition, and genuine concerns for our habitat that the competition represented. As an educator, it's irresistible for me to lecture and teach in international faculties of architecture and engineering, engaging with students on the topics of our human habitat. That cocktail, where science and technology meet with design and communication to seek accessible solutions for real-world energy problems, is magnetic and powerful.

What was your role in SDE14 in Versailles, France? What made that event successful?

I was the lead bilingual communication strategist; in other words, a hybrid filter for message and tone. My job was to communicate the who, what, where, when, why, and how of Solar Decathlon, which required a synthesized position for broad audiences. Accomplishing this in a cross-cultural environment was an additional challenge. I aimed to capture the international flavour of Solar Decathlon in a European context. The success of SDE14 hinged on several important factors:

- SDE14 was a multicultural, multilingual celebration of opportunity. Decathletes from 16 countries created an outstanding Solar Village, which was exhilarating for all involved, and for all audiences.
- The emphasis on communication was embraced through a) the Speed Peer Review™, a series of public, live-streamed, rapid-fire project pitches from the teams; b) culture days for each country to showcase their cultural heritage for visitors; c) the diplomats, ambassadors, and other international dignitaries that came to tour the Solar Village. These actions have since stimulated an on-going dialogue within and beyond the SDE community.
- The UDTA Contest (Urban Density, Transportation and Affordability) was a new initiative that focused on the challenges we face in the European urban environment. This contest urged teams to explore housing beyond single-family dwellings (for example, additions on top of existing multi-family housing blocks) and other topics related to urban living, such as mobility and shared services.

No European nation stepped forward to host SDE16. What steps did you take to "carry the torch" for the competition in Europe? Who or what inspired you to keep it going?

Peter Russell, Pascal Rollet, Sergio Vega, and I were steadfast in pursuing an SDE agenda. We formed a temporary ad-hoc committee referred to as the *Solar Decathlon Europe Secretariat*. We lobbied the European Commission (EC) to push the agenda and make the case for SDE at all levels of government. This resulted in an EC "tender project" to consolidate best practices for SDE and stimulate awareness in Europe, where resource-responsible habitation is a priority, with SDE as a way in. We also connected with key SDE legacy stakeholders through topic-targeted meetings, in hopes that we could build a European-wide consortium of ministerial representatives. Ultimately, financing the continuity required creative, entrepreneurial, and intrepid thinking.

Richard King encouraged Peter Russell and me to secure U.S. DOE endorsement to create a neutral, non-institutional, non-profit business entity to provide continuity for Solar Decathlon in Europe. With this impetus, Peter Russell and I founded the Energy Endeavour Foundation (EEF) to govern SDE through a business model that involves European-wide *Calls for Cities* and corresponding international *Calls for Teams*. Prospective Host Cities bid to host SDE; once designated, they pay fees to the EEF for transfer of strategic tools, best practices, and project-specific knowledge. Since 2017, the EEF has been the driving-force for SDE, acting as a custodian of the brand and rules, providing the structure and framework for SDE, and ensuring the quality and continuity of SDE from one edition to the next. The inspiration to foster this agenda has always been, and remains, SDE teams.

What is your current role with the Energy Endeavour Foundation? Why is the EEF important, and who is on the EEF team to support SDE competitions?

I am Co-Founder and Director of the Energy Endeavour Foundation, working with a small team of dedicated officers to perpetuate SDE and its evolution. Faced with the prospect of hosting SDE, various European cities experience a steep learning curve. Our goal is to guide, assist, educate, and transfer best practices to new host city colleagues as they embark on the complex SDE project. Ultimately, we are guardians of SDE values, setting the tone for this unique hands-on experience. Decathletes learn best, and teachings resonate, when they are mentored by leaders who communicate and represent a consistent, optimistic message of curiosity, safety, solidarity, and professionalism.

The EEF was, and is, important for this effort, as European leadership for SDE is vital. We are champions of the SDE building-in-situ format, where Decathletes come together to construct houses simultaneously under demanding, time-sensitive circumstances. Today, the EEF is a driving force behind SDE, perpetuating the European chapter within the worldwide Solar Decathlon movement. We recently completed a tender with support from the European Commission to collect data on projects related to the development of "smart" cities and buildings. The objective is to ensure high-quality support of SDE and the *European Innovation Partnership on Smart Cities and Communities* (EIP-SCC).

Claudio Montero is our in-house SDE Rules Official. He has unparalleled organizational experience with Solar Decathlon editions in Europe and the Middle East. Sergio Vega, who helped lead the first two SDE editions in Madrid (SDE10 and SDE12), is on the EEF board of directors. The EEF team includes educators, architects, engineers, building scientists, science communication experts, designers, brand strategists, finance experts, managers, and sustainability advocates. We represent nine countries (China, Romania, Italy, South Africa, Colombia, Canada, France, the Netherlands, Spain) and communicate through eight languages. All have SDE experience, and some are previous SDE Decathletes. Collectively, EEF team members have experienced, contributed to, advised, organised, or competed in 16 Solar Decathlon events from various SD chapters worldwide. The EEF also draws on the SDE Council of Experts, comprised of former SDE organisers, advisors, members of the now-retired Secretariat, and other supporters.

SDE19 was held in July 2019 in Budapest, Hungary. Share your reflections about that event.

The result of the EEF's first European *Call for Cities*, SDE19 in Budapest put Solar Decathlon Europe back on the map after a five-year gap. It was an opportunity to gauge interest, consolidate understanding of stakeholders, and review best practices from previous SDE editions. This edition moved us toward combining several energy strategies into potential solutions, rather than focusing on one main energy innovation. With SDE19, the concept of designing single-family homes shifted toward designing shared spaces with connected living in an urban context. Experimentation with a new contest called "Neighbourhood Integration and Impact" paved the way for SDE21 in Germany, when Solar Decathlon Europe became determinedly urban.

After postponement due to the SARS COVID-19 global pandemic, Wuppertal, Germany hosted SDE21-22 in June 2022. What changes did you observe in this competition?

The unforeseen delay presented unexpected challenges to the EEF, local Host City organisers, and SDE teams. Despite this, the competition was a smashing success. Teams demonstrated notable SDE values and qualities, going beyond all expectations with their stamina, dedication, and attention to detail. Excellent craftsmanship was visible in the clever detailing of their houses, as well as strong solidarity among them as a group of teams.

The 2021-22 competition focused on the challenges facing many cities in Europe: how to densify the city in a resource- and socially-responsible manner. Teams proposed designs for multi-unit housing with existing building stock in one of three urban situations: building on top, filling gaps in city blocks, or adding extensions. Each SDE contest was tailored to reflect the over-arching theme of "going urban."

Share your thoughts about the future of Solar Decathlon Europe and how global reactions to climate change have impacted this unique competition.

I think that Solar Decathlon Europe has gained momentum and is now more vital than ever. There is considerable interest from prospective European cities to host SDE25. Host Cities understand that the impact of SDE can be far-reaching and resonant. Host Cities contribute to the evolution of SDE by reflecting current European (and even worldwide) issues. By addressing these pressing topics, SDE Host Cities assist with the evolution of this university-level competition. These diverse European cities foster dialogue that encourages broad economic cooperation and international collaboration. This strengthens the relevance and vibrancy of SDE, an action-driven platform for the health of our communities and the planet.

Adapting to climate change requires innovation on many levels. Tackling this global challenge depends on our choices for materials, building practices, ways of living, use of personal and common resources, and mobility systems. Solar Decathlon Europe showcases the talent that's essential for addressing this challenge. Decathletes embody the concerns of our time. They are ingenious and responsible, while representing a sense of urgency. SDE fosters this steadfast spirit as an extraordinary platform to empower the next generation of clean-energy leaders.

Any additional thoughts?

The Energy Endeavour Foundation recently launched the EEF Empowerment Fund, an entrepreneurship initiative for young women graduates pursuing careers in the built environment. This initiative is largely inspired by our efforts to perpetuate the SDE competition, where women are front and centre in all phases of the design, build, and operation of the complex SDE project. We champion a world where women have a vital stake in the equitable reconfiguration of our built environment. We support economic and societal development in which arts and science contribute to each other, an arena where creativity, communication, interpretive and cross-cultural skills are mutually interwoven with science, technology, mathematics, and engineering. We envision a future in which hybrid, women-driven, entrepreneurial business initiatives are propelled into action.

SDE is a creative, industrious professional opportunity. In many ways, this is in the DNA of the EEF. SDE was the raison d'être behind the EEF, which is an example of multidisciplinary poetry in motion. When I witness the true grit of "Yes, I can do this, too!" among our program officers who are truly inspired by Decathletes, I see how we are all transformed by the SDE experience. It is a privilege to keep on creating opportunities in such a climate of profound optimism.

Solar Decathlon Europe 2019

After an unexpected 5-year break in European editions of the Solar Decathlon, the city of Szentendre, Hungary, located on the Danube River near Budapest, hosted SDE 2019. The event took place in

July 12–28 at the EMI Szentendre Science and Technology Industrial Park. This showcase of innovative houses was also extended for a two-week period after the competition ended, in order to allow more visitors to tour the site and learn about sustainable housing. Dr. Karoly Matolesy, an architect, deputy director at EMI, and honorary professor at TU Budapest, was the director of the competition, and Csaba Szikra, a mechanical engineer who was teaching at Budapest University of Technology and Economics, served as the competition manager.

Entrance to SDE 2019 and welcome tent at the EMI Szentendre Industrial Park.

The 2019 competition took an interesting turn, focusing on the renovation of existing buildings, such as multi-family houses, roof-top apartments, and townhouses (also called terraced houses). The basic idea was to demonstrate that older buildings could be refurbished and modernized to become more energy efficient and sustainable while relying on solar energy for power. Retrofitting and upgrading older buildings in densely packed urban areas throughout Europe is crucial because of precious little space available for construction of new energy-efficient buildings. The 10 contests for the competition in Hungary, shown below, were just slightly different from earlier editions of SDE.

Juried

- Architecture
- Engineering and Construction
- Energy Efficiency
- Communication and Social Awareness
- Urban Neighborhood Integration and Impact
- Innovation and Viability
- Circularity and Sustainability

Measured

- Comfort Conditions
- House Functioning
- Energy Balance

Decathletes from eight nations on three continents represented 10 participating teams at SDE 2019.

- Belgium
  - Ghent University
- France
  - Ecole Nationale Supérieure d'Architecture et de Paysage de Lille
- Hungary and Algeria
  - University of Miskolc, University of Pecs, and Saad Dahlad University of Blida (Algeria)
  - Budapest University of Technology and Economics
- Netherlands
  - Delft University of Technology
- Romania
  - Technical University of Civil Engineering Bucharest
- Spain
  - Universidad de Sevilla
  - Universidad Politécnica Catalunya
  - Universitat Politècnica de València
- Thailand
  - King Mongkut's University of Technology Thonburi

Over4 from the Technical University of Civil Engineering in Bucharest, Romania, won third place. This team developed their renovation project for a five-story apartment building that represented the late 1900s style found in many Romanian cities.

This prototype from team *Over4* represented a rooftop unit that could also be built as a free-standing dwelling.

The front entrance of the *MOR* team house from TU Delft, which took second place in the competition.

The *MOR* team design strategy was to renovate underperforming office buildings into net-positive, multipurpose buildings.

Habiter2030 from the Ecole Nationale Superieure d' Architecture et de Paysage in Lille, France, along with other partner schools and universities, took first place overall. This prototype represented a renovation of 1930s houses in urban areas of northern France.

Hungarian Nest+ house from the University of Miskolc, University of Pecs, and the University of Blida featured a Venturi plate at the top that helped with passive ventilation in summer (hot air flows up and out of the house).

In the winter, the mirrors in the rooftop Venturi plate reflect sunlight into the *Hungarian Nest+* house.

The back side of *Proyecto 3.1* from the University of Seville, Spain; this was an unusual design with open interior community space surrounded by pods that extended out from the main structure for private use, such as sleeping, studying, etc.

The front doorway of University of Seville's unique prototype house.

Special recognition was announced for several teams at the final awards ceremony. The *KMUTT* team from Thailand received the "Public Choice Award" for their *Resilient Nest* project. The team from Ghent University in Belgium received the "Tungsram Smart House Award" in honor of their *Mobble* project, and the *Koeb* team from Budapest University of Technology and Economics won the "Speed Peer Review Award" for an excellent summary presentation about their house. A consortium of universities from Pécs and Miskolc in Hungary and Blida in Algeria (Team SOMEshine) received the "ÉVOSZ Award" from the National Association of Construction Companies in Hungary, as well as the "Community Award."

The "Most Podium Awards" trophy went to the *MOR* team from TU Delft after their appearance in the winner's circle for eight of the 10 contests at SDE 2019. They received the following awards:

- **First place:** Energy Efficiency, Innovation and Viability, Communications and Social Awareness
- **Second place:** Urban Neighborhood Integration and Impact, Engineering Design and Construction, Residential House Functionality, Circularity and Sustainability
- **Third place:** Comfort Conditions

In 2019, plans were also well underway for the next European edition of Solar Decathlon to be held in Wuppertal, Germany.

However, the unexpected (and unwelcome) appearance of COVID-19 in the winter of 2020 threw a monkey-wrench into any efforts for large public gatherings. All across the globe that year, many nations announced restrictions on how many people could be together in one place. In addition, most events that attracted sizeable crowds, such as the Olympics (in Japan) and World Expo (in the UAE), were completely off the table, and they were postponed for at least a year. As a result of the unpredictable spread of COVID-19 around the world, SDE 2021 was delayed by a year and rescheduled for June 2022.

Despite multiple challenges during the pandemic, including cancellation of in-person classes at most universities for a year or more, teams selected for SDE 2021 never lost sight of their ambitious goals for the Solar Decathlon. Resourceful faculty mentors found new ways for students to proceed with digital learning at a distance and included lively online discussions as part of regular coursework. SDE teams maintained strong communications with each other and persevered with their design-and-build projects for the next competition. Unforeseen obstacles that emerged during this global health crisis did not diminish the passion of Decathletes so strongly committed to showcasing their innovative, net-zero-energy dwellings that would tread more lightly on the earth.

Solar Decathlon Europe 2021/22

Following a 12-month postponement, the fifth edition of SDE took place on June 10–26, 2022, in Wuppertal, a city in northwest Germany along the Wupper River, a tributary of the mighty Rhine. The city is home to four institutions of higher education, as well as the Wuppertal Institute for Climate, Environment, and Energy. SDE 21/22 carved out a unique path for Decathletes by asking them to consider solutions for Wuppertal's historic Mirke District, adjacent to the chosen site for the competition. This district includes an old railway station (Mirker Bahnhof) and numerous buildings constructed in the early 20th century and post-World War II period, some of which are no longer in good condition. Private landlords own many of the residential buildings, and this ownership fragmentation creates distinct challenges for urban revitalization. Recent local efforts to "transform" this area from the bottom-up led to greater interest in renewal, and SDE 21/22 became an excellent complement to this emerging trend.

Aerial photograph of the competition site for SDE 2021/22 in Wuppertal, Germany. Prototype houses were assembled along one main street, next to a central bike path/pedestrian walkway that runs by an old railway station, which provided easy access for visitors.

This view of the western end of the Solar Village shows prototype houses from Taipei, the Czech Republic, and Pecs, Hungary.

The historic Mirke District adjacent to the SDE 21/22 site in Wuppertal, Germany.

This view of the eastern end of the Solar Village includes prototype houses from Bucharest, Romania; Rosenheim, Germany; and Grenoble, France.

The competition site featured shaded rest areas for visitors made of recycled materials.

Solar Decathlon teams for this competition were invited to develop designs for sustainable housing that represented one of three possible solutions for the Mirke District in Wuppertal:
- Renovation and extension of existing buildings to make them more attractive and energy efficient.
- Closing gaps between existing buildings (vacant lots) to provide more living space and demonstrate renewable energy strategies.

- Renovation by adding more stories on top of an existing building to increase residential space and/or community areas that also function as rooftop sites for solar panels.

Another option was for teams to select a specific urban setting in their own country and then design a prototype house that represented one of the three approaches noted above. The emphasis on sustainable solutions for cities was in response to Europe's densely populated urban areas that created ever-increasing resource demands, while the threat of climate change became more urgent. This brilliant strategy for SDE 21/22 gave the teams a unique purpose: to suggest innovative solutions for current real-life situations that would benefit immediately from fresh, forward-looking ideas applicable and available right now.

Competition Director for SDE 21/22 was Wuppertal University Professor Karsten Voss of the Faculty of Architecture and Civil Engineering. Dr. Voss has been engaged with previous Solar Decathlon Europe competitions and has deep expertise in net-zero-energy buildings, equilibrium buildings, and carbon neutral cities. Dr. Daniel Lorberg was Director of the SDE project, and Marion Whitefield served as Head of Communications for this event.

An exciting concurrent development was the establishment of the Living Lab NRW (North Rhine-Westphalia) in Wuppertal to conduct research related to sustainable, energy-efficient construction and connect with universities in the region to promote R&D. Eight of the competition houses at SDE 21/22 were selected to remain onsite and become part of an ongoing exhibition and living laboratory where hands-on investigations, data monitoring, and data analysis will contribute to improved understanding of the state-of-the-art construction that embraces renewables and sustainability concepts. This commendable plan for studies and outreach, which includes guided tours and special events, will enable greater public engagement and learning for people of all ages. A huge win for everyone!

The German Federal Ministry of Economic Affairs and Energy also provided funding for a knowledge platform (online database) to gather and organize information and data from other Solar Decathlon competitions and living labs around the world. This effort is part of Annex 74, under the umbrella of the Energy in Buildings and Communities Program at the International Energy Agency. This central storage platform maintains copies of publications, manuals, drawings, scores, and other reports, which is a valuable asset for reference and research purposes. Collecting, preserving, and making these data accessible for relevant use is of enormous value for countless reasons, now and in the future. Hopefully, this supports scientific studies that enhance understanding of sustainable building practices and materials, energy efficiency, and architectural design for dwellings powered by renewable energy. Dr. Karsten Voss (University of Wuppertal) and Dr. Sergio Vega (Polytechnic University of Madrid) helped establish Annex 74, and they will continue to lead this rich knowledge platform.

The fifth edition of SDE involved 16 teams that represented universities from 10 countries. Unfortunately, some teams selected were unable to participate because of travel difficulties and shipping constraints due to the COVID-19 pandemic. In the list of competitors cited below, the "LL" designation indicates that this team's prototype house will remain onsite in Wuppertal as part of the Living Lab project.

- **Czech Republic:** Czech Technical University (LL)
- **France:** École Nationale Supérieure d'Architecture de Grenoble (LL)
- **Germany:**
    - o   FH Aachen, University of Applied Sciences
    - o   Biberach University of Applied Science
    - o   Düsseldorf University of Applied Sciences (LL)
    - o   Hochschule für Technik Stuttgart
    - o   Karlsruhe Institute of Technology
    - o   Rosenheim Technical University of Applied Sciences
    - o   Technical University of Applied Sciences Lübeck and Istanbul Technical University, Turkey
- **Hungary:** University of Pécs (LL)
- **The Netherlands:**
    - o   Delft University of Technology (LL)
    - o   Eindhoven University of Technology
- **Romania:** Research Institute in Bucharest (EFdeN)

- **Spain:** Universitat Politècnica de Valencia (LL)
- **Sweden:** Chalmers Technical University (LL)
- **Taiwan:** National Yang Ming Chiao Tung University (LL)

As in every Solar Decathlon, selected panels of distinguished professionals were invited to visit and evaluate the prototype dwellings during the onsite competition. Teams were well prepared for jury tours and devised creative ways of highlighting special features of their houses. In addition to excellent summary presentations, teams set up hands-on displays of sustainable materials used in construction, organized photos to highlight their design-and-build process, offered short explanatory videos, and even performed short skits to tell their story. Over time, Decathletes have learned that clear and convincing communications are essential for boosting scores. Most teams include students with expertise in the fields of marketing and public relations to ensure that messaging and delivery of information about their projects is top notch. SDE also requires that communications for the competition be available in English, and therefore, teams have extra steps to translate oral and written materials from their native language into English—no small feat. What these extraordinary students accomplish is impressive.

The 10 contests for SDE 21/22 embodied only a few changes from the previous edition, with the majority of points earned through contests that are judged by professional panels of experts.

Juried

- Architecture
- Engineering and Construction
- Communication, Education, and Social Awareness
- Urban Mobility
- Affordability and Viability
- Sustainability
- Innovation

Measured

- Energy Performance
- Comfort
- House Functioning

Europeans Take Up the Charge | 351

Level Up from Rosenheim University was designed as an elegant top floor for aging mid-20th century apartment buildings. Prefabricated modules for easy transport and construction made it possible to create different floor plans for these units, which are connected to community spaces such as gardens, greenhouses, studios, and work areas.

The attractive *EFdeN* house from Ion Mincu University of Architecture and Urbanism in Bucharest, Romania, had bifacial PV panels that produced solar energy with cells on the top and bottom; the system stored excess energy in repurposed batteries that came from electric cars. House walls were made of traditional straw panels insulated with wool. This prototype was planned as a model of affordable housing for single people in a community setting. This stunning prototype won the "People's Choice Award."

Stuttgart University's *coLLab* prototype was designed for use in student dormitories. It featured a wooden grid placed on top of an existing structure, with interior units that had functional walls containing built-in furniture and equipment, such as a kitchen table, desk space, seating, and bed that could be pulled out and unfolded for daily use.

The exterior walls of the *coLLab* prototype had living plants growing between solar cells that provided shade and power, which contributed to its dazzling appearance.

Aachen University designed their *Local+* prototype to "close gaps" between buildings in the Mirke District. By using small, flexible, movable "CUBEs" for personal retreats (sleeping) positioned within a larger central space containing a kitchen, pantry, and bathroom for shared use, this model encouraged social interaction. The team used exterior walls to communicate more information about the house.

SDE 21/22 teams developed impressive displays of sustainable materials used in fabricating their prototypes to educate visitors about the value of these options.

A non-competitive aspect of SDE in Wuppertal was the scheduling of specific dates dedicated to "culture and country days" during the onsite event. This provided an opportunity for teams to showcase their own unique heritage and traditions, including games, costumes, songs, food, and other activities of general interest. With such a multinational collection of teams engaged in this highly intense design-build-and-operate competition, the chance to celebrate and share your own cultural roots and customs with others was a welcome change of pace. Students thoroughly enjoyed it, and visitors had lots of fun learning about and experiencing traditions from other regions.

Reflections from Linus Knappe, who was communications manager for Team MIMO from Düsseldorf, Germany, at SDE 21/22. Linus is studying communication design at the University of Applied Sciences Düsseldorf, and he also teaches courses on web design, new media, and photography at Alanus University of Arts and Social Sciences.

What are you are currently engaged with?

I am studying communication design at the University of Applied Sciences in Düsseldorf, Germany, where I am also co-leader of the editorial team for the university magazine *FETZ* and part of a group that is creating a poster archive for the university. In addition, I'm doing contract work for companies, non-profit organizations, and local projects, plus designing a new book and website, to be published soon.

Your team won first place in the Communications Contest. What was your communications strategy, and why was it effective?

Probably the most important aspect of our communication strategy was to give it a lot of attention. In the end, all the ideas we developed in architecture, engineering, and social planning can only make a difference if other people get to know and understand them: whether architects, politicians, scientists or citizens. We have taken care to communicate clearly, so that people of all ages and backgrounds can understand our messages. We have even developed special media for children.

Even though social media played a big role in reaching people and getting the word out about the project, the website was the heart of our communication. The website allowed us to communicate our issues in detail, with all the necessary media and in a way that perfectly suited our content, free from social media structures and algorithms. In addition, the website makes the project accessible to many people: Multiple media can be used to explain the content in multiple complexities; texts can be searched, resized and contrasted, automatically read aloud, and automatically translated. For this reason, we have chosen to use the website as a digital guide that enriches the public tours of the prototype houses.

The attention paid to communication was also reflected in the size of the communication team: several people from different faculties and with different specializations worked together. At its core, the communication team consisted of designers, social researchers, and architects, who communicated with the other departments depending on the topic, creating balanced and socially conscious communication. Quite a lot of communication took place within team communication.

Finally, our media presence was consistently designed, which gave our team a distinctive appearance and helped our team members to identify with the team.

How did your experiences with Solar Decathlon Europe influence your academic, professional, and personal life?

I learned a lot while collaborating with talented people like photographers, animators, and social researchers. I had the opportunity to create a corporate design that had to work in many different contexts: Web, print, animation, visitor guides or team apparel. It was insightful to see how the corporate design evolved and adapted in all these mediums.

SDE21/22 has sparked many discussions about the future of cohabitation. The project has brought the issues to life and made them tangible. For students, it is a great opportunity to learn in a practical project and make their ideas heard.

When the scores were compiled from all 10 contests, the leader of the pack was *RoofKIT*. Decathletes from the Karlsruhe Institute of Technology in Germany took first place overall with a 13-point lead over the second-place winner, team *VIRTUe* from Eindhoven University of Technology in the Netherlands. For third place in the competition, two teams tied with the same exact score, just six points behind *VIRTUe*. The teams that tied for third place were: *SUM* from Delft University (Netherlands) and *AuRA* from the Grenoble National School of Architecture (France). The local team from the University of Applied Science in Dusseldorf (Germany) took fourth place with their *MIMO* prototype. What a celebration! The teams had put their all into this effort, and they were thrilled to make it to the finish line. In fact, everyone was a winner, and the collective exuberance was palpable at the final awards ceremony.

The *MIMO* team from Dusseldorf University of Applied Sciences in Germany took fourth-place honors. Their prototype was designed for additional stories on an old warehouse in Wuppertal.

The façade for *MIMO* was a climate shell of horizontal glass slats around the dwelling that allowed fresh-air ventilation between the slats, which also contained solar cells to generate electricity. The roof collected rainwater that can be stored in a cistern, and a rooftop greenhouse provided garden areas for residents to grow their own food.

The team from Dusseldorf, pictured here with Richard King and fellow CESA jury members, Jakob Schoof and Dr. Karin Stieldorf, won first place in the Communications, Education, and Social Awareness Contest (CESA). One of their clever ideas for marketing and public relations was creating a character called "little MIMO" to represent their prototype: Minimal Impact Maximum Output. Visiting children were asked to find five hidden MIMOs (a yellow sun with a smiley face, arms, and legs) in places throughout the house where they could learn something.

Surprise! There is a little MIMO behind a closet door for storage under the staircase.

SUM from Delft University made a bold statement on the side of their house to explain how their present-day dwelling connects the past and future. This prototype is based on a case study of the Bouwlust District of tenement flats in the Hague, Netherlands. The Delft team finished in a third-place tie with the team from Grenoble, France.

The Delft team prototype responded to two national problems in the Netherlands: (1) an exponentially growing housing shortage, and (2) the need for energy-neutral building infrastructure. Their *SUM* design suggested new public functions for the ground floor, such as a maker space and restaurant, and new apartments on the top floor with PV panels on the roof and integrated into the façade of the building.

Europeans Take Up the Charge | 359

Team *VIRTUe* from Eindhoven University in the Netherlands placed second overall in the competition. Their exemplary communications strategy centered on their concept of a ripple effect, like a drop of water on a pond. Their project hoped to create a ripple of influence to encourage the replication of more sustainable buildings.

Team *VIRTUe* utilized spaces effectively, both inside and outside their prototype, for messaging and educational purposes.

The *VIRTUe* prototype featured wood as the primary construction material and contained three distinct layers: fixed (load-bearing), flexible (façades), and free (furniture). Solar and solar-thermal panels were fabricated in a color that blended with the exterior wood for a sleek appearance. Each modular unit offered two apartments plus a common area with major appliances for shared use by residents, and the roof space was dedicated to gardens and social activities.

Team AuRA from the Grenoble School of Architecture in France were in a two-way tie for third place overall. Their design was intended as a roof extension on an aging hotel at a ski resort in the mountainous Vercors region of the French Alps.

Europeans Take Up the Charge | **361**

The AuRA team plans to work with the local community to refurbish the hotel as new apartments for sustainable housing powered by photovoltaics on the roof. Their goal is to revitalize the area and attract new permanent residents.

The AuRA team constructed their prototype with bio-based, geo-based, and recycled materials and set up a terrific demonstration space where visitors could learn about them and get hands-on practice applying them to an exterior wall of the dwelling.

CESA jury member Dr. Karin Stieldorf enjoying the bright, comfortable space, including a hammock, sink, and beverage storage, on the top level of the AuRA house.

Melissa King with AuRA Decathletes in the open communal space created for studying, socializing, and relaxing on the upper floor of their prototype.

RoofKIT from the team at Karlsruhe Institute of Technology, Germany, was thrilled to win first place at SDE 21/22. They designed a prefabricated, modular, lightweight prototype for rooftops in the Mirke District of Wuppertal to show how a housing unit could be supported on a typical roof.

The *RoofKIT* team demonstrated how living spaces could be flexible to meet the diverse needs of residents, while showcasing their beautifully crafted, highly efficient unit that aimed to promote a sense of togetherness.

The first-place winners from the Karlsruhe Institute of Technology were ecstatic to take top honors!

Reflections from Karsten Voss, Professor for building physics and technical services at the School of Architecture and Civil Engineering at the University of Wuppertal, Germany, and Competition Director of SDE 2021/22. A mechanical engineer with a PhD from EPFL in Lausanne, Switzerland, Dr. Voss has also been a researcher at the Fraunhofer Institute for Solar Energy Systems and expert consultant for the International Energy Agency. He has written numerous scientific articles and books on energy-efficient buildings and building performance. He was faculty advisor for Team Wuppertal at SDE 2010 and continues to be actively involved with IEA Annex 74.

How and when did you first get involved with Solar Decathlon Europe? What motivated you to become part of this effort?

My involvement started in 2008 with the application of Team Wuppertal for the first European edition of the Solar Decathlon in Madrid, Spain in 2010. The fascination for me was the intensive interaction of research and education with an interdisciplinary approach.

SDE 2021/22 took place on the heels of the COVID-19 pandemic. How did this timing impact the event?

Despite or even because of its Corona-related postponement of nine months, SDE 21/22 came at exactly the right time. In a situation of maximum uncertainty about our future energy supply, the event and all the university teams provided food for thought on how we can build our way out of the current energy, resource, and affordability crisis in architecture. The focus was on three main sustainability strategies: sufficiency (reduction of living space per person), efficiency (economy in the use of energy and resources), and consistency (consistent recycling of the materials and products used).

Some of the student-built houses from SDE 2021/22 are part of a new research study which you helped to plan and organize. What are the primary objectives of that study?

Eight of the 16 SDE demonstration prototypes will remain on-site for up to five years. They will form the so-called "living lab NRW," a specialized communications network to promote climate-neutral buildings and sustainable living in cities of the future. The project includes research within a PhD network from regional universities, as well as a practice-oriented place of learning for students, trainees, and the public.

Over the years, you have taken on different roles with SDE, from faculty advisor to competition director. Based on your experience, why and how has this competition been beneficial for European nations?

The main competition focus is well in line with the EU energy policy and namely the energy performance directive (EPBD). The "nearly zero energy building" was introduced in 2010, and the "climate neutral building stock 2050" becomes the current aim. It is obvious that the main task in Europe is on dense urban living and further development of the existing building stock. The correlating development and expansion of the SDE profile started with the "Declaration of Madrid" signed by all faulty advisors at the first European Solar Decathlon in 2010. It then took us twelve years until Solar Decathlon Europe became the first fully urban edition at Wuppertal, Germany in June 2022.

Europeans embraced the vision of Solar Decathlon, but they wisely adapted the competition to match the geophysical settings of their continent and the human habitats in their landscape. With an eye to the current constraints on limited available land and predominance of urban environments, as well as their propensity to focus on a sense of community and social awareness among diverse populations, the Europeans molded the 10 contests to better fit their needs. The visionary leaders noted in this chapter inspired the remarkable development of multiple European editions of the

Solar Decathlon. Their strong commitment to sustainability and emphasis on innovation in the *renewal of existing building stock* led to well-founded adjustments in some aspects of the competition. The capacity for flexible implementation of Solar Decathlon is a hallmark of its success as a creative endeavor over many years in different places.

Communications, Education, and Social Awareness (CESA) jury members Jakob Schoof, Dr. Karin Stieldorf, and Richard King at SDE 21/22 in Wuppertal, Germany.

Richard King sharing his perspectives about the power of communication in a presentation at the CESA awards ceremony. Pictured onscreen is the U.S. Capitol with the inaugural Solar Decathlon competition site in the foreground on the National Mall in Washington, DC.

Reflections from Professor Peter Russell, a Canadian-born architect and academic leader who took part in Solar Decathlon Europe 2012 as a Faculty Advisor; he has been involved with SDE ever since. He has led the faculties of architecture at the RWTH Aachen University, Germany, and TU Delft, the Netherlands. Peter is currently Dean of the Institute for Future Human Habitat Studies at the Tsinghua Shenzhen International Graduate School, China.

You've been engaged with Solar Decathlon in different locations and in varied roles (faculty advisor, support staff, organizer). Share your key take-aways from being part of the competition.

My first contact with Solar Decathlon was seeing the results of Solar Decathlon Europe 2010 in Madrid. Some of us at the RWTH Aachen University in Germany thought to ourselves, "We can do that!" We applied to and were accepted for SDE 2012. My role was as a faculty advisor. Later, I became involved in SDE 2014 as an observer for the German government. I also provided technical direction for the Speed Peer Review and Award Ceremonies. After that event, a group of organizers (Louise Holloway, Sergio Vega, Pascal Rollet and me) explored ways to secure a stable platform to perpetuate Solar Decathlon Europe. As a result of those efforts, Louise and I established the Energy Endeavour Foundation (EEF), the governing body of SDE endorsed by the U.S. DOE.

On a personal note, the experience as faculty advisor was one of the most rewarding periods in my time as an educator. The thrill of seeing Decathletes complete their house was emotional. At the most recent SDE 2021 event in Wuppertal, it was rewarding to witness and feel the exciting atmosphere. Professionally, the SDE pedigree has helped establish my reputation as a leading educator and secured my position as Dean at the Institute for Future Human Habitat Studies of the Tsinghua Shenzhen International Graduate School in Shenzhen, China.

What have you observed regarding how students have responded to SD competitions?

Students who decide to become Decathletes are a group of soon-to-be professionals who share common traits: ambition, curiosity, and courage. They lack experience, but SD gives them loads of that! The motto of the founder of the Olympics, Pierre de Coubertin, said, *"The important thing in life is not the triumph but the struggle, the essential thing is not to have conquered but to have fought well."* This holds true for Solar Decathlon. As in all competitions, there is a fierce competitive spirit, but Decathletes understand that the real value is not the competition or a prize, but the experience shared with other student teams.

You have attended SD events across the globe. How would you characterize the cultural influences on the competition in various locations?

Each edition of Solar Decathlon has featured two important aspects. The first has to do with the challenges facing the vernacular architecture, the local building culture, and the regional climate. This is reflected in the ten contests and the emphasis of the juried contests. The second aspect is cultural. This is reflected in how the public engages with the houses at the solar village. In some cases, the solar village becomes a festival of construction. In others, it is more technically oriented, and in some, it is carefully curated.

Both aspects are important, and the impact of SD is dependent on both. In the first place, SD addresses the pressing challenges of the regional construction industry and in the second place, SD is able to reach the broader public. As such, the diversity of different SD chapters simply reflects the broader diversity of humanity and this makes each chapter of SD effective and impactful.

Climate change has become an urgent problem. How has the architectural design of SD houses been a unique way of responding to this issue?

The architectural designs found in Solar Decathlon prototypes are as diverse as the students creating the designs. What is special is that students continue the 'pursuit for beauty' while responding to the performance demands of resource responsibility. The diversity of designs and students' creative solutions demonstrate that we can build and live responsibly without sacrificing comfort, quality, or beauty.

As a professor of architecture and faculty advisor for SD teams, what are some design ideas that have emerged as viable solutions to housing challenges?

In the wide range of designs for SD competitions, two trends stand out: spatial efficiency and resource optimization. Neither has to do with energy production (i.e., photovoltaics) per se, but these trends affect the overall energy balance of a home. First, constraints on competition house size (circa 800–1000 sq. ft.) have meant that the ambitions of designers must focus on creative use of flexible spatial arrangements. This results in innovative ways that make more efficient use of space, which affects the overall impact of the structure on energy use and the embodied energy in the materials.

The second trend is recognition of material-use as a huge factor in the CO_2 footprint of a building. New tools to assess this impact have enabled students to make interesting choices to recycle, upcycle, re-use, and re-purpose building components and materials. The most recent winner of SDE 21/22 (*RoofKIT* from Karlsruhe, Germany) embraced these principles in its winning design.

How do you envision Solar Decathlon competitions in the years to come?

Firstly, a true 'Solar Decathlon' is entirely dependent on the tangible reality of a 'Solar Village' and public exhibition. Creating design challenges can stimulate a semester's worth of ideas, but the real value of the SD experience is realizing the houses and doing that together, simultaneously, with other teams. Sure, this involves shipping and travel, but the total impact on visitors (virtual and physical) is far greater than any other way of spreading the word on sustainability in the built-environment to the general public.

Another point is that SD houses are prototypes built by students. Just as a racing boat may not be a vessel for a comfortable afternoon sail, these dwellings are not necessarily ready for market. Nonetheless, the competition serves as a test-bed for many ideas, from a scientific point of view (*does this actually work?*) and a test of the market (*do people like or understand or want this?*). Ideally, professionals will take the best of these tests and create new products and services. This might be Decathletes or the companies they work with.

Solar Decathlon has been called the 'America's Cup of architectural competitions' (Pascal Rollet). This is an apt comparison, as SD combines the spirit of intense competition, use of the latest technologies, and a spirit of adventure, ambition, and confidence. To demonstrate to the wider public that the planet is in good hands, there is no better place to observe this than in a Solar Village at a Solar Decathlon on-site event.

Chapter 7

Newfound Momentum

Solar Decathlon 2013

After learning about European competitions that spanned 12 years, we move our attention back to the United States where Richard King was searching for a location for Solar Decathlon 2013 (SD 2013). The highly successful 2011 event in West Potomac Park was a one-time arrangement, so a different site was needed for the next competition. The U.S. Department of Energy (DOE) issued a solicitation on October 18, 2011 to seek a new host and partnership agreement with city, state, and local governments for the next U.S. DOE Solar Decathlon. Several applications were submitted, and three venues looked promising: Kansas City, Missouri; San Diego, California; and the Orange County Great Park (OCGP) in Irvine, California. To evaluate the possibilities and determine which site best met the necessary criteria, Richard made onsite visits to all three places. There were pros and cons for each, but one spot stood out: the OCGP in Southern California.

Originally known as Irvine Ranch, the Great Park was once a rich agricultural region that encompassed 110,000 acres. After the attack on Pearl Harbor in 1941, the United States needed a new Marine Corps air base in the western part of the country, so that

---

Solar Decathlon: Building a Renewable Future
Melissa DiGennaro King and Richard James King
Copyright © 2024 Jenny Stanford Publishing Pte. Ltd.
ISBN 978-981-5129-47-2 (Paperback), 978-981-5129-13-7 (Hardcover), 978-1-003-47759-4 (eBook)
www.jennystanford.com

land was purchased to build El Toro training base for pilots, flight crews, and military personnel. After the Marine Corps Air Station El Toro officially closed in 1999, Orange County voters declared their support for a new park and nature preserve to be created on that huge tract of land. When Richard King visited the Great Park, he was impressed with the expansive acreage and mile-long runways that would enable a solar village to be constructed on hardscape. Paved surfaces make it easier and less expensive to build houses, assemble big tents, and set up walkways, in contrast to previous construction of a solar village on grassy areas prone to wear and tear and muddy ground. In addition, facilities offered at the Great Park included a historical hangar to accommodate large gatherings, plus ample parking, picnic areas, public rest rooms, and proximity to metropolitan areas in Southern California. Without doubt, it was a winning site!

The Marine Corps Air Station El Toro, which boasted mile-long runways, was later developed as the Orange County Great Park in Irvine, California. It was an excellent venue for Solar Decathlon 2013.

The Orange County Great Park with the historic airplane hangar was an outstanding location for SD 2013.

In January 2012, the U.S. DOE announced that the Great Park had been selected as the host venue for SD 2013. Government leaders in Orange County and the city of Irvine were delighted, and they offered generous support for the event, such as marketing, park security, online media coverage with daily updates, and co-funding. At subsequent meetings in Irvine, California, Richard King and staff members of the OCGP hatched a brand-new idea: expand the educational value of Solar Decathlon by including an Energy Exposition (XPO) for the public. The XPO would feature interactive exhibits, children's activities, and learning opportunities for visitors of all ages. The XPO would also offer the SunShot Innovation Pavilion, an informative trade show that aimed to connect the visiting public with clean energy services and products available at that time.

Visitors to the 2013 Solar Village appreciated the pleasant, sunny weather every day of the competition.

The Energy XPO included engaging, interactive children's activities.

On "school days" at SD 2013, buses brought students to the Solar Village so they could tour the houses and learn about clean energy solutions.

Richard recalls the planning phase with these words: "I was excited about the possibilities with this outstanding location in California. It was quite a transformation to begin planning a west-coast edition of the Solar Decathlon where we could tap into a different audience. The addition of an XPO meant we could extend and diversify our outreach, capturing a new set of visitors out west and educating thousands more people about clean energy and sustainability. However, I realized there was a steep learning curve for Solar Decathlon Organizers. We all had to figure out how to manage logistics and challenges at the new site, how to work with park staffers and local government, how to orchestrate hospitality for teams and visitors, and how to set up public transportation to the Great Park. Plus, diving into plans for an XPO represented a whole new dimension of effort, but I was enthused about broadening Solar Decathlon's exposure and intensifying its educational value."

Members of the Board of Directors for the Foundation of the Great Park were especially supportive of the Solar Decathlon. They were active in the local community, and their spirited enthusiasm for this unique competition led to increased sponsorship, financial contributions, and heightened neighborhood engagement. For example, Janet and Walkie Ray were so thrilled about SD 2013 that they organized formal and informal gatherings in Orange County to spread the good news and encourage attendance at the event.

To solicit teams for SD 2013, the U.S. DOE issued a formal request for proposals on July 22, 2011. At that point, the location for the event was not yet clear, but that did not deter universities from sending in applications. By November, more than two dozen universities from the United States and abroad had responded to the RFP. Team proposals were evaluated on the following criteria:

- Technical innovation and design
- Fundraising and team support
- Organization and project planning
- Curriculum integration
- Conceptual design
- Special considerations, such as diversity, geographic location, etc.

The evaluation group, comprising U.S. DOE and NREL staff members, was more impressed than ever with the overall quality of university proposals. With each successive competition, proposal content became more sophisticated, innovation was more prominent, and "smart" technology emerged front and center. For this sixth edition of the Solar Decathlon, students from returning universities who had participated in prior SD competitions could stand on the shoulders of giants. In other words, those students benefitted from the first-hand expertise gained at previous SD events.

In January 2012, U.S. DOE Secretary Steven Chu announced the 20 teams chosen to participate in SD 2013. The selected competitors included nine returning teams and 11 new teams representing four countries, 14 states, and Washington, DC. In the press release from the U.S. DOE on January 25, Dr. Chu expressed these words: "The heart of the Solar Decathlon lies within the creative students and the passion and energy they bring to this important design challenge. I feel confident the competitors announced today will produce the most exciting competition held to date. We look forward to working with this new class of energy leaders."

Team projects for SD 2013 showcased a wide range of approaches to architectural design, engineering solutions, and applications of innovative technology. University teams zeroed in on certain types of building structures that would be optimal for specific geographic locations and climates, such as the coastal desert climate of San Diego, California. They researched locally available, sustainable materials and factored in life-cycle costs. They identified target markets in urban, suburban, and rural regions with an eye for more impactful communications with potential homebuyers. In essence, over time, Decathletes had learned to be more holistic in planning their net-zero-energy houses by considering a wide variety of factors during the initial design phase that included input from a highly diverse group with different backgrounds. As a result, what they produced became laser focused on solar houses that would be more energy efficient, sustainable, appropriate, attractive, affordable, meaningful, and *acceptable for a designated locale*. Hence, this led to an increasingly eclectic collection of ingenious designs that were engineered with strategic effectiveness.

The field of competitors for SD 2013 included six teams from the west, five from the east, four from the middle of the country, and four international teams.

- Arizona State University and The University of New Mexico (Tempe, Arizona, and Albuquerque, New Mexico)
- Czech Technical University (Prague, Czech Republic)
- Middlebury College (Middlebury, Vermont)
- Missouri University of Science and Technology (Rolla, Missouri)
- Norwich University (Northfield, Vermont)
- Queens University, Carleton University, and Algonquin College (Kingston and Ottawa, Ontario, Canada)
- Santa Clara University (Santa Clara, California)
- Southern California Institute of Architecture and California Institute of Technology (Los Angeles, California)
- Stanford University (Palo Alto, California)
- Stevens Institute of Technology (Hoboken, New Jersey)
- The Catholic University of America, George Washington University, and American University (Washington, DC)
- The University of North Carolina at Charlotte (Charlotte, North Carolina)
- The University of Texas at El Paso and El Paso Community College (El Paso, Texas)
- University of Calgary (Calgary, Alberta, Canada)

- University of Louisville, Ball State University and University of Kentucky (Louisville, Kentucky; Muncie, Indiana; and Lexington, Kentucky)
- University of Nevada Las Vegas (Las Vegas, Nevada)
- University of Southern California (Los Angeles, California)
- Vienna University of Technology (Vienna, Austria)
- West Virginia University (Morgantown, West Virginia)

All teams came together in January 2012 for an orientation workshop inside the historic hangar at the OCGP. Decathletes had a tour of the site, and each team visited the assigned lot where their house would be located in the Solar Village. That gathering was a fantastic opportunity for teams to get acquainted, while growing their understanding of the new venue and upcoming competition. Richard King had discovered the tremendous value of bonding within and among participating teams, and he believed this helped create a friendly competition built upon genuine community respect.

In addition, offering a stimulating workshop midway through the planning phase proved to be an excellent way to elevate Decathlete motivation. By then, teams were knee deep in design details and probably facing unexpected constraints due to time, cost, and engineering. Bringing them together helped teams realize they were not alone in managing many of these challenges. Again, leadership surfaced as a key factor for Solar Decathlon teams. When they met 19 months before the opening of SD 2013, leaders had already emerged. The Organizers watched in amazement as projects unfolded and leadership became a decisive element in overall success.

The 10 contests for SD 2013 were basically the same as SD 2011, except for some adjustment in how points were determined. There were five juried contests (subjective) and five measured contests (objective), each one worth 100 points, which created strong balance for overall scoring.

Juried

- Architecture—100 points
- Engineering—100 points
- Market Appeal—100 points
- Communications—100 points
- Affordability—100 points

Monitored

- Comfort Zone—100 points
- Hot Water—100 points
- Appliances—100 points
- Home Entertainment—100 points
- Energy Balance—100 points
    - A bidirectional meter on every house determines the net energy produced and consumed during the competition.

Jury Members for Solar Decathlon 2013. Photo credit: Jason Flakes, U.S. Department of Energy Solar Decathlon.

Distinguished leaders from the field committed their time and talent to serve as jurors for the 2013 competition. Most of them noted that they came away from this experience feeling they had gained much more through their participation than they had given. The following experts participated in SD 2013 jury teams.

- **Architecture:** Richard Swett, Amy Gardner, Victor Olgyay
- **Engineering:** Kent Peterson, Brad Oberg, Mark Thornbloom
- **Market Appeal:** Susan Aiello, Brian Baker, Steve Glenn
- **Communications:** Ariel Schwartz, Mark Walhimer, Haily Zaki

- **Affordability:** External professional cost estimators determined the approximate final construction cost of each entry in the competition.
  o Teams that build their houses for $250,000 or less would be awarded all 100 points for this contest. A sliding scale was applied to determine points awarded for houses that cost between $250,001 and $599,000. Any house that cost more than $600,000 to build earned no points in this contest.

Learn more about these experts at:
https://www.solardecathlon.gov/past/2013/juries.html

Melissa and Richard King meet up with friends and former Decathletes Gregory Sachs, David Schieren, Cristina Zancani, Joe Simon, and Mike Wassmer at the 2013 event. The welcome banner features Bosch, Cisco, Edison International, Schneider Electric, and Wells Fargo, major sponsors of that competition.

Healthy sponsorship for SD 2013 boosted the quality of the event. Many were returning sponsors, but the new location attracted valuable newcomers to the field. Thirty-five sponsors contributed funds, volunteers, outreach, a microgrid for the Solar Village, sensor equipment for data collection, design software support, uniforms for SD organizers, food, drinking water, buses for public transport, and virtual tours for the website. Those charitable partnerships were instrumental in the remarkable achievements of Decathlete teams and resounding triumph of that sixth edition of the competition and accompanying Energy XPO.

Team Highlights

After the assembly phase, teams were rocking, rolling, and ready to go full-speed ahead. The competition got underway on October 3, 2013, and southern California weather came through in flying colors with plentiful sunshine in the Solar Village every day. In a clear demonstration that the SoCal climate is ideal for solar dwellings, every team in the 2013 competition earned all 100 points for the Energy Balance Contest. For the first time in the history of SD events, every house had a positive net balance of power, making as much or more energy than needed for daily functions in their houses. This was a strong testimony for the visiting public, as it proved that solar-powered homes are both viable and reliable in that region.

Decathletes from the Catholic University, George Washington, University, and American University aimed to design a home for renewal and regeneration. Team Capitol DC created *Harvest Home* for returning military veterans, and every aspect of the dwelling considered their physical, emotional, and social needs. This outdoor table with built-in garden containing fresh vegetables and herbs was quite ingenious.

The decks and landscaping with edible and native plants of the *Harvest Home* were meant to calm the mind, body, and spirit, while also providing healthy nourishment.

The sleek, beautiful house from Team Capitol DC incorporated biomedical sensors that could monitor and analyze physical therapy data, and the smart home management system could adapt to the inhabitant's behavior and lifestyle.

Newfound Momentum | 383

Phoenix House from the University of Kentucky, Ball State University, and the University of Louisville was built to demonstrate resilience in response to a natural disaster. The structure is made of reclaimed materials that combine traditional aesthetics with contemporary style. Prefabricated modules could be easily assembled, and structural insulated panels (SIPs) in the walls and roof systems were durable, resilient, and efficient.

Start.Home from Stanford University included modular elements that could be added, subtracted, or reconfigured for maximum adaptability. The CORE mechanical room, shown in the center of this photo, provided power control for the house.

Stanford Decathletes were one of the three teams at SD 2013 to earn all 100 points in the Affordability Contest. Their classy, stylish interior made an immediate impression on visitors. The redwood siding and hardwood floors made of Douglas fir came from old houses in the California Bay area.

The wide glass doorways of *Start.Home* opened up completely for a spacious, light-filled experience. Photo credit: Jason Flakes, U.S. Department of Energy Solar Decathlon.

The delightful deck with artistic pergola spilled out from the main living area, providing a comfortable and welcoming outdoor space.

Derek Ouyang is an educator and researcher at Stanford University and Executive Director of City Systems, a nonprofit consultancy. In both endeavors, he focuses on community-engaged, data-driven projects to shape a more sustainable, resilient, and equitable future for urban systems. He was project manager of Stanford University's entry in the 2013 Solar Decathlon in California.

I first started thinking about sustainability when I was an undergraduate at Stanford University. We'd been accepted into the 2013 Solar Decathlon, and I was faced with the juggling act of being chosen to lead over a hundred students, convincing Stanford to let our team build a solar house on campus, building that house, keeping it together, then breaking it apart to deliver on trucks to the competition site in Southern California; and all the while maintaining a good GPA, making the most of the Stanford experience, finding a girlfriend, and getting enough sleep. *This* was a problem of sustainability.

Let's just say some balls were dropped. Sustainability eventually meant just the sustainability of the house—namely its net-zero-energy performance. This was also a new notion to me. Growing up, I associated home with many formative values, like family and education and creativity, but never the value of energy. Every time I turned on a light switch or sink faucet as a kid, it was a miracle of biblical proportions, and in all due respect to the Creator, I often left them on. It wasn't until I attended Stanford that I began to understand the concept of ecological capacity, and just how over the limits we are—in large part due to our unsustainable resource consumption patterns. A group of us (students) wondered how we might reshape the homebuilding industry to better honor the obligations we have to future generations, and discovered that every two years, teams of college students were already sharing their design visions with the world.

Through Stanford University's first entry into the Solar Decathlon, my team sought the unique opportunity to showcase some of our ideas, such as: (1) a CORE bathroom-and-kitchen module that delivers resource-saving technologies without dictating the overall design of the home, which can be mass-manufactured in the future; (2) passive architectural design principles like in-wall thermal storage, natural ventilation, and daylighting; and (3) human-centered design touches like an eco-feedback light switch, knee-powered bathroom sink, and wall-mounted, framed, kinetic artwork that features undulating fans representing the past twenty-four hours of electricity production and consumption, like an "energy diary." Our central message was that, altogether, these innovations empower sustainable people, not just sustainable homes.

What sustained us through that Solar Decathlon marathon? More than coffee: our relationships, forged over two years. Just as a house becomes a home only through its telling of human stories, the competition became a community, first on campuses around the world, and then suddenly all of us together on a former airstrip at the Orange County Great Park. Well, seven days into the Decathlon my team, like every other team, was exhausted from non-stop public tours, technical trials, and jury evaluations. We wanted nothing else but to spend the evening *living* in our solar home. Invites were dispatched, and pretty soon a bunch of us were piled on living room sofas watching *Star Trek*

2 on our large flat-screen TV with the 9.1 surround sound system bass turned all the way up. The next day, it rained. Might I remind the reader that the tenth contest in the competition is *literally a* net-zero-energy contest, where every kilowatt-hour of energy preciously consumed during the week must be, in return, harvested from the sun through photovoltaic technology, in order to receive 100 points. The next evening, our meters showed consumption had dropped to net negative! I must admit that we appealed to more than just science for sunshine on the last day. Luckily, the clouds cleared, and we did just fine in that contest, and in the competition overall. Would I go back to that couch and recklessly splurge electrons with friends all over again? Yep, in a heartbeat!

Now, a decade down the road, what are my thoughts about sustainability? The memories to cherish, of course. Work-life balance is still a tricky balancing act, though I'm getting enough sleep these days. But when it comes to net-zero-energy homes, to be honest, I'm not so focused on them anymore. After the Solar Decathlon I started to wonder whether we, as students, should have explored more than just single-family homes, and have argued for a more holistic approach to sustainability that tackles suburbanism, car-dependence, and deeper social inequities. I have traveled around the world and discovered vast disparities in access to basic needs. I began to rethink our CORE as an off-grid building block for emergent cities.

Then I came back to the San Francisco Bay Area and discovered the poverty within our Silicon Valley bubble, so easily ignored before. I began to rethink our CORE as a plug-in accessory dwelling unit (ADU) to expand affordable housing. I have watched floods, wildfires, and pandemics impact vulnerable communities, and realize we need to think about not just mitigation of greenhouse gas emissions, but adaptation of our buildings and neighborhoods to the shocks and stresses of an already changing environment. In hindsight, I realize that Solar Decathlon was always about pushing boundaries, not setting them. Most of all, what that experience gave my fellow students and me was a sustained drive for meaning, for impact, for change. That light switch has—and will—always stay on.

For more inspiration, tune into Derek's powerful TED Talk delivered in June 2013, a few months before the California event: https://www.youtube.com/watch?v=mX2rzFL9hL4&t=1s

Winning Teams

The scores in SD 2013 were surprisingly close throughout the competition. Every house in the Solar Village was well engineered, and the overall performance was remarkably consistent. By October 12, the last day of the competition, the range separating competitors from the lowest to the highest total score was only 177 points. In addition, the range for the top five teams was only 19 points, and just six points separated the top three teams. What a razor-thin margin! The final awards ceremony to announce the winners was held inside Historic Hangar 244. Exuberance was high when the top three teams were called out: Czech Technical University in third place, University of Nevada at Las Vegas in second place, and Vienna University of Technology in first place overall.

The *AIR House* from the Czech Technical University was designed for empty-nesters who are 50+ years old as a preretirement weekend home and later a permanent retirement residence. AIR stands for Affordable, Innovative, and Recyclable. The house featured an attractive wood façade and wooden canopy overhead with solar panels, plus wood fiber insulation for acoustic and thermal control.

The outside terrace of the *AIR House* added more living space for cooking, relaxing, entertaining, gardening, and storage.

Decathletes from the Czech Republic are ecstatic about their third-place finish. Photo credit: Stefano Paltera, U.S. Department of Energy Solar Decathlon.

The *DesertSol* from the University of Nevada Las Vegas was designed as a highly efficient vacation retreat that could withstand harsh desert conditions.

Exquisite interior design for the UNLV dwelling showcased a sophisticated combination of modern simplicity with a rustic feel and natural colors highlighted with brightly colored accents. *DesertSol* won second place overall at the 2013 competition.

The *DesertSol* team from UNLV celebrates their second-place victory. Photo credit: Stefano Paltera, U.S. Department of Energy Solar Decathlon.

The elegant *LISI House* from the Vienna University of Technology won first place at SD 2013. Hundreds of visitors waited in long lines for a glimpse inside this gorgeous dwelling.

Exterior curtains surrounding the Team Austria house opened up for fresh air and light-filled ambience.

Team Austria sharing their joy onstage with Richard King and Mayor of Irvine Steven Choi in being awarded first place overall at SD 2013.

The People's Choice Award for SD 2013 went to *Urban Eden* from the University of North Carolina at Charlotte. The team made a strong statement about sustainability and the vital connections between outdoor and indoor spaces for an urban dwelling. The use of geopolymer cement on the exterior, instead of conventional concrete, reduced the carbon footprint of the home by 90% and was an effective sound barrier.

Urban Eden had vertical exterior gardens that provided ornamental and edible plants. The large glass walls could open to allow fresh air and cooling breezes inside.

Team Ontario was honored to win first place in the Engineering Contest at SD 2013. They wanted their house, called *ECHO*, to change the way consumers choose to live and to "echo" a message of sustainability into the future. Photo credit: Stefano Paltera, U.S. Department of Energy Solar Decathlon.

The overarching atmosphere at Solar Decathlon 2013 was casual, upbeat, and happy. The weather was perfect, innovative competition houses performed flawlessly, and the XPO added delightful entertainment interwoven with educational opportunities. There was something for everyone, and smiles were displayed in abundance. The shift from Washington, DC to Southern California highlighted the influence of location and climate on the dynamics of culture.

From October 3–13, approximately 64,000 visitors experienced SD 2013, with an average 16,000 house tours per team, which totals 300,000 house visits. There were 627 million media impressions, including 1700 online articles, 180 television stories, 137 radio segments, and more than 350 articles in print. Media exposure spanned PBS Nightly News, CNBC, *The LA Times*, *San Diego Times*, NBC News, FOX, *The Huffington Post*, *USA Today*, and *The OC Register*. The SunShot Innovation Pavilion reported over 10,000 visitors during this period of time. Thanks to the impressive ingenuity of Decathlete teams, the due diligence of event organizers, and the immeasurable support from the terrific sponsors, thousands of people benefitted from their exposure to SD 2013 at the OCGP.

California continues to be a leader in solar technology applications, and SoCal turned out to be an ideal spot for this showcase of innovative homes powered by sunlight. SD 2013 was a smashing success, so it made sense to stage another competition at the same venue.

Solar Decathlon 2015

The U.S. DOE was committed to a seventh edition of the Solar Decathlon, and Richard King was enthusiastic about another event at the OCGP. After receiving plenty of positive feedback about the 2013 experience, he looked forward to returning for a repeat performance. In response to the call for university proposals to enter the competition, 20 university teams were selected to participate in 2015. Faculty advisors were energized, and a new group of supercharged Decathletes geared up to begin their design-and-build projects.

About 18 months before the 2015 competition was set to begin, Richard discovered that the OCGP runway that accommodated SD 2013 was no longer available. Plans were afloat in Orange County to redevelop that land, which necessitated the removal of paved airstrips. With that significant change, the next solar village would have to be relocated. As an alternative site, park officials suggested use of an existing parking lot next to the Palm Court in the Great Park. Hence, that became the new home for SD 2015. Dimensions of the space and the required north–south orientation for solar rooftops meant that the village would be arranged in a diamond-like configuration. Rather than a single main street with sun-powered houses on each side, the village would consist of four intersecting pedestrian streets with houses along each side of those walkways.

Unfortunately, the new location at the OCGP did not offer sufficient space for an energy exposition, so this edition of Solar Decathlon did not include an XPO. Visitors could still take advantage of rides in the huge Great Park balloon for excellent views of the Solar Village. They could also visit the carousel, an art gallery that showcased William Pereira's master plan for UC Irvine, and the Farm and Food Lab that focused on growing edible plants.

The new site (in the foreground) was adjacent to the Palm Court (background) where repurposed military buildings are now a complex that features a multidisciplinary public arts exhibition and learning center.

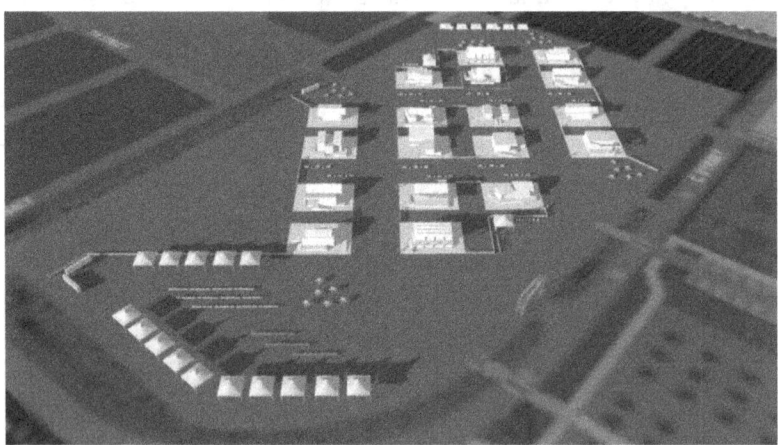

Layout for the 2015 Solar Village in the OCGP parking lot resembled a diamond shape with four intersecting paved pedestrian walkways flanked by solar houses.

Of the initial 20 teams chosen to participate, six were unable to complete their projects, so the final list of competing teams totaled

14. Participants for SD 2015 included six university teams from eastern states, four from the Midwest, and four from California. Three of those teams collaborated with international universities in Germany, Italy, and Panama. Nine of the competitors were newcomers to Solar Decathlon, but their exuberance and appetite to excel gave them a leg up from the get-go.

- California Polytechnic State University at San Louis Obispo
- California State University, Sacramento
- Clemson University, South Carolina
- Crowder College and Drury University, Missouri
- Missouri University of Science and Technology
- New York City College of Technology
- State University of New York at Alfred College of Technology and Alfred University, New York State
- Stevens Institute of Technology, New Jersey
- The University of Texas at Austin and Technische Universitaet Muenchen, Germany
- The State University of New York at Buffalo
- University of California, Davis
- University of California, Irvine; Saddleback College, Chapman University, and Irvine Valley College
- West Virginia University and University of Roma Tor Vergata, Italy
- Western New England University, Universidad Tecnológica de Panamá, and Universidad Tecnológica Centroamericana

For SD 2015, a Commuting Contest and Home Life Contest became part of the set of 10 contests (see details below). Five contests were juried (subjective), and five were measured (objective). Each contest was worth 100 points; a total of 1000 points would be a perfect score in the competition.

Juried

- Architecture—100 points
- Engineering—100 points
- Market Appeal—100 points

- Communications—100 points
- Affordability—100 points

Monitored

- Comfort Zone—100 points
- Appliances—100 points
- Commuting—100 points
   - On eight separate trips, teams had to drive 25 miles or more in two hours or less; they also had to recharge their electric vehicles using their own PV system.
- Home Life—100 points
   - Each team's solar system had to provide sufficient power for electric lights, hot water, and home electronics (such as a television and computer), as well as meal preparation for two dinner gatherings, and a movie night with fellow Decathletes.
- Energy Balance—100 points
   - A bidirectional meter on every house determines the net energy produced and consumed during the competition.

SD Organizers recruited another top-notch group of jurors to evaluate the quality of every entry, in order to determine the winners of each of the juried contests. The list included the following professionals:

- **Architecture:** Alastair Reilly, Ashara Nelson, Ann Edminster
- **Engineering:** Michael Brandemuehl, Cynthia Cruickshank, Ginger Scoggins
- **Market Appeal:** Brian Baker, Loraine Fowlow, Annette Stelmack
- **Communications:** Carolynne Harris, Macie Melendez, Mark Walhimer
- **Affordability:** External estimators

For more information about individuals on the jury panels, visit https://www.solardecathlon.gov/2015/competition-juries.html.

Stefano Paltera, expert photographer for many Solar Decathlon competitions, taking a group photo of Jury Members for SD 2015.

The competition spanned 10 days, from October 8 to 17, with plentiful sunshine. However, daytime temperatures were surprisingly hot for that time of the year, with several days in the upper 90s. Most teams had to rely on air-conditioning systems to maintain indoor comfort, and this increased their use of electrical energy. This real-life scenario was a serious test for these solar-powered homes. For some teams, that heatwave affected the overall energy balance of their houses during the competition.

Media engagement at SD 2015 added interesting new dimensions for public outreach. Irvine City Television (ICTV) supported the event with a "Daily Minute" broadcast from the OCGP Solar Village. Every day of the competition, a film crew arrived onsite to confer with Richard King about timely activities taking place. After sharing some background, they would then decide what and who to highlight, and off they went to a specific spot. Following the filming, a short video would then be posted on the SD website. Upbeat media exposure, which also included live broadcasts from local television and radio channels proved to be quite effective in enticing more visitors to attend the event.

Decathletes from Cal Poly San Luis Obispo jumping for joy outside their solar house for the Daily Minute film crew.

ICTV filming Decathletes from the New York City College of Technology in front of their house called *DURA*, which stands for Diverse, Urban, Resilient, Adaptable.

The youngest volunteer tour guide was a walking encyclopedia about Solar Decathlon 2015. He quickly became an onsite celebrity and was featured in the Daily Minute.

SD Director Richard King participating in a live interview with Los Angeles NBC affiliate Channel 4 about the 2015 competition. Photo Credit: Thomas Kelsey, U.S. Department of Energy Solar Decathlon.

Team Highlights

SD 2015 teams

The Clemson University team from South Carolina used an innovative approach to create their house, called *Indigo Pine*. They boasted about sending computer-aided design specifications via email to California, where those parts were cut locally and delivered. The team reassembled the house foundation through the night.

The framing system for *Indigo Pine* was reminiscent of Lego® building blocks, with every part numbered for reassembly like a 3D puzzle. Using a movable scaffold on wheels, Clemson Decathletes put interlocking parts of their house together securely in a tab-and-slot configuration.

The exterior siding on the outer shell of the Clemson house had perforations to allow air flow between the inner and outer shells, which allowed warm air to move upward and away from the house to help keep it cool.

When completed, *Indigo Pine* (center of photo) featured a spacious, shady porch for gatherings in the fresh air. Their goal was to provide digital plans for a solar house that could be sent anywhere and then be constructed relatively quickly by homeowners. Clemson won second place in the Architecture and Communications Contests and third place in Market Appeal.

The team from California State University Sacramento created a home to reflect styles typical of neighborhoods around the state Capitol Building, such as Craftsman Bungalows. Students from Sac State designed and built ingenious solar skylights that were installed above a large open porch on one side of their beautiful house. *Reflect Home* would become part of a living lab at the Technology Optimization Research Center, where students learn about net-zero design and sustainable construction.

Decathletes from Sacramento State University talking with U.S. Secretary of Energy Dr. Ernest Moniz inside their *Reflect Home* during competition week. Photo credit: Thomas Kelsey, U.S. Department of Energy Solar Decathlon.

West Virginia University partnered with the University of Roma Tor Vergata to make *STILE* (Sustainable Technologies Integrated in a Learning Experience), which is the Italian spelling of style. The team combined architectural elements from ancient Rome with traditional elements of the Appalachian Mountain region. The distinctive arched rooftop of *STILE* was made of recycled steel shipping containers.

Decathletes from West Virginia and Rome, Italy, demonstrating outstanding teamwork as they move a large glass window into position for their *STILE* house.

The wall of glass windows on the southern side of the *STILE* house created a light-filled interior, while the solar panels mounted on the arch above offered partial shade to help cool the house.

Alfred University, a school with less than 2000 students on the southern tier of NYS, was one of the first coeducational institutions in the United States. Those students teamed up with the State University of New York at Albany and Alfred College of Technology for a Solar Decathlon project. Their two-bedroom solar *Alf House* aimed for simplicity, efficiency, and affordability. A family in Los Angeles purchased *Alf House* for use as a vacation home after the competition.

Team Orange County, the local favorite, included Decathletes from the University of California at Irvine, Chapman University, Irvine Valley College, and Saddleback College. They built *Casa del Sol* as a three-bedroom, two-bathroom house with adjustable coverings for outdoor spaces to help control the temperature. Decathletes were especially proud of their greywater recycling system, which used rainwater and wastewater for irrigation of landscape plants.

Team Orange County sharing a happy moment onstage with U.S. Department of Energy Secretary Dr. Ernest Moniz. They were excited about winning second place in the Engineering Contest. Photo credit: Thomas Kelsey, U.S. Department of Energy Solar Decathlon.

Art Boyt, Faculty Advisor for the Crowder College Team, working alongside his students at SD 2015. Art is a solar trailblazer whose early engagement with sun-powered activities included a 1984 cross-country adventure in the Solar Phoenix, a rudimentary solar car.

Students from local schools in Orange County visited the disaster-resilient house from Crowder College and Drury University called *ShelteR*[3] (shelter cubed). Photo credit: Thomas Kelsey, U.S. Department of Energy Solar Decathlon.

Winning Teams

Drawing on three words: Interactive, Intuitive, Integrated, *Inhouse* Decathletes from California Polytechnic State University, San Louis Obispo, wanted to connect systems of the house with residents in energy-affordable ways. Bifacial solar panels created a translucent roof above a spacious porch.

Inhouse had private and public "wings" surrounding a mechanical core. The pubic areas could be opened up onto an exterior area that doubled the living space. The Cal Poly team was thrilled to capture third place overall at the 2015 competition. They also took second place in Market Appeal, third place in Architecture, and fourth place in Engineering.

Decathletes from the State University of New York at Buffalo created a house for their climatic region of widely varying seasonal temperatures. A flexible design with an attractive overhead canopy shaded the structure and decreased cooling loads, as well as functioning like a trellis for plants.

GRoW Home from SUNY Buffalo emphasized sustainability through attention to home-grown edibles. The enclosed "Growlarium," where vegetables and herbs could be nurtured year-round, was part greenhouse and part solarium.

Enthusiastic Decathletes from SUNY Buffalo, first-timers at a Solar Decathlon, surprised everyone by taking second place overall at the competition. This team won fifth place or better in all 10 of the contests for SD 2015. Photo credit: Thomas Kelsey, U.S. Department of Energy Solar Decathlon.

The *SURE HOUSE*, SUstainable and REsilient, from the Stevens Institute of Technology on the Mid-Atlantic shoreline near New York City, wanted to make a statement about surviving in a changing, more extreme climate. Following Hurricane Sandy in 2012, the students were keenly aware of the need to build super-strong dwellings that could withstand unexpected storm surges.

SURE HOUSE had a storm-resistant exterior shell with encapsulated thick insulation to manage temperature swings. The entire structure could be elevated to resist periodic flooding.

Stevens Decathletes were flying high after the announcement of their first-place win overall at SD 2015. Their clever use of an "islanding" photovoltaic system could produce power even after a storm that damages or disconnects the house from a utility grid. This made it possible for *SURE House* to maintain its own power and also share power with exterior USBs for neighbors to charge their electronic devices. Photo credit: Thomas Kelsey, U.S. Department of Energy Solar Decathlon.

Richard King and Dr. David Danielson, Assistant Secretary for Energy Efficiency and Renewable Energy, sharing the stage with Stevens Decathletes at the final awards ceremony. In addition to first place overall, Stevens came in first in Architecture, Engineering, Market Appeal, Communications, Appliances, Home Life, and Commuting. What an impressive performance across the board! Photo credit: Thomas Kelsey, U.S. Department of Energy Solar Decathlon.

Reflections from Carolynne Harris, professional museum planner, who specializes in the implementation of new museums, renovations, expansions, and large exhibitions. After employment at the Smithsonian Institution and Fernbank Museum of Natural History, Carolynne started her own firm. She worked with Solar Decathlon Organizers from 2005 to 2011 and served as a Jury Member at SD 2015 in Orange County, CA.

Following my initial efforts with SD 2005, I helped plan the signage, branding, and educational displays for SD 2009. When SD 2011 moved off the National Mall to West Potomac Park, I had an opportunity to reconceive the Solar Village to be more reflective of an actual neighborhood with streets and houses that backed up to each other. Organizers had learned a lot about visitor interests, flow, and team needs, and we envisioned an even bigger imprint on this smaller location adjacent to the FDR Memorial. The ground-breaking ceremony at that new site was set to open just a few months before SD 2011.

I was excited about this refresh of the experience! With graphic designers from Studio Ammons, our group redesigned the gateway entrance, sponsor, house identification, and educational signage to fit into this more compact space on baseball fields. We included QR codes, which were fairly new in 2011, so visitors could access deeper information about the event and energy efficiency. We also allowed more space for teams to present information to visitors while they queued up.

As we started the installation, torrential rain fell - for days. That was quite different from SD 2005, when we could barely drill into the soil for signs because it was so hot and dry; here it was soup. The hard plastic walkways the National Park Service required only served to hold pools of mud underneath, which squished and squirted up when walked on. Keep in mind that we were at sea level next to the Anacostia River and Tidal Basin, so there was no drainage. Our innovative QR codes? Raindrops prevented cellphones from reading them. I remember laughing with our NREL Communications colleagues out of sheer exhaustion from dealing with those conditions.

Nevertheless, thousands of people showed up, waiting in line with umbrellas and rainboots. They enjoyed the event, with the average visitor staying long enough to tour at least two or three houses. Decathletes weathered (literally) conditions never expected with indominable energy and purpose to build houses and showcase energy efficient designs. Once again, the competition showed the public cutting-edge solar technology and beautiful livable design. The response was positive, and that Solar Decathlon was a great success, despite being wet and muddy.

SD moved to a new location in Orange County, California for SD 2013 and 2015. I was honored to serve on the 2015 Communications Jury. Our task was to evaluate how well teams told the story of their houses and how effectively they used communication elements to excite, engage, and educate the public. Each team developed a website, brochure, video tour, house tour exhibit panels, and on-site tours. I was thrilled to be on the 'other side' of the competition and really dig into these projects. Every year they were more sophisticated and interesting.

Solar Decathlon teams made their houses approachable and described how their solutions were appropriate for residential communities that would occupy them. Some entries were intended for mass production. For example, Stevens Institute of Technology (SIT) from New Jersey, responded to the destruction of Hurricane Sandy and developed ways to address flooding, hurricane winds, and storm preparation. They showed empathy for their community and interpreted technical information clearly. During the tour for jury members, they explained their branding and had strong messaging and sophisticated metrics on communication. The SIT team integrated social media into their website in real time, which was unique at the time. Different team members hosted the tour to explain each area of the house, which was engaging. They ended up being the overall winning team of SD 2015!

Clemson University from South Carolina won second place in the Communications Contest. Their branding, tour approach, and general vibe were consistent with the 'homey' theme of their house and reflective of their university and community culture of "southern charm."

SD 2015 was a great capstone to my Solar Decathlon experiences. I will always be impressed with the growth and impact of this project as it progressed from teams giving tours, to hosting dinner gatherings and running errands in an electric vehicle, to producing a suite of communication and educational elements for public consumption. Over the years I have remarked to friends and colleagues about the number of students who had amazing experiences and received fantastic exposure to companies interested in innovators like them. The sheer number of public visitors exposed first-hand to the availability and beauty of energy-efficient designs, as well as how everyday decisions can make a difference in energy savings, was remarkable. The camaraderie engendered among everyone who participated reflected the leadership, focus, and intention of Solar Decathlon. It's a purpose-driven project that, in my career in museums and public engagement, will always remind me of the power of informal education and science interpretation. Challenging projects with strong missions can create team spirit, friendships, and gratification that lasts for decades.

The second edition of the California competition marked exactly 15 years since the very first Solar Decathlon call for proposals was issued in October 2000. Richard reflected on what an exhilarating ride he had experienced as his vision flourished. After a successful launch of the inaugural Solar Decathlon, the competition had expanded from east coast to west coast and beyond the shores of the United States. Emerging international interest in the event beckoned a new phase. Richard realized that his visionary idea had transitioned "from being here to being everywhere" (*What do you do with an idea?*" by Kobi Yamada, Compendium, Inc.).

With a distinct nod to innovation and transformative technology, Solar Decathlon continued to push the envelope to address climate change. Organizers of Solar Decathlon events around the globe arrived at the Solar Village in Irvine, California, to observe and learn more about managing this unique competition. Several members of the Solar Decathlon Europe team came to SD 2015, as well as leaders of new international competitions on the horizon. The U.S. DOE willingly shared with other nations SD competition rules, scoring details, and other essential aspects of program implementation. This led to cooperative planning efforts, and as a result, collegial friendships developed under the Solar Decathlon umbrella. This spontaneous and notable collaboration among bold pioneers helped to mobilize grand efforts to initiate new competitions on four continents.

Special guests at the final awards ceremony for SD 2015 included (left to right): Waleed Salman, EVP of Strategy and Business Development for Dubai Electricity and Water (DEWA); Dr. Peter Russell, Professor of Architecture; Louise Holloway, Director of the Energy Endeavour Foundation; Richard King; His Excellency Saeed Mohammed Ahmad Al Tayer, Vice Chairman of DSCE and MD&CEO of DEWA; Dr. Hongxi Yin, Chairman of SD China Organizing Committee; Michel Orlhac, Vice President for the Global Marketing division of Schneider Electric.

Chapter 8

International Expansion

As Solar Decathlon Europe took flight, other places around the world seized upon the opportunity to formulate and render their own versions of this unique program. Pioneering leaders in different regions of the world realized that the educational value and workforce development strengths of this competition could help propel their university students toward successful professional careers. In addition, Decathletes who participated in these events became remarkable change agents capable of promoting transformational thinking, applied practice, and government policy from the ground up. Forward-looking pathfinders also recognized the potential for Solar Decathlon to move the needle with the expansion of renewable energy and growth of sustainable building practices. For many nations, dependence on fossil fuels was untenable, unhealthy, and costly in the long run, especially if they lacked their own resources. Plus, affordable housing had become a global issue that begged for innovative solutions, particularly in developing countries.

Solar Decathlon was a tangible, realistic program that offered hope and promise in the fight against climate change. After more than a decade of success, it had a proven track record, as well as systematic methods of operation that could be modified and replicated. The flexible model established by the U.S. Department of

---

Solar Decathlon: Building a Renewable Future
Melissa DiGennaro King and Richard James King
Copyright © 2024 Jenny Stanford Publishing Pte. Ltd.
ISBN 978-981-5129-47-2 (Paperback), 978-981-5129-13-7 (Hardcover), 978-1-003-47759-4 (eBook)
www.jennystanford.com

Energy (DOE) made it possible for other locales outside the United States to use the trademarked name of "Solar Decathlon" alongside a national, regional, or continental identifier, such as "Solar Decathlon Europe." After a formal memorandum of understanding (MOU) was signed by the United States and its shared partner, individual contests and competition rules could be revised or updated for best fit to the new location. That was the beauty of the program: Solar Decathlon had a clear vision and a firm foundation, but it allowed for reasonable modification over time and in various corners of the world.

Richard King was delighted with expanded horizons for his visionary idea. In his words, "I wholeheartedly welcomed inquiries from other countries about hosting their own Solar Decathlon and offered encouragement for those new SD initiatives. Fortunately, my role at the U.S. DOE had evolved to include guidance for those who would lead the development of new SD competitions. Fortuitously, I became a mentor for competition directors in other places around the world, and together, we created a bold new vision for our collective dream. These international relationships forged with commitment to a shared passion taught me much more than I had ever imagined. I gained newfound appreciation for the significance of cultural background, architectural heritage, and a sense of place on residential dwellings. A home reflects your roots. A home contributes to who you are. A home represents your personal identity. A home is your safe refuge."

With worldwide recognition and growth, the power of Solar Decathlon was unleashed. Richard King thought about that wonderful book titled *What Do You Do with an Idea* by Kobi Yamada (c. 2013 by Compendium Inc.). The child in that beautiful story suddenly discovers that his idea has "changed right before my very eyes. It spread its wings, took flight, and burst into the sky." Although international expansion was beyond the scope of Richard's initial idea, he was absolutely thrilled to see his vision soar to new heights.

Throughout this period of international expansion, Solar Decathlon was standing on the shoulders of giants. Each new iteration of the competition relied on the tenacious resolve of champions who believed in the mission of this unique challenge for university students. Just as a handful of committed individuals led the charge to create SD Europe, more unsung heroes emerged to shepherd the founding of Solar Decathlon competitions in new locations on three more continents. Of course, the rich flavor of diversity was on full display as Decathlon dwellings reflected the strong influence of climate, geography, local resources, and cultural heritage on architectural design.

On the heels of the second edition of SD Europe in 2012, another fresh set of 10 contests was brewing on the other side of the globe. We begin with the remarkable story of Solar Decathlon China.

The Beginning: SD China 2013

In 2007, Applied Materials, a world leader in materials engineering with headquarters in Silicon Valley, California, sponsored the savvy Solar Decathlon team from Santa Clara University, which took third place at SD 2007. After that experience, the company well known for producing semiconductors quickly realized the benefits of this competition for workforce development. They sponsored the second Santa Clara team at SD 2009 and also decided to become a top-level sponsor of that competition. Their generous support for the event contributed significantly to successful public outreach that year. Applied Materials donated $250,000 to produce and deliver daily updates about the Solar Decathlon that were aired on a local network in the nation's capital. Twice a day, during the 6 pm and 11 pm news, viewers in the Washington metro area got updates on the competition and team standings. Thanks to Applied Materials, just about everyone in the DC area seemed to know about Solar Decathlon 2009.

Success often begets success. China was a huge market for the semiconductor materials that Applied Materials developed, and the company was actively engaged with the business community in China. They looked ahead and forecast positive outcomes that would emerge from a Solar Decathlon competition held in China. To explore that possibility, they reached out to Peking University, the National Energy Administration for the People's Republic of China, and the U.S. DOE. By 2010, there was enough support to make formal plans.

An MOU was signed in Washington, DC on January 18, 2011 to authorize plans for Solar Decathlon China 2013. The timing coincided with an agreement between President Obama and President Hu to encourage mutual cooperation and public–private partnerships in order to deploy clean energy strategies in the United States and the People's Republic of China. The intent was to stimulate economic growth and fight climate change. Key players who rallied for the establishment of SD China included Mark Walker and Charlie Gay (Applied Materials), Michel Orlhac (Schneider Electric), Mark Ginsberg and Richard King (U.S. DOE), and Pingrong "PR" Yu (Peking University).

Peking University is one of China's premier educational and research institutions with extensive international collaboration experience in green energy innovation and development. Its Center of Solar Energy in the College of Engineering provided valuable expertise in organizing the inaugural Solar Decathlon China. A notable champion of SD China was Pingrong Yu, a research professor at Peking University. Professor Yu, known to many as "PR," was chosen to become the Director of SD China 2013. He was excited but also wise enough to know that he needed support from others who understood the competition "from inside out." Brittany Williams, a member of the University of Maryland's Solar Decathlon team in 2007 and 2011, was a perfect fit, and she was willing to live in China during the planning and execution of the event. Brittany accepted an invitation to become the competition manager, and her life took a bold step forward.

Pingrong Yu also recruited from within China to fill some important positions on the organizing staff. A key person was Yuan Tian, who had experience directing China's Global Future Leader Assistantship Program (GFL), a language and cultural exchange program for international university students. Yuan, who was looking for work in the field of education, was intrigued with this unique program called SD China. She got down to business right away as General Director of Event Management, and her exceptional organizational skills proved to be an asset for the organizers.

The organizing team for SD China 2013 had their hands full, but they drew upon helpful guidance from Richard King and others who were seasoned experts. Perhaps the most useful preparation for the organizers from China was their trip to Washington, DC to shadow U.S. organizers of Solar Decathlon 2011. By observing the competition director, manager, and other key leaders firsthand during the onsite event on the National Mall, the group from China gained valuable understanding about how to run a Solar Decathlon competition.

PR needed a site for SD China, so his team issued a call for cities. The city of Datong in Shanxi Province (northwest of Beijing), known for its rich coal deposits and many coal mines, responded with an impressive proposal. The mayor wanted Solar Decathlon to be a representative symbol of his efforts to rebuild and transform Datong into a model "green city." Datong was selected as the host city. Without doubt, incredible effort was dedicated to China's first Solar Decathlon over the next couple of years. Their boots-on-the-ground approach was a marvel to watch as the chosen site was converted into a beautifully designed space that included new buildings and roadways for the competition. The city of Datong created an impressive Solar Village, complete with a hotel called the Sun Palace where hundreds of Decathletes could stay and facilities to accommodate thousands of visitors. To top that, the site was located in the middle of an expansive new cultural center where a museum, theater, library, and sports stadium were being built for the city's residents. The full-scale development of an entirely new site that offered free lodging and meals to Decathlete teams, as well as other services, was a first for Solar Decathlon events. What a splendid achievement for the People's Republic of China!

In Richard King's words, "Solar Decathlon took a dramatic step forward with China's guidance and direction. They were the first in the world to design and develop a brand-new site for the competition, complete with housing for participants and concrete pads to build on. In the United States and Europe, SD events took place on open fields or in parking lots. Everything was temporary, and the sites were free, since resources and available, appropriate land were often insufficient for a large-scale venue. China was also the first nation to offer teams the opportunity to leave their houses at the solar village (permanently). China wanted to keep the prototype dwellings onsite for education and demonstration purposes. Teams loved that idea!

It was a lot of work and cost to disassemble competition houses and ship them back home. More importantly, this plan provided an added benefit for the Chinese academic community. The 22 houses from this competition provided a set of leading-edge designs from around the world as learning tools that would yield valuable data for subsequent analysis."

Richard King arrived in Beijing and toured Tiananmen Square on October 24, 2012, before traveling to Datong, China, to participate in a workshop with teams selected for SD China 2013.

Reflections from Pingrong Yu, a venture capitalist and entrepreneur who lives in Silicon Valley, is the Founder and Managing Partner of Yu Galaxy and Co-Founder and Managing Partner of SV Tech Ventures; his investment teams have invested in over 100 startups in a variety of industries. PR has also been a research professor at Peking University, a longtime industry mentor for Schwarzman College, and an external advisor for research and innovation at the University of Colorado at Boulder. He earned a doctorate in physical chemistry from CU-Boulder and has published more than 30 patents and peer-reviewed papers, which have been cited more than 5000 times. PR served as chairman for SD China 2013.

You were at NREL doing research on solar energy in 2000–2005. How did that lead to your involvement with Solar Decathlon competitions? Talk about your role in getting the first Solar Decathlon China off the ground.

Because NREL was helping to manage SD, I learned about the competition early on. After NREL, I went to Silicon Valley and joined a solar energy startup called Innovalight, which was eventually acquired by DuPont. In 2007 I started my own solar energy/energy efficiency company called Optony in Silicon Valley, and later expanded the business to China. While I was an entrepreneur, I also took a part-time professorship position at Peking University. In 2010, I started to discuss a potential Solar Decathlon competition for Asia with senior folks from Peking University, industry leaders such as Applied Materials, and of course, the father of SD, Richard King. Given my background from NREL, industry, and academia, I knew the potential impact an SD could have on research, industry, and education. In getting the first SD in Asia off the ground, my role was identical as that of a startup CEO: setting the vision, getting necessary resources (including funding), assembling a team, figuring out ways to solve challenges, and executing all the details to ensure success.

You served as competition director and chairman of the organizing committee for the first SD China in Datong, China in 2013. What were the greatest challenges of that event? Attendance at SD China 2013 was impressive. How did people find out about the competition, and why do you think they were so eager to experience a Solar Decathlon?

I've built a few successful companies, however, running the first SD China was more challenging than all of them. One of the major challenges early on was to raise enough funding and get adequate in-kind support to run the competition. To solve that challenge, we adopted a page from the playbook of the Olympics: to have cities compete for the host city. I've personally visited several cities in China, tried to sell them the vision of SD, and convinced them to submit proposals. The host city provided a significant portion of the funding and in-kind contribution for the competition.

For SD China 2013 we had about 250,000 visitors to our competition site, and millions of people learned about SD from media sources such as TV, newspapers, and web media. These are the results of focused marketing over the course of three years. We leveraged our network and marketing resources from industry, academia, and governments from the beginning of the program in 2011 to the final competition in 2013. We used each training session or special event to grow our outreach. We even ran advertising at Datong Airport and on taxies in the final weeks in 2013.

Houses are relatable to the general public. We've been repeating the message successfully: SD houses are not only energy efficient, but also beautiful and comfortable. The public wanted to check it out. They were also curious how teams from different cultures and countries would implement these elements differently.

How did your participation in SD China influence your professional career? How did that engagement change your perspectives about the future of renewable energy and innovative building design?

Solar Decathlon was the first major non-profit work I've done, yet it was also one of the most rewarding. For most of my career, I was a researcher, entrepreneur, and venture capitalist. After the SD experience, I decided to do more non-profit and high-impact work in the future.

Despite the influence of SD and much more active discussions about climate change than when I started renewable energy research more than two decades ago, I still believe there is a long way to go to reach wide commercial adoption of renewable energy combined with innovative building design. This is an important component to reduce our carbon footprint.

How has the Solar Decathlon influenced change?

The Solar Decathlon competition we organized was a successful collaboration between the U.S. and China, with eleven other countries. To reduce our carbon footprint, we need global collaboration. SD was a wonderful platform for people from different countries to share their best ideas and technologies.

Datong, a traditional energy city, was given a valuable opportunity to learn about new energy.

The greatest value of the Solar Decathlon was in the people who experienced it. During the course of a few years, many young students told me that the competition had changed their lives. Yes, it has also changed mine.

What other comments or insight would you like to share about the Solar Decathlon?

SD is of great impact to humanity, and I'm glad that Melissa and Richard are writing a book to let a lot more people know about it.

Datong was more than ready for the first SD China event in 2013, which took place from August 2–12. To everyone's surprise, public attendance was overwhelming and broke records with an estimated total number of visitors at 250,000. In fact, people were so eager to enter the competition site that they knocked down fences after standing in long lines to get in. Their intense curiosity about these innovative model dwellings was rather astonishing, but welcomed by participating Decathletes who were happy to showcase their applied ingenuity for this design-and-build competition.

The 10 contests for SD China 2013 were very similar to the contests for SD 2011 in the United States; each contest was worth 100 points.

Juried

- Architecture
- Engineering
- Market Appeal
- Communications
- Solar Application

Monitored

- Comfort Zone
- Hot Water
- Appliances
- Home Entertainment
- Energy Balance—Schneider Electric provided a bidirectional meter on every house onsite to determine the total net energy produced and consumed during the competition.

SD China 2013 featured broad international participation, including 22 teams that represented 35 universities from 13 countries on five continents. The following teams participated:
- Abbaspour University of Technology (Iran)
- Alfred State College, Guilin University of Technology, and Alfred University (China, United States)

- American University in Cairo (Egypt)
- Beijing Jiaotong University and Bern University of Applied Sciences (China, Switzerland)
- Chalmers University of Technology (Sweden)
- Inner Mongolia University of Technology (China)
- London Metropolitan University and Guangzhou Academy of Fine Arts (England, China)
- Middle East Technical University (Turkey)
- National University of Singapore (Singapore)
- New Jersey Institute of Technology and Harbin Institute of Technology (United States, China)
- Peking University and University of Illinois at Urbana-Champaign (China, United States)
- Shandong Jianzhu University (China)
- Shanghai Jiaotong University (China)
- Southeast University (China)
- South China University of Technology and Huazhong University of Science and Technology (China)
- Tel Aviv University, Shenkar College of Engineering, and Neri Bloomfield School of Design, College of Management Academic Studies (Israel)
- Tsinghua University and Florida International University (China, United States)
- University Teknologi of Malaysia (Malaysia)
- University of Wollongong (Australia)
- Worcester Polytechnic Institute, Ghent University, and Polytechnic Institute of New York University (Belgium, United States)
- Xiamen University (China)
- Xi'an University of Architecture and Technology (China)

To come up with a slogan for SD China, there was a competition that yielded 800 entries. The winning slogan was a wonderful expression of what international Solar Decathlon events are all about: "Brighten the Future, Harmonize the World." SD China was awesome! The event took on the flavor of an Olympic competition in both size and complexion. News coverage was extensive with more than three million media impressions recorded.

International Expansion | **427**

Richard King with SD China 2013 jury members and organizers at a special dinner. Competition Director Pingrong Yu (called PR) is seated in the front row (on left end), and Competition Manager Brittany Williams is standing in the back row (second from left).

The City of Datong built an impressive site for the competition, complete with a hotel called the Sun Palace (shown in the background) for hundreds of Decathletes. They created a beautifully designed village for 22 competition houses with facilities to accommodate thousands of visitors.

The opening ceremony on August 2, 2013, was held outdoors in front of the Sun Palace hotel.

For the opening ceremony, a huge stage was erected on the front steps of the Sun Palace hotel. Each team was invited onstage to be introduced.

International Expansion | 429

SD China 2013 was a huge success, with an estimated 250,000 visitors in attendance. This aerial view shows how the Chinese patiently waited in long lines to see the houses. Team HelioMet from the London Metropolitan University (black façade) appears in the foreground of this photo.

At times, the Solar Village was very crowded, and many visitors used umbrellas for shade in the hot August sun. The scene resembles a colorful canopy that moved with the people.

There were observation towers at each end of the Solar Village that offered spectacular views of the prototype houses and people waiting in lines to look inside.

This interesting design came from Team Green Sun from Inner Mongolia University of Technology. The shape echoed a traditional "yurt" used by Mongolian nomads in the steppes of Central Asia.

International Expansion | 431

The expansive solar façade made it possible for this house to produce plentiful solar energy.

At the end of the competition, the final scores showed that the top three teams came from three different continents. Chalmers University of Technology from Sweden came in third place, South China University of Technology from the People's Republic of China came in second place, and the University of Wollongong from Australia took top honors with their first-place win overall.

Team Sweden placed third overall in the 2013 competition. Their unique design included a large curved solar roof that covered the entire house.

Team SCUT South China University of Technology placed second overall in the 2013 competition.

The second place team SCUT getting ready for all-team photo after the final awards ceremony.

Team UOW from the University of Wollongong in Australia took first place at SD China 2013. Their entry was a complete retrofit of a typical house in Australia. The team took an existing house, updated it with solar panels and energy-efficient features, and expanded and modernized the kitchen.

The Australian team's superb renovations impressed the judges. They had to disassemble the entire house before the competition to ship it to the site in China.

Brittany L. Williams, AIA LEED AP BD+C, is a registered architect and educator. As a practicing architect at an award-winning firm, Gardner Architects LLC, Williams focuses on a detail-oriented, multidisciplinary approach to the synthesis of sustainable active and passive design strategies at the residential scale. She is also a Clinical Assistant Professor in the School of Architecture, Planning, and Preservation at the University of Maryland.

You have been a Decathlete, as well as a faculty advisor for Solar Decathlon teams, and now you are a practicing architect. Talk about what you observed in housing prototypes at SD China 2013.

SD China 2013 demonstrated a wide variety of construction types, architectural styles, and sustainable building technologies. Even during the construction phase of the competition, a full range of assembly techniques and construction materials were on display: Some houses were entirely prefabricated in the university's home country and then craned into place; some entries were panelized; and some houses were entirely site-built from traditional construction materials in just a few days. There was even one house that was made of large foam pieces assembled with the use of a crane. The architectural styles represented the widely diverse universities and nationalities competing. It was transformative to see how teams adapted their local vernacular architectural traditions to modern, innovative housing types that integrated sustainable building technologies. The entries certainly pushed the envelope in terms of what's possible for housing. Many of the solutions seem innovative years later, and their architectural impact can still be felt.

How did your work as competition manager at SD China impact your future professional life?

The experience highlighted and reinforced the need for the implementation of global, multidisciplinary, innovative, and time-tested solutions to sustain the built environment into the future. With teams from 13 different countries, it was fascinating to see the range of architectural solutions that represented many different dwelling types while also seeing similarities to strive for the goal of a more sustainable future. Despite regional differences in building culture and construction, the passionate commitment to environmental stewardship in the built environment from so many different countries, universities, and students enacted was truly inspiring. That enhanced my own drive and desire to pursue the integration of sustainable strategies in the built environment in my role as an architect and educator.

What other thoughts would you like to share about SD China 2013?

Being the only non-native Mandarin speaker on the SD China 2013 organizing team was a fascinating experience. While the official language of the competition was English (my native language), everyone still spoke Mandarin in the office. After all, we were in Beijing. One frustrating experience with the language barrier ultimately proved to be one of the most enlightening moments of the competition for me. I was coordinating some site-related issues with a Chinese organizer who was a civil engineer. He spoke significantly more English than I spoke Mandarin, but he wasn't fluent, and the few phrases I'd learned in Mandarin weren't going to cut it. We were having trouble communicating, and despite help from several bilingual speakers, we were unable to communicate effectively enough to solve the site coordination issues. The translators couldn't understand our conversations that were laden with construction and design-related jargon. Finally, we had a breakthrough. We stopped trying to use spoken language and started drawing. We passed a pen back and forth and worked through the issues silently, using universal drawing conventions. Solving those problems by relying on visual representation underscored the entire point of the competition: to create a sustainable built environment, we need to collaborate beyond specific languages and cultures.

SD China 2018

SD China 2013 was a huge success, but the Chinese government did not immediately commit to a second competition. Then in 2015, approval was granted and government funding secured for the second Solar Decathlon China. This time organizational responsibility was assigned to the China Overseas Development Association (CODA), with support from the central government and local authorities. A new Organizing Committee was established for SD China, and planning for another competition commenced. Yuan Tian was invited to help manage SD China 2018, and she was excited about this opportunity. The theme for the competition was *Light up Your Talent*.

A new host city was needed, and the selection process began. Again, cities were encouraged to submit proposals. The city of Dezhou in China's Northwestern Shandong Province, which was strategically located as a transportation center, was pleased to be selected as the host site. Referred to as "China's Solar Valley," Dezhou is home to many research labs and over 100 solar companies. This urban area is also known for the ubiquitous presence of solar panels on rooftops.

Richard King was invited to become an honorary member of the Steering Committee for SD China, and he eagerly accepted this position. CODA quickly assembled additional experts who had appropriate experience. With participation from representatives of Dezhou, the organizing team set out to design a much larger, more impressive venue, building on lessons learned from 2013.

Richard recalls, "As I listened to their ideas, I became more and more impressed with the incredible scale and vision emerging from planning sessions. In addition to creating a full-scale modern site with magnificent facilities, they built a large reflecting pool with a huge fountain to attract visitors. Quite impressive!"

Dr. Hongxi Yin, Associate Professor of Architecture at Sam Fox School of Design and Visual Arts at Washington University in St. Louis, focuses on design-build as an educational tool for decarbonization. He was faculty lead for Team WashU at Solar Decathlon 2017 and SD China 2018. Dr. Yin and Ms. Xiaoqing Qin provided the conceptual and schematic design for the 2018 SD China competition campus in Dezhou City, Shangdong Province, China. Dr. Yin describes the site plan in detail below. Inspired by the original agricultural context of Dezhou, the "green village" concept for Solar Decathlon China 2018 created a collaborative, user-friendly, sustainable competition site for 20 collegiate teams. A Central Loop encircled the entire competition site, with south-facing Solar Decathlon houses located along that road. The competition complex in the middle of the Central Loop included team gathering space, full-size auditorium, and office compound for competition organizers. Considering the hot-humid climate and rainy season during the assembly phase, the site design maximized the opportunity to plant shade trees, install permeable paving for runoff reduction, and create rain gardens for water infiltration. The 40-acre core zone was surrounded by an 80-acre service entrepreneurship park to the south, a 30-acre service zone to the east, and a 20-acre waterfront park to the north.

During the assembly phase, teams worked simultaneously in pleasant surroundings with gathering spaces for meetings, collaboration, and rest. After opening to the public, it became a vibrant space with a spacious corridor connecting the solar prototypes where visitors could easily walk and explore the demonstration houses. A green buffer space at the edge of the campus separated it from noisy urban streets and nearby parking lots. Meandering paths in the green buffer space offered an alternative route for visitors to look at the solar houses from different angles. Every day during the competition, thousands of people came to see this unique exhibition.

For three months after the 2018 competition, the campus became a net-zero energy, net-zero carbon emission, net-zero-water-waste park to provide education, entertainment, and exhibition functions. The greenhouse garden on the west end provided water recycling and LED farming demonstrations for the public. The student cafeteria with zigzag roof configuration and 10,000 square feet of roof-mounted solar panels maximized the use of daylighting, as did the large auditorium nearby. Green roof technology was also displayed on the building that contained office space. The design offered plenty of open space for a variety of activities.

Map of the competition campus for SD China 2018. Photo credit: Hongxi Yin, Xiaoqing Qin.

The SD China Steering Committee wrote new rules for the 2018 competition, and recruitment for university teams began in 2016.

Solar Decathlon was gaining popularity in China, so there was no shortage of interest. SD China also caught the interest of many international universities that were eager to participate. For the second SD China competition, 23 teams from 34 universities that represented 11 countries on three continents were selected.

- South China University of Technology and Polytechnic University of Turin (China, Italy)
- Tsinghua University (China)
- Southeast University and Technical University of Braunschweig (China, Germany)
- Shandong University, Xiamen University, National Institute of Applied Sciences of Rennes, University of Rennes 1/Superior School of Engineering of Rennes, University of Rennes 2/Institute of Management and Urbanism of Rennes, High School Joliot Curie of Rennes, Technical School of Compagnons du Devoir of Rennes, European Academy of Art in Brittany (EESAB), National School of Architecture of Brittany (China, France)
- Beijing Jiaotong University (China)
- The University of Hong Kong (China)
- College of Management Academic Studies (COMAS) Afeka College (Israel)
- Istanbul Technical University, Istanbul Kültür University, Yildiz Technical University (Turkey)
- McGill University and Concordia University (Canada)
- New Jersey Institute of Technology and Fujian University of Technology (United States, China)
- Indian Institute of Technology Bombay (India)
- Shenyang Institute of Engineering (China)
- Shanghai Jiaotong University and University of Illinois at Urbana-Champaign (China, United States)
- Hunan University (China)
- Seoul National University, Sung Kyun Kwan University, AJOU University (Korea)
- Shanghai University of Engineering Science (China)
- Tongji University and Technical University of Darmstadt

(China, Germany)
- University of Toronto, Ryerson University, Seneca College (Canada)
- University of Nottingham, Ningbo, China (China)
- Washington University in St. Louis, Missouri (United States)
- Xi'an University of Architecture and Technology (China)
- Xi'an Jiaotong University and Western New England University (China, United States)
- Yantai University and Illinois Institute of Technology (China, United States)

As shown below, the 10 contests for SD China 2018 were slightly different from the contests in SD China 2013. Each contest was worth 100 points.

Juried

- Architecture
- Engineering
- Market Appeal
- Communications
- Innovation

Monitored

- Comfort Zone
- Appliances
- Home Life
- Commuting—Drive an EV 24.8 miles (40 km) in 1 hour or less four times during competition week.
- Energy Balance

Richard King reflects on his first impressions of SD China 2018. "I arrived in Dezhou, China on Monday, July 30, 2018, and after a good night's sleep, I awoke at 6:00 AM and went to the site. AMAZING! The teams were busy working to finish assembling their houses by noon, the 'stop work time' for final electrical, safety, mechanical, and landscaping inspections. Following 20 days for the assembly period, most teams were ready. As I walked around the village with

Yuan Tian, I was impressed with the site layout and the size and complexion of the houses. SD China rules allowed for larger, two-story houses, but this makes assembly more complex. A second story added height risks, more materials, and additional construction. Building solid, permanent foundations also required substantially more work up front, both in staff time and cost. As Yuan Tian noted, 'We are building a small city'."

"I was just amazed as I walked around the solar campus for the first time. The exhibition houses showcased a highly diverse set of designs. I was struck by the magnitude of both the competition site and the industrial innovation center. All-in-all, it was the largest, most spectacular Solar Decathlon ever held."

The SD China 2018 Opening Ceremony on Friday, August 3, was extraordinary. The Chinese delivered a magnificent ceremony, including a light show, videos, and VIP speeches. Teams were excited, and with more than 700 people and 60 media outlets from all over China in attendance, word of the event spread quickly. Over the course of the 15-day competition, more than 500,000 visitors and representatives from 370 business enterprises came to the solar campus. This event appeared to break all existing records for attendance at Solar Decathlon competitions.

Richard King arrived at the SD China 2018 competition campus on July 31, 2018.

International Expansion | 441

The SD China 2018 campus is a 40-acre site with a 30-acre service zone on the east (see the large suspended buildings in the background) that housed a cafeteria and auditorium for award ceremonies. Competition houses were positioned in an oval around a central core with office space for staff and air-conditioned facilities for the Decathletes.

There were prepared spaces, or lots, for each team to build on a solid foundation. This photo was taken from an observation tower in the center of the campus.

Several prototype houses along the roadway that encircled the solar village. The road was closed during the competition to allow pedestrians to visit the site.

Many people came to see the new innovation center, with its colored lights and beautiful water fountain, next to the solar campus. The competition houses were open until 9:00 PM every evening, and there were long lines of people waiting to get inside.

International Expansion | **443**

The SD China 2018 site was large enough to accommodate tens of thousands of visitors.

Team B&R's *Low House* from the University of Hong Kong and Beijing University of Civil Engineering and Architecture had extensive landscaping with water features and a bridge.

Decathletes from *Low House* created attractive landscaping around the dwelling, including walkways encircling a pond.

Team PKU from Peking University, China, built tree-like pedestals positioned around the house to support solar panels.

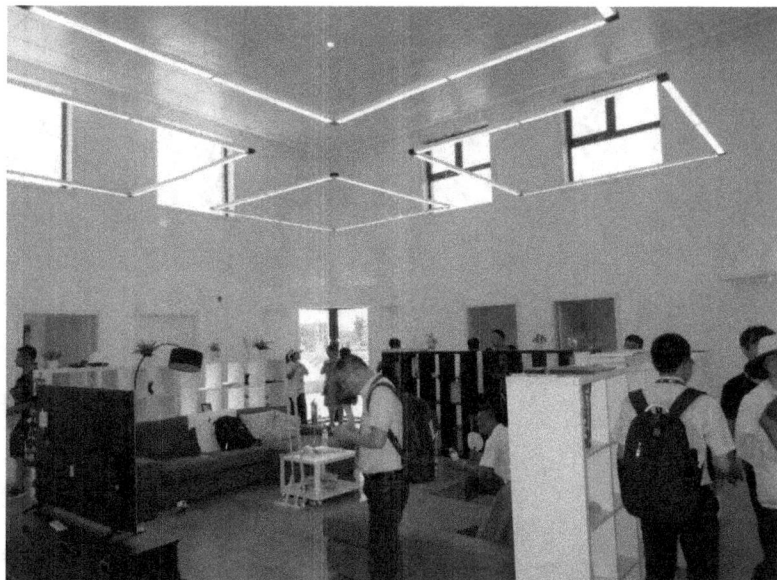

This interior view of Peking University's house shows the large, open living area with a high ceiling. Windows in an elevated position provided ample daylighting while allowing privacy.

Team XAUAT from Xi'an University of Architecture and Technology designed a traditional house that reflected the typical Chinese custom with three generations under one roof. They created a comfortable, green home for rural living.

View of the living room of Team XAUAT's house with a staircase to second-floor bedrooms. The team used many low-cost materials to make the home affordable.

Team TJU-TUDA's *Energy PLUS Home 4.0* from Tongji University, China, and the Technical University of Darmstadt, Germany.

Team TJU-TUDA's *Energy PLUS Home 4.0* had a large solar array on a flat roof. There were two ways to access the second floor: one way was to go outside and climb a staircase at the rear of the house.

The other way to reach the upper level in the Team TJU-TUDA house was to use the climbing wall in the living room.

The team from Washington University in St. Louis (WUSTL) in Missouri (USA), created a stunning 3D printed prototype called *Lotus House*, the first Solar Decathlon house to be made with this process. Their project mimicked the most significant flower in Chinese culture: the lotus flower.

The team from WUSTL focused on reducing waste at the construction site, reducing CO_2 generated during the process, and conserving resources. All walls and furniture in *Lotus House* used 3D manufacturing.

This loveseat was designed and created using 3D fabrication.

Side view of the *Lotus House* shows its unique shape, which invited further exploration.

Adding to the drama of SD China 2018, this turned out to be a tight competition, with scores for the top teams being very close. For example, two teams tied for third place because they were separated by only 0.02 of a point. The two third-place winners were (a) *C-House* from Team TUBSEU (China and Germany) and (b) *Nature-Between*, the house from Team JIA+ (China and France). *WHAO House* from Tsinghua University in China captured second place, and *Long-Plan* from South China University of Technology and Politecnico di Torino in Italy won first place overall.

Team JIA+, comprising seven different universities from China and France, placed third overall. The *Nature-Between House* used natural materials to synchronize indoor spaces with the outdoors.

Team TUBSEU from Southeast University, China, and the Technical University of Braunschweig, Germany, won the other third-place award. The *C-House*, called the "cube," was a high-tech dwelling with impeccable engineering. The design included solar panels mounted on the east and west sides of the house, in addition to rooftop modules.

International Expansion | 451

C-House had an open interior with a central core containing all the mechanical and plumbing equipment. A unique feature was an electric vehicle charge port on the inside wall, so they could park an EV inside the house during charging.

C-House placed their kitchen along the rear of the central core, where heat and humidity could be managed effectively. Bedrooms were on the upper level.

WHAO House from Tsinghua University in China took second place. This prototype was constructed in three modular sections connected by open courtyard areas.

Team SCUT-POLITO from South China University of Technology and Politecnico di Torino, Italy, won first place overall at SD China 2018 with their *Long-Plan*, a townhouse designed for urban living.

International Expansion | 453

Long-Plan, with simple lines and beautiful architecture, would fit easily into urban areas where row houses were popular.

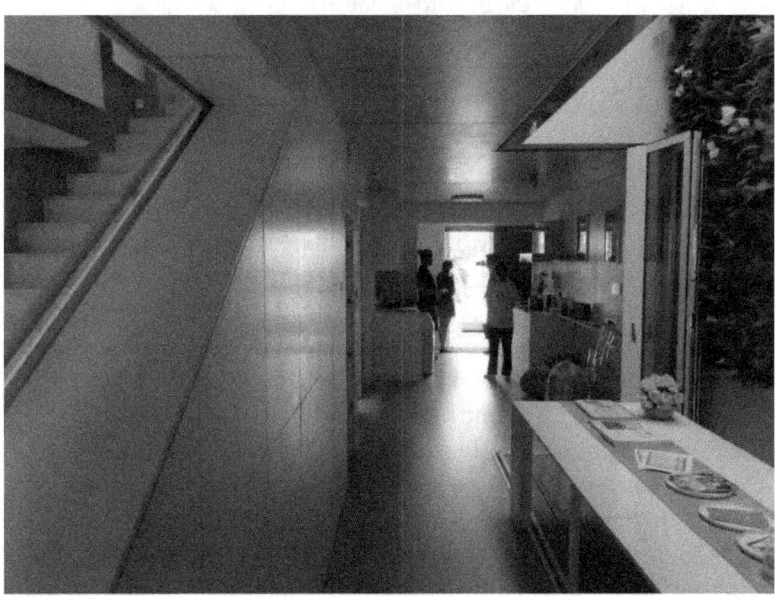

Long-Plan was well insulated and meticulously crafted.

A green wall inside the *Long-Plan* dwelling provided clean air and comfort from natural elements, which can be hard to find in an urban setting.

SD China 2022

Soon after the closing ceremony at SD China 2018, the organizers issued a call for cities to host the next SD China event, scheduled for October 2021. The city of Zhangjiakou, located in Hebei Province in northern China, was selected as the host site. Along with Beijing, Zhangjiakou was chosen to cohost the 2022 Winter Olympics. The organizers planned to use competition prototypes as housing for Olympic athletes and staff during the February 2022 winter games. That was a great idea, but the COVID-19 pandemic led to a year's delay of the competition (following the Winter Olympics).

As it turned out, the decision to postpone SD China until August 2022 was made after many teams had already assembled their houses at the event site in Zhangjiakou. Unfortunately, those competition prototypes had to remain unoccupied and locked up there all through the winter and Spring of 2022. However, the delay did not dampen the spirits of the Decathletes, and they were eager and determined to compete that summer.

Aerial view of snow-covered prototype houses at the 2022 SD China site during the winter of 2021–22. Photo credit: Solar Decathlon China Team.

After two-and-a-half years of preparation, the organizers and participating teams for SD China 2022 were finally ready to begin. The competition took place from August 8–14 with a triple challenge for all competitors, who were asked to focus on (a) *sustainable development*, (b) *smart connections*, and (c) *human health*.

Aerial view of the SD China 2022 event site. Photo credit: Solar Decathlon China Team.

Decathletes from 16 enthusiastic teams represented 29 universities in 10 countries around the world. For over half of those teams, the Solar Decathlon project was a collaborative effort that involved more than one university or nation.

- Beijing Jiaotong University and Loughborough University (China, United Kingdom)
- Chongqing University (China)
- China University of Mining and Technology and AGH University of Science and Technology (China, Poland)
- Technical University of Denmark (Denmark)
- Soochow University (China)
- Dalian University of Technology (China)
- Harbin Institute of Technology (China)
- Hefei University of Technology and University of Lille (China, France)
- Southeast University, Sanming University, Swiss Federal Institute of Technology Zurich (China, Switzerland)
- Shenzhen University and RMIT University (China, Australia)
- Zhejiang Normal University, Shenyang Jianzhu University, Chemnitz University of Technology (China, Germany)
- Xi'an University of Architecture and Technology and Southwest Minzu University (China)
- Tsinghua University (China)
- Tianjin University, Tianjin Chengjian University, The Oslo School of Architecture and Design (China, Norway)
- Xi'an Jiaotong University (China)
- Xi'an Jiaotong-Liverpool University, Zhejiang University/ University of Illinois at Urbana-Champaign Institute, Thomas Jefferson University (China, United States)

The houses on display at the 2022 event were stunning in design, as well as performance. Various shapes and styles demonstrated modularization and integration of photovoltaics into the façade of the buildings, as well as on rooftops. These climate-responsive designs were intended as prototypes for both urban renewal and rural construction. The grand plan was that the management of all 16 Solar Decathlon houses would be transferred to the host city of Zhangjiakou at the end of the competition, so they could be used for long-term public education and demonstration purposes. Zhangjiakou had agreed to maintain the SD China houses as "living labs" to serve as models of sustainable development for the region. Planned follow-up activities for those houses included academic seminars, industry fairs, and extended research projects. In addition, teams from SD China were granted access to their houses for one year after the competition, so they could monitor house performance and collect data for analysis.

The 10 contests for SD China 2022 included the following:

Juried

- Architecture
- Engineering
- Energy
- Communications
- Market Potential

Measured

- Indoor Environment
- Renewable Heating and Cooling
- Home Life
- Interactive Experience
- Energy Self-Sufficiency

When the scores of all 10 contests were compiled at the end of the competition, the top three teams overall for SD China 2022 were as follows: third place—*Steppe Ark* from Tsinghua University; second place—*Solar Ark 3.0* from Southeast University, the Swiss Federal Institute of Technology Zurich, and Sanming University; and first place—*R-Cell* from Tianjin University, the Oslo School of Architecture and Design, and Tianjin Chengjian University. The winning teams were thrilled, and the entire solar campus erupted with excitement about this incredible undertaking that defied the unforeseen challenges of COVID-19 to become a huge success story.

Prototype *Steppe Ark* from Tsinghua University came in third place overall. Photo credit: Solar Decathlon China Team.

Prototype *Solar Ark 3.0* from Southeast University, Swiss Federal Institute of Technology Zurich, and Sanming University came in second place overall. Photo credit: Solar Decathlon China Team.

Stunning interior of *Solar Ark 3.0* prototype. Photo credit: Solar Decathlon China Team.

R-Cell prototype from Tianjin University, the Oslo School of Architecture and Design, and Tianjin Chengjian University came in first place overall. Photo credit: Solar Decathlon China Team.

Beautiful interior of *R-Cell* prototype. Photo credit: Solar Decathlon China Team.

All-team photo for SD China 2022. Photo credit: Solar Decathlon China Team.

 *Reflections from Yuan Tian, Executive Deputy Director of SD China, who supervises the competition rules and educational outreach programs. She has worked for SD China for nearly a decade and is one of the core members who organized the 2013, 2018, and 2022 events. Prior to SD China, Yuan Tian worked for the China Education Association for International Exchange affiliated with the China Ministry of Education. She holds two master's degrees from the United States and China.*

How and when did you hear about the Solar Decathlon? Why did you want to get involved?

I first found out about the Solar Decathlon when a professional headhunter who was doing recruitment for the SD China Organizing Team contacted me about being part of the group. I was attracted to the project because it's the first educational program I had ever heard about that can actually create a real situation for the learners.

What helpful guidance did you receive when planning SD China competitions?

The guidance from the SD team in the U.S. was quite helpful, including job shadowing in Washington, DC during the 2011 SD event and the follow-up online meetings with Richard King, Cecile Warner, and Sara Farrar-Nagy. We also received technical support and on-site services from experienced SD organizers from the U.S., such as Mike Wassmer, Joseph Simon, Tom Meyers, and Brittany Williams.

SD China competitions were held in Datong (2013) and Dezhou (2018). What are your best memories from those experiences, and what made those events successful?

My best memories from those two SD China competitions are mainly about SD China participants, including the students and interns. It's rewarding to see most of them become part of the mainstream in academic fields, the green building industry, and the renewable energy sector. The extensive international participation and strong support from the host city in China made the competitions a big success.

What is most rewarding about Solar Decathlon competitions?

I think it's that almost every participant told us how much they benefitted from the experience, no matter what their role was: organizer, student, faculty mentor, jury member, enterpriser, or the public. It's like a kaleidoscope, reflecting different perspectives about what is the ideal life in the future.

Share your thoughts about the impact of SD China on student Decathletes.

The student Decathletes have been given the most autonomy, compared to any other previous learning scenarios. Most of them succeeded in showing incredible creativity and resistance to pressure in all the processes of making decisions and coping with unexpected situations.

How has your involvement with Solar Decathlon influenced your professional career and life?

The greatest influence is that Solar Decathlon has become my goal for professional development and will continue to be so in the next five to ten years. It helps me break the boundaries across my original major as a linguist and expand into other fields like architecture and energy. With SD China, learning became continuous and constant for me. The past 11 years with SD China happened to coincide with my life being transformed from a single person to a mother of two children. It affords me lessons that merit what I can and should do in order to support my children to face all the uncertainties of the future. Climate change is one of them, for sure.

SD Latin America and Caribbean 2015

Marvelous new developments sometimes catch us off guard or take us by surprise, and that was certainly true for the inception of Solar Decathlon Latin America and Caribbean. Richard King was keenly interested in hatching a new event in South America, and it was with delightful astonishment that he discovered our neighbors to the south were gearing up to make that dream a reality. He had lively conversations with a few individuals who were interested in launching a Solar Decathlon competition in South America, but time had lapsed between those discussions.

During the inaugural SD event on the National Mall, Richard met Marcela Huertas from Colombia. She was genuinely enthused about the competition, and Richard applauded the notion of a new Solar Decathlon in the Southern Hemisphere. Several years went by, and he did not hear more about any specific plans. However, face-to-face verbal communication drives a lot of day-to-day business operations in Latin America, so if you do not happen to see or speak with colleagues in person or by phone, you might not know about emerging news. Fast forward to 2013, and Marcela was back in touch with Richard to share a fantastic update: Colombia was eager to host Solar Decathlon Latin America and Caribbean (SD LAC)!

The story of SD LAC begins with Dr. Rodrigo Guerrero, mayor of Santiago de Cali. He traveled to Washington, DC in February 2014, where he met with Richard King. The mayor, a huge fan of the Solar Decathlon, was deeply committed to getting SD LAC off the ground in his city and pledged his full support. Cali, an urban metropolitan area of more than 3 million people with a fast-growing economy, is located in the Valle del Cauca, southwest of Bogotá, the nation's capital. Hosting a Solar Decathlon event was an opportunity for Cali to shine as it showcased sustainable housing prototypes, demonstrated how solar works, and promoted a diverse clean energy portfolio.

SD LAC Competition Director Carlos Rodriguez-Marin with government officials: Amilkar Acosta, Dimitri Zaninovich, Maria Eugenia de Guerrero (Mayor's wife), and Mayor of Cali Dr. Rodrigo Guerrero.

On Monday, March 10, 2014, the formal signing ceremony took place at the U.S. DOE to authorize the Solar Decathlon expansion in Colombia, South America in 2015. Colombia's Minister of Energy, Amilkar Acosta, Mayor Guerrero, and the Colombian Ambassador to the United States, Luis Villegas, were in attendance to celebrate this significant step forward. Thanks to the fervent resolve of Marcela Huertas and Rodrigo Guerrero to make this happen, SD LAC sprang to life!

Reflections from Marcela Huertas Figueroa, Founder and CEO of Green Center Capital, a financial boutique with expertise in executing projects to build sustainable, smart cities. Previously, she worked at the World Bank as Chief of Staff for President James D. Wolfensohn and for Juan Manuel Santos, former President of Colombia and Nobel Peace Prize winner. She was manager of the Inter-American Development Bank Fund and senior advisor on renewable energy for the Minister of Mining and Energy. Marcela studied Political Science at Los Andes University in Colombia, Urban Planning at New York University, Finance at ESADE University in Spain, and Executive Sr. Management at Harvard University.

How did you get involved with Solar Decathlon Latin America and Caribbean?

I am responsible for initiating interest in bringing Solar Decathlon to Latin America and Caribbean. It all began when I had the wonderful opportunity to visit the first Solar Decathlon in Washington, DC in the fall of 2002. At that time, I was working for the World Bank as an urban planner specializing in renewable energy. After visiting the competition, my dream became to take it to my region, and more specifically, to launch it from Cali, Colombia, the city where I was born.

I was motivated to bring SD to Latin America and the Caribbean for the following three reasons:

- Promote innovative projects in developing countries with photovoltaic energy;
- Engage universities throughout the region to conduct R&D, in order to contribute to climate change mitigation; and

- Explore the possibility to build "Green Neighborhoods" for low-income communities.

Who was instrumental in getting the 2015 competition organized in Cali?

The following two people made Solar Decathlon a reality for Latin America and the Caribbean:

Richard King: He kindly accepted my invitation to have lunch near the Department of Energy after my visit to the Solar Village. He had a wonderful disposition to listen to my dream and goal for a Solar Decathlon in Latin America and the Caribbean. We decided to give it a try.

Mayor Rodrigo Guerrero: Solar Decathlon LAC would have never been possible without the leadership, vision, and political will of Rodrigo Guerrero. In 2015, Dr. Guerrero was the Mayor of Cali. He believed in the importance of implementing projects to promote the use of renewable energy in a city with lots of sun, water, and wind.

What made SD LAC 2015 and SD LAC 2019 successful? How have these competitions had an impact in Colombia and Latin America?

What made the competition a complete success was the strong alliance of the public, private, and academic sectors. There was a homogeneous belief in the project, its impact, and potential to contribute to a better quality of life for society as a whole.

However, in Colombia, we do not always focus on long-term goals, so we sometimes fail to invest our energy in the right ways. For example, the development of sustainable neighborhoods to improve the quality of life for our citizens, among other economic and social benefits, has not happened on a large scale. To date, our region has not yet made a significant contribution to climate change mitigation.

How have those Solar Decathlon events influenced education, design, the sustainable building industry, and public awareness?

Both Solar Decathlon competitions have indeed influenced education, design, and public awareness in positive ways. Although the number of "green buildings and houses" has not reached or surpassed the number of regular buildings in our region, we hope the sustainable building industry will make more progress in the future. The lack of technical knowledge and the high price of equipment, such as photovoltaic panels, affects the business models. I really hope many more Solar

Decathlons will take place in Latin America and the Caribbean, and that they can also evolve in this part of the world.

The inaugural SD LAC was held on the grounds of Universidad del Valle in Santiago de Cali, Colombia, from December 4–15, 2015. For the first time ever, the competition was held in a tropical savanna climate where the average annual temperature is approximately 76°F (24.5°C), and intermittent rain showers are common. Colombia, the second most biodiverse country in the world, is located near the equator (3° north latitude) and has no major seasonal variation. It is an emerging tourist destination that offers a vast array of geographical and cultural treasures. Cali, the site for SD LAC, was no stranger to competitions. Cali had previously hosted well-known sporting venues, such as the Pan-American games in 1971, World Games in 2013, and World Youth Championships in Athletics in 2015.

Carlos Rodriguez-Marin was tapped to be Director of this new competition. An industrial engineer with 25 years of experience in management of large city projects, including the 2013 World Games, plus two decades of university teaching in finance and business innovation, Carlos had great insight and the appropriate business acumen to lead the project. His fundraising efforts resulted in close to a million dollars of sponsorship for SD LAC. Also, a heap of credit goes to Carlos for galvanizing a skilled team of organizers who masterminded event communications and onsite logistics in a short period of time. From the initial recruitment of teams, to design and assembly of the solar village, to media coverage and jury panel selection, Carlos and his crew gave a virtuoso performance. He also hired a highly skilled competition manager, Edwin Lopez-Bouzas, to provide invaluable support for this great new adventure in Cali, Colombia. Leaning on available SD expertise, Carlos connected with Edwin Rodriguez-Ubinas (SDE Competition Manager), Joe Simon (at NREL), and Richard King (at U.S. DOE). Their guidance helped him hit the ground running for SD LAC.

Carolyn Field, wonderful onsite assistant and superb translator for Richard King at the 2015 event, sharing a happy moment with Competition Director Carlos Rodriguez-Marin.

A local government representative with Competition Manager Edwin Lopez-Bouzas and Richard King.

Fifteen teams representing eight nations participated in the competition, which local journalists referred to as the "Olympics of Solar Energy."

- London Metropolitan University, *Heliomet* (United Kingdom)
- Pontificia Universidad Bolivariana, *Yarumo* (Colombia)
- Universidad del Valle, *Wiwa* (Colombia)
- Pontificia Universidad Javeriana de Bogotá, *PEI* (Colombia)
- Pontificia Universidad Javeriana de Cali and Universidad ICESI and Sena Valle del Cauca, *Alero* (Colombia)
- Sena Valle del Cauca, *Vrissa* (Colombia)
- Universidad de los Andes, *Huerto* (Colombia)
- Universidad de Sevilla and Universidad Santiago de Cali, *Aura* (Spain, Colombia)
- Universidad la Salle and Hochschule Ostwestfalen-Lippe, *Habitec* (Colombia, Germany)
- Universidad Nacional de Ingeniería del Perú, *Ayni* (Perú)
- Universidad ORT Uruguay, *Casa Uruguaya* (Uruguay)
- Universidad San Buenaventura and Universidad Autónoma de Occidente, *Mihouse* (Colombia)
- Instituto Tecnológico y de Estudios Superiores de Monterrey, Campus Querétero, *Kuxtal* (Mexico)
- Universidad Nacional de Colombia sede Medillin, *Unsolar* (Colombia)
- Universidad Tecnológica de Panamá and Western New England University, *Casa Smart* (Panamá, United States)

The teams were required to choose a specific urban community where their prototype dwelling would be located. The challenge was to engage in intentional, purposeful design that would meet the daily needs of people living in an identified location. In addition, the houses had to be affordable for those in that target community. Decathletes created scale models of their houses, plus realistic renderings of the setting of those prototypes in the target neighborhoods chosen.

The 10 contests for SD LAC 2015 were as follows:

Juried

- Architecture
- Engineering and Construction
- Communications and Marketing
- Sustainability
- Urban Design and Affordability
- Innovation

Measured

- Energy Efficiency
- House Functionality
- Comfort Conditions
- Electrical Balance

An outstanding aspect of the competition was the incredible resourcefulness of student teams. They demonstrated clever ways of using and reusing existing materials and devised innovative equipment and materials that incorporated local resources. Ever mindful of the need for affordable, energy-efficient housing solutions for Latin America as a developing region, Decathletes doubled down on how to "do more with less." The outcomes were impressive!

True to their reputation as "happy people who enjoy social gatherings," SD LAC organizers included entertainment in the "Villa Solar." They provided a large stage, open to the public, where singers, musicians, and dancers could perform during competition week. Decathletes and others in the audience joined in the fun, which contributed to an atmosphere of delightful merriment as the event unfolded. Any residents nearby who did not already know about SD LAC were quickly brought up to speed with the colorful lights and joyful music.

Casa Habitawlum from Universided de la Salle de Bogota, Colombia, and University of Applied Science from eastern North Rhine-Westphalia, Germany, is a great example of various exterior shading devices to keep the house cool on hot days.

Casa Yarumo from Universidad Pontifica Bolivariana de Medillin, Colombia, included beneficial rooftop gardens for growing food for residents of the house.

The exterior walls of bamboo in *Cultural Machine* from Universidad Javeriana de Bogota, Colombia, opened up completely on all four sides to allow fresh air and cooling breezes into the structure.

The interior of *Cultural Machine* from Bogota shows generous flexibility in use of space, including beds that could be raised or lowered and movable shades that enabled privacy and protection from outdoor elements as needed.

SOL-ID from London Metropolitan University had a unique shape to help shade the house from direct sunlight.

In a strong nod to sustainability, unique handmade bricks made of discarded cocoa shells were used in the construction of the *SOL-ID* house.

Casa Smart from Universidad Tecnologica de Panama and Western New England University in the United States was constructed from reused shipping containers. Notice the green box-like frames around the windows to reduce heat from direct sunlight inside the dwelling.

Curious visitors eager to peek inside *El Armatodo de UnSolar* from Universidad Nacional de Colombia in Medillin waited patiently in lines to tour the house.

A Decathlete from *UnSolar* showing the special equipment they used to make their own bricks by hand.

Natural elements, such as bamboo, were transformed into attractive decorative features for *Casa UnSolar*.

UnSolar Decathletes developed a practical use for plastic bottles—as furniture!

Students at *Mas Huertas Mas Casa* from Universidad de los Andes, Colombia, demonstrated how to grow plants in different ways. They took advantage of recycled containers as pots and reused 70% of greywater from the house for watering their plants.

A *Mas Huertas Mas Casa* Decathlete shows his strong commitment to horticulture with "green hands."

This colorful hammock made of recycled plastic bottlecaps is a great example of how teams made creative use of available materials.

Green walls with living plants were displayed prominently on the exterior of *Mihouse* from Universidades Autonoma y San Buenaventura, Colombia.

Over the course of the competition from December 4–15, teams jockeyed for the top position as scores from each of the contests were released. In the end, Decathletes from the Universidad de Sevilla

(Spain) and Universidad Santiago de Cali finished in third place. The team from Pontificia Universidad Javeriana de Cali and Universidad ICESI (Colombia) won the second-place award, and the Universidad ORT in Uruguay took top honors with first place overall.

La *Casa Aura* from Universidad Santiago de Cali, Colombia, and Universidad de Sevilla, Spain, won third place overall in the competition. The exterior shades could be opened manually to welcome breezes that cooled the deck and interior of the dwelling.

Strikingly beautiful, *Casa Alero* from Universidad Icesi y Javeriana was awarded second place at SD LAC 2015. The bamboo shades on all sides could be partially or fully open or closed, depending on weather conditions and time of day.

Distinctly elegant and highly energy efficient, *Casa Uruguaya* from Universidad ORT, Uruguay, took first-place honors at the 2015 competition. The large deck with deep overhanging roof offered a comfortable spot for relaxing or dining.

The interior of *Casa Uruguaya* made clever use of every bit of space, while keeping the living area open and brightly lit.

Consistent, outstanding performance of their well-crafted *Casa Uruguaya* led to big smiles from this hard-working team, pictured here inside their dwelling with Melissa and Richard King.

The closing ceremony was an upbeat celebration with the vibrant exuberance of Decathletes on full display as the winning teams were announced. To everyone's surprise, students were so excited with their achievements that, during the ceremony, the temporary stage collapsed beneath their enthusiastic jumping. In a fitting tribute to everyone's dedicated efforts and huge success, the incoming mayor of Cali pledged his support for a second event in Cali, Colombia.

Following that remarkable success story, the second MOU was signed by representatives from the U.S. DOE, the Colombian Ministry of Mining and Energy, and the Department of National Planning on November 21, 2016. This document authorized the second edition of SD LAC.

Onstage at the final awards ceremony for SD LAC 2015, Mayor Rodrigo Guerrero, Richard King, and the incoming mayor pledge their support for another SD LAC in Cali, Colombia.

Reflections from Carlos Rodriguez-Marin, an electrical engineer from the Universidad Autónoma de Occidente and telecommunications manager for EMCALI, worked on communications technology projects with public and private companies for 25 years. He was Competition Director for SD LAC, and later served as director of ERT, a government group from the Valle del Cauca.

Share your vision to inspire greater response to climate change in Latin America.

I'm excited about promoting sustainable housing and making it available and affordable for everyone. My vision is that every family and every home, no matter where they live, would have this opportunity. We know now that climate change is real, and we all feel the negative effects of our warming planet. In Cali, we were fortunate to learn about a university competition called the Solar Decathlon. There was interest in bringing this event to South America, and several key leaders in Colombia rallied around the idea. Their determined efforts made Solar Decathlon Latin America and Caribbean (SD LAC) a reality. I was so happy to be part of this movement, and we had our first successful event in December 2015 at a site next to the Universidad del Valle in the city of Cali, Colombia.

It was so fantastic that we decided to host a second edition of Solar Decathlon LAC in December 2019 at the same location. Both these amazing events involved model homes that were small in size (e.g., 80m^2) but bursting with big ideas. Those solar-powered houses, designed and built entirely by students, were constructed with sustainable materials and had greywater recycling systems for use with daily tasks, such as laundry, cleaning, and watering plants. Some of the homes even had their own vegetable gardens or fruit orchards where they could produce food for residents. Watching so many visitors come to explore and tour these unique housing prototypes during the competition made me realize that people are really interested in learning more.

Solar Decathlon Latin America and Caribbean was a great opportunity for the public to realize that sustainable housing begins with each one of us. If you understand more, you can transform your own habits, and you can then plan for a more energy-efficient, sustainable home. Environmentally-friendly materials and systems are available right now to help us conserve water and other resources, but we have to change our attitudes and daily habits. We have to say to ourselves, "I am responsible for the water and energy that I consume, as well as the waste that I produce. If I don't take responsibility, no one else will."

Solar Decathlon LAC got me thinking about expanding on this idea. I wanted to share how outreach and education can help transform the way we build homes and how we go about our daily lives. With support from the regional and local governments of all 42 municipalities in the Valle del Cauca and the Universidad del Valle foundation, we were able to begin a new project. This model is based on five important pillars: materials and design, water and energy, comfort for everyone, waste reduction, and food security.

Our plan is to reach out to local communities to teach them how every single person can make a difference. We cannot continue to think that, "If I don't do it, someone else will." Instead, we need to help people understand that every person is responsible for their own actions and the effects of those actions. To encourage transformation worldwide and halt climate change, we must show people how to change their own behaviors. People should say, "It starts right here, with me."

SD Latin America and Caribbean 2019

Due to unexpected political and socioeconomic situations in Colombia, plans to host another competition took a bit longer than anticipated, but new teams were ready to go four years later. The 2019 event would take place at the same location on the grounds of Universidad del Valle. Once again, Carlos Rodriguez was leading the charge. That fall, Richard King received an invitation from Carlos, asking if he would return to Cali for SD LAC 2019 as a jury member for the Innovation Contest. Without hesitation, Richard said "Of course, I wouldn't miss it!"

Leader of the Architecture Jury panel (in wheelchair) gave an inspiring presentation at SD LAC 2019 about the challenges of living with a physical disability. His remarks aimed to raise awareness about the importance of universal design.

With brilliant foresight, SD LAC 2019 organizers chose a compelling theme for this competition: *A home for everyone*. No prior Solar Decathlon had focused on universal design, defined as designs that function well for everyone. Carlos realized the significance of this concept and embraced it for the next edition of SD LAC. He asked the participating teams to ensure that their Decathlon prototypes were accessible and comfortable for all people, in terms of the physical arrangements, as well as accommodating for daily functions, such as communication for those who are vision-impaired, blind, hard-of-hearing, deaf, etc. This was a huge leap forward, as this theme recognized the need for architects and the construction industry to incorporate universal design into new or retrofitted dwellings and buildings. Carlos made this statement: "The idea is to promote this concept at universities and academic institutions, to show that we are all equal."

SD LAC 2019 included 15 teams that represented seven nations:

- Pontificia Universidad Javeriana, *PEI Maquina Verde – El Arca* (Colombia)
- Pontificia Universidad Javeriana de Cali, Universidad Federal Santa Catarina, and Instituto Federal Santa Catarina, *Minga* (Colombia, Brazil)
- SENA, *Vrissa* (Colombia)
- Instituto Tecnológico y de Estudios Superiores de Occidente, *Tonal Casa Solar* (Mexico)
- Universidad de los Andes, *Puert Abierta* (Colombia)
- Universidad Nacional de Ingeniería, *Játi* (Perú)
- Universidad Santiago de Cali, *AEter* (Colombia)
- Universidad Arturo Prat, *Willkallpa* (Chile)
- Universidad del Magdalena, *Huru* (Colombia)
- Universidad Nacional de Colombia, *PV4* (Colombia)
- Universidad de la Salle and Hochschule Ostwestfalen-Lippe University, *InnoNativo* (Colombia and Germany)
- Federal University of Paraíba, *Casa Nordeste 1.0* (Brazil)
- Universidad del Valle, *Chamaleon House* (Colombia)
- Universidad de Sevilla, *AURA* (Spain)
- Universidad de San Buenaventura and Universidad Autónoma de Occidente, *Tuhouse* (Colombia)

The 10 contests for the 2019 competition echoed those from the first event.

Juried

- Architecture
- Engineering and Construction
- Communications and Marketing
- Sustainability

- Urban Design and Affordability
- Innovation

Measured

- Energy Efficiency
- House Functionality
- Comfort Conditions
- Electrical Balance

The onsite event took place in the first two weeks of December. Cali, Colombia, was lit up with Christmas decorations, which added a festive flavor to the arduous challenges that the teams had faced in preparing for such a demanding competition. The Decathletes rose to the occasion with this brand-new set of unique model houses. The jury panels comprising experts in the field were notably impressed.

Carlos Rodriguez-Marin (second from left) with jury members for SD LAC 2019, which included Richard and Melissa King.

Richard and Melissa King served as jury members for the Innovation Contest, along with Walter Torres, an esteemed expert in innovation from Argentina. All three were blown away with the ingenuity on display. For example, the Decathletes utilized locally available resources in novel ways to make their own materials that enhanced the sustainability of their prototype houses. To respect the need for affordability in Latin America and the developing world, the teams did not necessarily strive for costly hi-tech solutions. Instead, they often embraced simplicity, relying on new approaches or creative combinations of existing materials to incorporate into prototype dwellings. Look closely at the photos from SD LAC 2019 to see examples of clever, low-cost elegance.

As in the inaugural edition, model houses were attractive dwellings well-suited for a warm, tropical climate. With less concern about the need to keep houses heated and greater emphasis on keeping interior spaces sufficiently cool and comfortable, the designs included plentiful shading, as well as clever options to encourage refreshing breezes. To ensure that these "smart houses" were affordable and energy efficient, manual operation of some equipment was prioritized over mechanized operation that relied on electricity. These Decathletes approached the design task with a mindset that was laser focused on sustainability and conservation. In other words, "How can we do more with less?"

Inno Nativa from Universidad de la Salle (Colombia) and Hochschule Ostwestfalen-Lippe University (Germany) featured luxurious green walls and a variety of shading devices, both vertical and horizontal, made of bamboo and wood.

Team *AEter* from Universidad Santiago de Cali, Colombia, used rice husks to make their own bricks, which were lightweight and easy to snap together, just like Lego® blocks, for relatively quick construction.

This clever idea for a suspended bed that could be tucked up into the ceiling when not in use allowed space for other daily functions.

Flexible storage in this house came in the form of sturdy wooden boxes that could be moved around easily and combined with woven baskets to tuck away clothes or other items.

This gorgeous handcrafted wooden fixture in the ceiling was designed intentionally for acoustic purposes; it absorbed and softened harsh sounds inside the house for a more pleasant environment.

Huru from Universidad del Magdalena, Colombia, won the Innovation Contest, which considered holistic approaches to systems, as well as the relationship between the dwelling and its intended neighborhood. The Huru team used only locally available natural resources for construction, such as stucco from guava trees, and materials that Decathletes had conceptualized and fabricated themselves.

To build *Huru*, the team made fireproof, waterproof blocks made of recycled paper mixed with cement, a completely new invention.

The teams eagerly awaited the announcement of the overall winners, which would be recognized at the final awards ceremony on December 15. The scores revealed that third place went to *Tuhouse* from the Universidad de San Buenaventura and Universidad Autónoma de Occidente in Colombia. Second place went to *PEI Maquina Verde* from Pontificia Universidad Javeriana in El Arca, Colombia. *Minga,* from Pontificia Universidad Javeriana de Cali, Universidad Federal Santa Catarina, and Instituto Federal Santa Catarina in Colombia and Brazil took top honors with their first-place win overall. That team created compelling, heartfelt communication materials to highlight their commitment to disaster resistance, recovery, and multi-unit dwellings for two generations within a family. That made quite an impression on the judges.

After the winners were announced at the awards ceremony, the entire solar village erupted with jubilant cheering. The Decathletes had worked diligently to design, build, and showcase their creations, and they were ecstatic to realize success. Finally, it was time to celebrate! Cali, well known for salsa dancing, lived up to its reputation of fancy, rhythmic footwork. As the celebratory music began, everyone in the outdoor tent was moving to the beat with huge smiles on their faces!

Tuhouse from Universidad de San Buenaventura and Universidad Autónoma de Occidente was made of concrete to ensure security and longevity and used bamboo for window screens. This forward-thinking team won third place overall at SD LAC 2019.

Tuhouse had a spacious green roof with productive vegetable gardens. They also grew plants on the exterior green walls. On this house, PV panels extended out from the roof as overhangs for decking and walkways below, to preserve roof space for gardening.

PEI Maquina Verde from Pontificia Universidad Javeriana, Colombia, was enchanting and exquisitely designed. Steel and glass siding was covered with unusual handmade shades fabricated from thick stems of locally harvested vines that created alluring appeal.

The cheerful, bright interior of *PEI* lent a sense of grandeur to this dwelling, which offered two levels for various daily functions. Use of space was flexible, and the entire house felt relaxed and comfortable.

The team from Pontificia Universidad Javeriana de Cali, Universidad Federal Santa Catarina, and Instituto Federal Santa Catarina that built *Minga* had studied a community in Buonaventura for 3 years to better understand their needs. These Decathletes embraced the concept of co-housing for two generations of residents (older and younger), and the house was ADA compliant.

Minga provided outdoor decking areas for family relaxation and social enjoyment.

PEI was overjoyed to win second place in the Innovation Contest, as well as the second place overall in the SD LAC 2019 competition.

The buoyant *Minga* team was thrilled to take first-place honors in the competition.

Carlos Rodriguez expressing his sincere gratitude to Regional Governor Dilian Francisca Toro for her steadfast support of SD LAC. Pictured here to accept the award is the Governor's sister, Jimena Toro, Social Manager for Valle del Cauca.

Special Tribute to Carlos Rodriguez-Marin

We are sad to share that Carlos Rodriguez passed away in September 2022, while this book was being written. Carlos fought a courageous battle with pancreatic cancer, but in the end, he was unable to overcome that dreadful foe. In tribute to such a wonderful person, we honor Carlos here, so that readers are introduced to our cheerful, generous friend who had a heart of gold. It was a privilege to be with Carlos, as he always seemed to radiate happiness in every direction. Easy going and amicable, he was a natural leader and dedicated mentor who inspired others to do their best. At the same time, Carlos knew how to buckle down and tackle the tough stuff with true grit and determination. No matter what the odds might be, Carlos would find a way forward through unwavering perseverance. Mindful of treating everyone with respect, his gracious manner made every individual feel comfortable in his presence. Gifted with abundant love from his beautiful family, Carlos gave every ounce of that love right back to those around him. He is survived by his loving wife, Sandra, and their two children, Santiago and Bella.

Carlos Eduardo Rodriguez-Marin

We are grateful to Carlos for helping to establish a powerful legacy with Solar Decathlon Latin America and the Caribbean. We are grateful to our dear friend Carlos for shining his light to teach others through his own fine example. Thank you, Carlos, for enriching our lives.

Melissa and Richard King

SD Middle East 2018

In July, a few weeks after the final awards ceremony for Solar Decathlon Europe 2014, Richard King received an invitation from a colleague involved with solar car racing. A new race called the Abu Dhabi Solar Challenge would take place in January 2015, just before the opening ceremony of the World Future Energy Summit in Abu Dhabi. Richard was asked to participate as a protest resolution juror for the race. Hans Tholstrup, his longtime friend and solar car aficionado from Australia, encouraged Richard to attend. Hans was president of the International Solar Car Federation (ISF) and saw this as an opportunity to bring together some early leaders of solar car racing: Dr. Freddy Sidler from Switzerland, Takahiro Iwata from Japan, and Richard King from the United States. Freddy Sidler's team from the Engineering School in Biel took first place at the 1990 World Solar Challenge (WSC) across Australia, and Takahiro Iwata's team from Honda R&D won the next WSC in 1993. In that same year, Richard King had served as competition director for Sunrayce '93 in the United States. Along with Hans Tholstrup, those pioneers helped catapult the vision that solar cars would accelerate research and development in clean energy as an essential path to safeguard the health of our planet. Without hesitation, Richard accepted the invitation and made plans for travel to Abu Dhabi in January. He was eager to expand the Solar Decathlon into the Middle East, and he was sure that SD prototypes customized for this region would make an impact.

Fortuitously, in fall 2014, Dr. Pedro Banda, Senior Specialist for R&D at the Dubai Electricity and Water Authority (DEWA), reached out to Richard King at the Department of Energy to inquire about organizing a new edition of Solar Decathlon in the Middle East. Richard was surprised and intrigued with the idea. He planned to be in Abu Dhabi for the January 2015 solar car race, so he arranged for a face-to-face meeting during his visit to the United Arab Emirates (UAE) with Waleed Salman, EVP of Strategy and Business Development for DEWA. Waleed was eager to move forward with the creation of a new event. During the discussion in his office at DEWA headquarters in Dubai, preliminary plans for Solar Decathlon Middle East (SDME) emerged.

Six months later, on June 17, 2015, the U.S. DOE, the Dubai Supreme Council of Energy (DSCE), and Dubai Electricity and Water Authority (DEWA) signed an MOU to collaborate in organizing SDME in Dubai, UAE. Describing this agreement, His Excellency Saeed Mohammed Ahmad Al Tayer, Vice Chairman of DSCE and MD&CEO of DEWA, stated, "We consider the Solar Decathlon a unique opportunity to generate incentives among Emirate students to use their innovative skills to design buildings that are energy-efficient and self-sufficient, contributing to national and regional growth."

In a press release about this news, Richard noted that "President Obama's Climate Action Plan had called on the U.S. to work with other nations to lead the world to a cleaner, more prosperous future." Solar Decathlon competitions held in different countries had become outstanding examples of productive, cooperative endeavors that offered innovative solutions for a more sustainable future. The MOU with Dubai referenced cooperation for two SD events in the Middle East. Richard King was excited that Waleed Salman and his team at DEWA planned to include SDME in the 2020 World Expo, set to be held in the UAE. That event had the potential to expose hundreds of thousands of visitors from around the globe to the remarkable ingenuity displayed in Solar Decathlon prototype houses, boosting public understanding of energy-efficient designs for dwellings and buildings worldwide.

The new competition needed an experienced leader with in-depth knowledge of how to manage a Solar Decathlon. In 2014, after disassembly of the solar village at SD Europe in Versailles, no European nation stepped forward to host SDE 2016. As a result, Edwin Rodriguez-Ubinas, a highly qualified leader who had served as Competition Manager for SDE competitions, was looking for a job. DEWA quickly realized that Edwin was a perfect match for the crucial role of Competition Director, and Edwin accepted their invitation to lead this new effort in the UAE.

It was costly and rather difficult to ship houses from various locations to the UAE, so DEWA set up monetary awards to help defray the cost of transportation and basic team travel. They provided funds for all SDME teams equivalent to more than 10 million UAE Dirham (AED, roughly 2.5 million USD). These awards included guaranteed money for each participating team, plus substantial prize money for the winning teams.

Team Place	Monetary Award
First	900,000 AED (~$250,000)
Second	800,000 AED
Third	700,000 AED
Fourth	550,000 AED
Fifth	450,000 AED
All others	400,000 AED

An official call for proposals was issued early in 2016, and SDME organizers were pleased with the response. The prize money attracted 22 universities from the Middle East and around the world. Those that answered the call targeted the need for sustainable living in desert regions, where extremely high temperatures, blowing sand, and dust create a challenging set of conditions. This location in the UAE represented a very different scenario from the humid, subtropical conditions of Washington, DC, the Mediterranean climate of Madrid, Spain, and the tropical, savanna conditions of Cali, Colombia. Therefore, model homes at SDME would need to feature various design elements and engineering solutions that reflected a desert environment. Another major goal of this competition was to recognize appropriate sociocultural contexts in creating and promoting innovations and technologies that would benefit those who live in the Middle East.

By mid-2018, some university teams reported that they were unable to complete their design-and-build projects. Seven teams withdrew, so the slate of contenders for the inaugural SDME was winnowed down to 15 teams representing 28 universities from 11 countries on four continents. Many of those teams were collaborative efforts that included Decathletes from several institutions. This international cross-fertilization of ideas was an important part of the process that nurtured multiple perspectives and enhanced mutual respect among diverse nations and regions. These are the teams for SDME 2018:

- Ajman University of Science and Technology (UAE)
- American University in Dubai (UAE)
- American University of Ras Alkhaimah (UAE)
- Eindhoven University of Technology (The Netherlands)
- Heriot-Watt University Dubai (UAE)
- Islamic Science University of Malaysia; University of Technology – Malaysia (Malaysia)
- Ion Mincu University of Architecture and Urbanism; Technical University of Civil Engineering, Bucharest; Birla Institute of Technology and Science, Pilani in Dubai; Polytechnic University of Bucharest (Romania, UAE)
- National Chiao Tung University (Taiwan)
- New York University Abu Dhabi (UAE)
- Sapienza University of Rome (Italy)
- University of Wollongong (Australia)
- University of Sharjah; University of Ferrara (UAE, Italy)
- University of Bordeaux; Amity University; An-Najah National University; Arts & Métiers Paris Tech; Bordeaux School of Architecture (France, UAE, Palestine)
- Virginia Tech (United States)
- King Saud University (Kingdom of Saudi Arabia)

The next step for SDME organizers was to write competition rules and begin the search for a good site. To assemble all the model houses in a solar village, five to seven acres of open land is needed. The search for an appropriate location yielded no suitable spot within the city limits of Dubai. Therefore, DEWA decided to hold the competition next to the Mohammed bin Rashid Al Maktoum Solar Park, a huge site of ground-mounted solar arrays. At the time, those solar arrays were capable of producing 700 MW of power (That system has since been expanded for even greater power output). This site provided a beautiful, spacious area in the desert, several miles away from Dubai. It would be more challenging to attract visitors there, but the site offered ample space to host a spectacular event.

International Expansion | 499

Melissa King at the entrance to Solar Decathlon Middle East 2018 in Dubai, UAE.

An elevated view of SDME 2018 "Solar Hai" (solar village), including infrastructure buildings that provided food, meeting rooms, prayer rooms, and a large auditorium.

Mohammed bin Rashid Al Maktoum Solar Park, one of the world's largest solar systems, next to the site for SDME 2018.

SDME's "Solar Hai" was surrounded by breathtaking desert landscapes.

Richard King with Abdullah Al Hammadi, Edwin Rodriguez-Ubinas, and Waleed Salman.

The 10 contests and their scoring values for SDME were as follows:

Juried

- Architecture (100)
- Engineering and Construction (100)
- Energy Efficiency (80)
- Sustainability (100)
- Communication (80)
- Innovation (80)

Measured

- Energy Management (140)
- Comfort Conditions (120)
- House Functioning (120)
- Sustainable Transportation (80)

In addition to those 10 contests, the teams could also win recognition in four other categories: Smart Solutions, Building Integrated Photovoltaics (BIPV), Interior Design, and the People's Choice Award.

SDME teams designed and built homes with substantial thermal mass (stone, concrete) that sat directly on the ground. This strategy can help maintain cooler temperatures on hot days when temperatures approach 110°F (43°C). Another feature typical of desert regions is a distinct entryway or vestibule for the dwelling. Many teams also built walls around the house for additional shading, privacy, and protection from natural elements, such as high winds and blowing sand.

SDME 2018 took place from November 14–29. The event included more than 600 students and university faculty, and media coverage of this outstanding competition was excellent. Jury members were impressed with the ingenious prototype designs for desert conditions, along with the notable innovation on display.

The final awards ceremony on November 28 was an elaborate, carefully orchestrated event that began with a narrated walk-through of the "Solar Hai" for His Excellency Saeed Mohammed Ahmad Al Tayer, Sheikh Hamdan bin Mohammed bin Rashid Al Maktoum (Crown Prince of Dubai), Waleed Salman (EVP at DEWA), and other officials. Richard King accompanied them, while SDME participants and the press watched their live-streamed tour from the auditorium. A few minutes later, the group joined Decathlete teams inside, and the final awards ceremony began. Teams erupted with excitement as winners were announced and prizes awarded.

Team Baitykool, which included students from France, the UAE, and Palestine (University of Bordeaux, Amity University, An-Najah National University, Arts & Métiers Paris Tech, Bordeaux School of Architecture) took third place overall. Team UOW from Wollongong, Australia, won the second prize, and Team FutureHAUS from Virginia Tech in the United States won the first-place award overall at SDME 2018.

Crown Prince of Dubai tours "Solar Hai" with His Excellency Al Tayer, Waleed Salman, and Richard King.

The house from Team *Baitykool* had frameless PV modules sandwiched between concrete walls, plus a shaded entryway on one side. The dwelling represented a strong rock that surrounded an interior oasis.

The façade of Team *Baitykool*'s house showcased building integrated photovoltaics (BIPV), along with thermal walls with holes that allowed light to enter the dwelling.

A green wall for hydroponics coupled with an aquaponic system in the *Baitykool* courtyard enabled a harmonious symbiotic environment that created a calming ambience. This open patio provided natural light and a retractable shading system that blocked sunlight on hot days.

Like other SD prototypes, this space inside the *Baitykool* house was multipurpose for sleeping, studying, and working; notice the attractive window shading and ingenious staircase integrated into storage cubbies.

View of the north side of UOW's *Desert Rose* house with DEWA's Innovation Center tower (under construction) in the background. This team designed an innovative "house for life" that could adapt to people's needs as they age.

The open floor plan inside *Desert Rose* was ADA accessible to accommodate for those who need assistance with everyday movement.

This demonstration exhibit form shows a cross section of exterior wall layers containing foam and batt insulation to achieve an energy-efficient dwelling with a thick thermal envelope.

Va Tech's *FutureHaus* was constructed of "cartridges" using advanced prefabrication methods, which enabled faster assembly, greater accuracy, reduced material waste, and less pollution. These modular elements are ready for plug-and-play assembly.

An outdoor patio around *FutureHaus* featured attractive screens that filtered light for shading and privacy. The ceiling cavity is lined with phase-change materials for load shifting to help balance energy demand.

FutureHaus was a marvel of technology and leading-edge design features, such as programmable height adjustments for kitchen and bathroom sinks and counters, as well as a closed-loop smart shower recycling system that purifies, reheats, and recirculates water while someone is taking a shower. The dwelling is a stunning showcase of Smart Built Environments (SBEs), which include sensors and multi-modal user interfaces that can improve people's lives.

The Virginia Tech team receiving the first-place award from the Crown Prince and His Excellency Al Tayer.

Reflections from Dr. Edwin Rodriguez-Ubinas, Principal Researcher for the Energy Efficiency Area at the Dubai Electricity and Water Authority R&D Center in the UAE. He was Competition Manager for SDME in Dubai (2018) and the Dubai Expo (2021). Edwin has also been a professor and researcher at the Technical University of Madrid. He is one of the founders of Solar Decathlon Europe (SDE), and served as SDE Competition Manager for several events: 2010 and 2012 (Spain), 2014 (France), and 2022 (Germany).

After working on Solar Decathlon Europe competitions, you accepted an offer to lead Solar Decathlon Middle East in Dubai. What motivated you to take a leading role in that first event in a different part of the world?

Starting a Solar Decathlon in a new region was a very attractive challenge for me. Having participated in the U.S. DOE Solar Decathlon and being one of the founders of Solar Decathlon Europe and its

Competition Manager for the first three editions, I felt I was ready to take on this new responsibility. Being part of the organizing team and leading the Solar Decathlon Middle East (SDME) competition surpassed all my expectations. It has been an excellent learning experience, and I enjoy my time in the UAE. I want to thank HE Saeed Al Tayer, MD and CEO of Dubai Electricity and Water Authority (DEWA), Mr. Waleed Salman, EVP of Business Development and Excellence, and Dr. Pedro Banda, Director of R&D, for trusting me and letting me lead the competition efforts of SDME.

What was most challenging about organizing the inaugural SDME competition?

Starting a Solar Decathlon competition in a new region is not easy, and it's even more difficult if you have never lived in this place, as was my case. My previous experience was in regions that I knew very well. After participating in U.S. DOE SD events and organizing SDE competitions in Spain, the transition to France was relatively straightforward. However, to effectively work on Solar Decathlon Middle East (SDME), I had to learn more about Middle Eastern culture, connect with new colleagues, and figure out how to approach universities to encourage their participation in SDME. Fortunately, from day one, I received full support from the top management and colleagues at the Dubai Electricity and Water Authority (DEWA), which organizes SDME. Additionally, the educational system in the UAE follows European and American best practices, and the UAE is quite advanced in terms of technology and other technical aspects. All of this made it easier to plan and run the SDME competition.

Another big challenge for the organization of SDME was finding a good site for the solar village. Unfortunately, an appropriate place was not found in or near downtown Dubai. Therefore, DEWA provided a huge site, infrastructure, and facilities to host a most impressive SD competition. The site is located in the Mohammed bin Rashid Al Maktoum Solar Park, the largest solar park in the world, and next to the DEWA R&D Center and the Innovation Center, an iconic building. However, since the site is not close to the city, a bigger effort was required to encourage the public to visit the solar village.

The keys to SDME's success include the Dubai Electricity and Water Authority's (DEWA) top management commitment, the fact that DEWA has all the expertise required for running SDME in-house, and the involvement of all divisions' employees. As in any big organization, effective coordination and communication were critical challenges for SDME. These were overcome with the help and commitment of DEWA managers and employees. In DEWA, I felt most welcome at all times, and my colleagues and superiors became good friends while working on SDME. Finally, I want to acknowledge the outstanding contribution of many young Emirates in developing SDME: Noura Hassan Ali Al Hammadi, Sarah Abdulmajeed Alzarouni, Habiba Ebrahim Ahmed Mohamed, Abdulla Al Suwaidi, and many others from all DEWA divisions.

What difficulties did you encounter in preparing for the next SDME competition, which took place during the 2022 World Expo in the UAE?

After the successful completion of the first Solar Decathlon in the Middle East, we were very excited about the second edition, especially due to the fact that it would take place during and in alignment with the World Expo Dubai 2020. We were even better prepared for the second edition, with a consolidated and experienced organization, lessons learned from the first edition, and a fantastic and fully equipped site. In the beginning, everything was great. We started the competition with 22 strong teams from universities around the world. Indeed, the first international workshop for SDME 2020 was the best one ever.

Then suddenly, the COVID-19 pandemic happened, affecting the whole world. Universities were closed, reopening only in an online format. Students could not work together, teams lost their sponsors, cargo costs had escalated, and this climate of uncertainty caused many teams to withdraw. The Dubai Expo 2020 and SDME 2020, as well as many other international events, were postponed for a year. The pandemic lasted longer than expected, and the competition was held under its restrictions. However, SDME teams do not give up and faced COVID-19 challenges with courage and determination. Similarly, DEWA and SDME organizers did not stop. We continued supporting the teams and maintaining communication among all stakeholders. The winning house of SDME 2018 from Virginia Tech was transported back to the UAE, assembled, and exhibited during the entire time of the Dubai Expo 2020. A participating house was exhibited at a World Expo for the first time in Solar Decathlon's history. Finally, eight teams assembled their outstanding projects at the SDME Solar Village. They participated in the final phase of the competition, showing the world that it is possible to turn adversity into success. The SDME held in 2021 was a big celebration and a song of hope in very difficult times for the whole world.

How has SDME made an impact in Dubai, the UAE, and the Middle East?

Solar Decathlon Middle East is part of the Dubai Electricity and Water Authority's initiatives to support energy transition efforts, demand-side management programs, and carbon neutrality actions of Dubai, the UAE, and the region. The project was supported by the Dubai Supreme Council of Energy and organized under the patronage of His Highness Sheikh Hamdan bin Mohammed bin Rashid Al Maktoum, Crown Prince of Dubai and Chairman of the Dubai Executive Council.

SDME has impacted many sectors of society, including academia, industry, professionals, and policymakers. Thousands of students have benefited from applied experiential learning, getting knowledge through doing, discovering, reflecting, and testing. These future professionals are better prepared to face expected urban growth, the required energy transition, and climate change challenges. Current professionals, policymakers, and the public have accessed and experienced sustainable solutions that respond to their lifestyles and the region's harsh climate. SDME teams designed, constructed, operated, and showcased a variety of ways to develop sustainable zero-energy buildings for the Middle East. The SMDE projects have also revalorized vernacular architecture, finding their inspiration in historical Middle Eastern buildings, reinterpreting traditional designs, or using advanced materials and technology to enhance the benefits of traditional regional solutions.

SD Middle East 2020/21

Under the patronage of His Highness Sheikh Hamdan bin Mohammed bin Rashid Al Maktoum, Crown Prince of Dubai and Chairman of Dubai Executive Council, the second edition of SDME was planned in conjunction with the 2020 World Expo in Dubai. DEWA was designated as the Official Sustainable Energy Partner of that event, and the organizers were enthusiastic about prospective global exposure and focus on clean energy. Seven interrelated pillars that reflected the goals of DEWA and the World Expo were the focus of the second SDME competition: Sustainability, Future, Innovation, Clean Energy, Mobility, Smart Solutions, and Happiness. The participating teams were asked to represent these seven pillars in their prototype houses. Edwin Rodriguez-Ubinas, who led the SDME team for the 2018 event, served as the Competition Director, with support from Claudio Montero, a highly skilled veteran of SDE competitions.

The COVID-19 pandemic led to the postponement of SDME until 2021, but once it was rescheduled, Decathletes were eager to showcase their creations. The following teams participated in this event, which many referred to as "the Olympics of Sustainable Building."

- *Team KU*: Khalifa University
- *Team TAWAZUN*: Manipal Academy of Higher Education Dubai Campus
- *Team HARMONY*: The British University in Dubai
- *Team Sharjah*: University of Sharjah
- *Team Desert Phoenix*: University of Louisville, American University in Dubai, American University in Sharjah, Higher Colleges of Technology
- *Team ESTEEM*: Heriot-Watt University
- *Team SCUT x CSCEC*: South China University of Technology

SDME 2021 took place from November 11–25 at the Mohammed bin Rashid Al Maktoum Solar Park, and the houses remained on display from December 5–11. This made it possible for more World Expo visitors to experience the SDME competition and tour the innovative prototypes. At the final awards ceremony, these teams were recognized as the overall winners: *Team Go Smart* from the University of Bahrain took third prize, *Team Sharjah* of the University of Sharjah came in second, and *Team SCUT x CCSIC* of South China University of Technology won first place.

In addition, *FutureHAUS*, the SDME 2018 winning house that made such an incredible impression on everyone who toured it, was invited to be on display at the 2021 Dubai World Expo for 6 months, courtesy of DEWA. To date, that special invitation was the greatest honor for any Solar Decathlon prototype. More than a million people from all over the world toured that dwelling during the exhibition. Joe Wheeler, *FutureHAUS* Program Director and Co-Director of the Center for Design Research at Virginia Tech, took the lead in organizing and managing the *FutureHAUS* exhibit. In his words, "Being part of the World Expo was totally awesome." Scale models of all the SDME 2021 houses were also on display at the World Expo, so it was fantastic exposure for the Solar Decathlon competition.

VirginiaTech's *FutureHAUS* on display at the 2021 Dubai World Expo; photo credit: Va Tech.

Student Decathletes who supported the *FutureHAUS* exhibition at the 2021 Dubai World Expo beaming with pride; photo credit: Va Tech.

SD Africa 2019

As the year 2016 began, Richard King reflected on the successful expansion of Solar Decathlon competitions from the United States into Europe (Spain and France), Asia (China), and South America (Colombia), with plans under way for an event in the Middle East (Dubai). Realizing that Africa was the second largest continent with a rapidly growing population, it seemed fitting to hold a new edition of the Solar Decathlon there. Then, out of the blue on February 8, Richard received an email from Souad Lalami, Head of the Energy Efficiency and Green Buildings Department at the Research Institute in Solar Energy and New Energies (Institut de Recherche en Energie Solaire et Energies Nouvelles, or IRESEN) in Benguerir, Morocco. In her message, Souad noted that IRESEN was planning a new program dedicated to green buildings and energy efficiency in buildings. As part of their strategy, IRESEN wanted students to be involved and take the lead. Souad knew about the Solar Decathlon and thought it was the perfect platform to combine the dual objectives of promoting green energy and energy-efficient buildings while strengthening educational programs.

Richard could hardly believe the opportune timing of this communication exchange. He shares this recollection: "I was ecstatic to connect with Souad Lalami at IRESEN. Morocco was on the leading edge of solar energy R&D in Africa, and their plentiful supply of year-round sunshine made their location ideal for large-scale photovoltaic projects. I relished the opportunity to initiate a new competition in Africa and was eager to learn more. I recognized some potential issues, such as how to transport SD prototypes from one place to another on a huge land mass that lacked cross-continental highways and railroads. However, I believed that this type of educational competition would actually yield the greatest benefits in the developing world. Perhaps Solar Decathlon Africa (SDA) would lead to significant gains for rising young leaders in African universities. In my mind, this was all systems go!"

Souad Lalami with Richard King at breakfast in Casablanca, Morocco.

For the next few months, Souad Lalami, Richard King, and the U.S. DOE stayed in close communication about laying the groundwork for SDA. Fortunately, Marrakesh in Morocco was the site of COP22, the United Nations conference held every year to address climate change. Dr. Ernest Moniz, U.S. Secretary of Energy at the time, would be participating. Arrangements were made for Dr. Moniz to meet with Badr Ikken while there to discuss Morocco's renewable energy programs and future projects.

Reflections from Souad Lalami, civil engineer and Environmental Safeguard and Sustainability Manager for Price Waterhouse Cooper and former World Bank employee, has professional expertise in renewable energy R&D, energy efficiency, and the finance industry. She founded her own consulting firm to help companies assess and mitigate their environmental impact. She was an early advocate for establishing Solar Decathlon Africa in Morocco and dedicated supporter of energy efficiency, smart mobility, "smart" cities, financial inclusion, and gender equity.

You were an early champion for Solar Decathlon Africa. How did you hear about this competition?

I first heard of Solar Decathlon in 2014 when SDE was held in France. I didn't have the opportunity to visit, but I felt amazed by the videos and pictures, the spirit behind the competition, and the impact it had on the participants. The students, visitors, and above all, the building industry, really got me thinking that we needed to plan an African version. Richard King was amazing. I got his email very easily and sent him an introductory message, saying that I loved the idea and wanted to give it a try in Morocco. He responded instantly, and that was the beginning of the adventure.

Why did you want to bring this idea to Africa?

First, I wanted the African edition to showcase African culture around the building industry. We have amazing traditional and very efficient materials to build homes that are able to keep you cool and fresh while the temperatures are high outside. On the other hand, I wanted to have an impact on the construction industry by bringing fresh, new ideas that can be tomorrow's way of building. Finally, as an engineering student who had studied in Morocco at one of the best public schools in Africa, I know that we lack opportunities to practice what we've learned. That confrontation with the real world, the one we dig into as soon as we get out of school, is important. This practical experience with Solar Decathlon has a unique impact on students and has certainly changed the way they perceive their future profession.

Talk about some of the problems in Africa related to energy, education, and housing. How did you envision SD Africa as a pathway to address those problems?

First, it's important to note that African countries are the ones struggling the most from climate change. It has had a huge impact on our people, as it can cause displacement, food insecurity, water scarcity, floods, and more. It's a priority in Africa to find ways to live that will address those challenges on an everyday basis. In the housing sector, among others, the challenge is to find passive and efficient ways to cool houses without extensive use of energy or water. The systems need to be low maintenance, affordable, and suitable for the local culture. Those were the primary problems we wanted to address through this new competition. One last point seemed important for us: making sure that Africa was represented through African students who know the struggles, as well as the traditional building methods, of African countries. This point was very important, as I personally wanted the students to take part in this practical and unique experience which would have an incredible impact on their careers.

What were your dreams for Solar Decathlon Africa? What did you hope SDA would achieve for Morocco and the other African countries?

IMPACT is what drives me, and I hope that by my small contribution of bringing this amazing competition to Morocco and creating an African version of Solar Decathlon, I succeeded in leaving a tiny fingerprint on the spirit of the students who participated, the companies who were involved, and moreover, the people who visited the on-site event. This project is definitely one of my dearest accomplishments. I feel so lucky that I've been able to make a contribution and so grateful for Richard King and other colleagues who have given such tremendous effort to make it happen.

Souad Lalami in a U.S. DOE Solar Decathlon hat.

On November 15, 2016 during COP22 the Moroccan Research Institute in Solar Energy and New Energies (IRESEN), the Moroccan Ministry of Energy, Mines, Water, and the Environment (MEMEE), and the U.S. DOE signed a Memorandum of Understanding (MOU). This MOU was official recognition of planned collaboration among these organizations to develop a new edition of the Solar Decathlon. SDA would take place in Morocco in 2019 to showcase local and regional architectural design while honoring the visionary example of the U.S. DOE Solar Decathlon.

Reflections from Badr Ikken, Executive President of Gi3 (Green Innov Industry Investment), which develops industrial units for green technologies. He has an engineering degree from TU-Berlin, where he worked in the Institute for Factory Management (IWF) as Researcher and Project Leader for Fraunhofer-IPK and CTO of Lunos-Raumluftsysteme. In 2010, he became R&D Director for the Moroccan Solar Agency (MASEN) and co-founded the Moroccan Research Institute for Solar Energy and New Energies (IRESEN) in 2011. There, he established the largest solar research and innovation platform in Africa (Green Energy Park), as well as the Green & Smart Building Park. Badr Ikken is VP of the greenH2 cluster in Morocco and VP of the green economy commission of the Moroccan Confederation of Enterprises.

You were at Solar Decathlon Europe in Versailles in 2014. How did you hear about this competition? What were your impressions of that event?

I was first informed and invited by a friend and partner, Vincent Jacques le Seigneur, who was the General Secretary of the National Institute of Solar Energy in France (INES) and involved in the co-organization of Solar Decathlon Europe 2014 in Versailles.

I was drawn, then inspired, by all these realizations: architecturally original, technologically impressive, ecologically sustainable, made and realized by young students. But mostly, I was impressed by their strength and determination.

What struck me as well, were the eyes of the young students, with whom I could exchange and which clearly showed fatigue, a heaviness of the tasks worthy of Hercules, but who were lit up with pride and happiness, just like kids that discover Disney World and would like to stop time to enjoy this experience forever!

We can't stay indifferent to this unique combination of technological prowess, team competition, recreational event, and above all, the joy and happiness of these young people. It was contagious and motivated me to engage myself to work on organizing the first edition of Solar Decathlon in Africa, precisely in my country, Morocco.

What are the challenges for energy and housing in Morocco and in Africa? How did Solar Decathlon convince you that it would be a great way to address those problems?

Morocco, such as the African continent, enjoys a climatic diversity that ranges from semi-arid and desert conditions to snowy mountain regions and temperate climates. This implicates the necessity of having habitats resistant to very high temperature gradients, but which remain financially affordable. The use of natural materials and ancestral techniques is relevant, but requires an evolution and adaptation to meet all current safety and comfort requirements, in order to have sustainable African construction. So, the subject is therefore cross-cutting and does not only concern architects and building engineers, but all the other expertise related to sustainable development, human development, social-economics, and environmental impacts.

Solar Decathlon integrates this state of mind and makes it possible to combine all these specialties and skills, as well as to accompany young students from different fields to develop and create real new technological solutions and implement them. Only the sky is the limit… and sometimes the financial limitations.

What motivated you to establish Solar Decathlon Africa in Morocco?

As part of the development of the young research center for which I was responsible, the Solar Energy and New Energies Research Institute (IRESEN), we had begun to develop and set up research platforms as bridges and innovation highways between universities and industries, with the aim of supporting R&D and accelerating the incubation of new innovative products.

The first applied R&D platform was the "Green Energy Park" in Benguerir. This facility integrated several solar power plants which constituted a veritable open-air laboratory, a kind of living lab. We were also able to use the platform to organize several events to attract more investors, new partners, researchers, and students, including the first solar car races, the Moroccan Solar Race Challenge, as well as innovation competitions for solar technologies like the Green Africa Innovation Booster.

The second platform, the Green & Smart Building Park, should have been dedicated to energy efficiency in the building, smart grids, and sustainable mobility, and therefore, we thought about how to implement a small green town as a living lab which would be inhabited by students from Mohammed VI Polytechnic University and used for testing and research activities.

Solar Decathlon Africa was going to fit perfectly into this program and make it unique at the continental level. We were going to be able to hit two targets with one bullet!

You met with U.S. Secretary of Energy Ernest Moniz at COP22 in Marrakech in November 2016 to sign a Memorandum of Understanding for Solar Decathlon Africa. How did you feel about that step forward?

For years, the Kingdom of Morocco has been strongly committed to the fight against global warming. COP22 in Marrakech was an excellent opportunity to present our progress in the field to the world, while supporting the international dynamic. I very much appreciated the exchange with U.S. Secretary Moniz because he is a politician who managed very strategic and important topics, that he had discussed in Marrakech. He was also humble, and he took the necessary time and attention to support our beautiful project, which would make the eyes of hundreds of young people shine and prove to them that they are capable of moving mountains and helping to protect our planet. He spoke with great interest about the Noor Solar Power Plant and recommended that I work to integrate Morocco into the Mission Innovation Alliance, initiated by President Obama and Mr. Bill Gates, that aimed at accelerating innovation by development of clean energies to limit global warming. Three years later, I was very proud to accompany my minister Aziz Rabbah to Vancouver to officially join this prestigious Alliance and be the first country to represent the African continent.

What key individuals or groups helped you get started with SDA in Morocco? How did you gain their support?

We approached Richard King, who was very friendly and open to supporting and accompanying us. This support was to be the effective start of Solar Decathlon Africa. At the national level, IRESEN, the Green Energy Park, and the German cooperation GIZ enabled us to finance the first awareness-raising and preparation activities. Then we approached our partners, the OCP Group and the Mohammed VI Polytechnic University, with whom we had developed the first R&D platform and with whom we planned to develop new ones. The chairman of the OCP group, Mostafa TERRAB, joined the project and assured us of his full support by granting substantial funding to allow us to move from the dream to the realization of this wonderful project. It was the same case for my other partners, in particular, Hicham El Habti, president of the Mohammed VI Polytechnic University, and Mohammed Benkamoun, co-president of the Green Energy Park.

We were able to request the High Patronage of His Majesty King Mohammed VI, who kindly granted it to us. This gave Solar Decathlon Africa real international positioning and prestige and enabled us to mobilize the remaining necessary funds. The Moroccan and U.S. Departments of Energy also supported us, and I would also like to mention the U.S. Embassy in Rabat and the local authority of Benguerir, as well as the region of Marrakech.

My colleagues then took over the project and poured their hearts and time into it. We were able to organize an exceptional event with a much smaller budget than the American and European Solar Decathlons, thanks to the commitment of my colleagues and our doctoral students, whom I could never thank enough. I would need several pages to praise them all, but, I would like to name a few people who spared no effort: Samir Idrissi, Oualid Mghazli, Souad Lalami, Abdellatif Ghennioui, Zakaria Naimi, Selma Tazi, and many other fellows from IRESEN, Green Energy Park, UM6P, and OCP.

Share your thoughts about the impact of Solar Decathlon Africa 2019. How did the competition benefit Morocco and Africa?

Beyond the organization of an unforgettable competition with a rich and varied program, Solar Decathlon Africa has enabled the establishment of one of the most important platforms worldwide for testing, research, and training dedicated to energy efficiency in buildings, smart grids, and sustainable mobility—the Green & Smart Building Park. This living lab developed by IRESEN and UM6P is now a real source of continental pride, which has enabled the establishment of several masters and certifying training courses such as the greenBEE Master (green building and energy efficiency), Master BIM, and certification trainings in sustainable construction.

Today, several ambitious research projects are carried out at the platform level, several conferences and workshops for the benefit of students of different African countries are organized, and new sectors are emerging. For example, the valorization of hemp for the construction sector (thermal and acoustic insulation), as well as the construction with compressed hay, and mud bricks and sandbags have been initiated. Several advances have been made as part of the organization of the Solar Decathlon, and it is truly a source of pride.

What are the greatest accomplishments of Solar Decathlon Africa?

Solar Decathlon Africa was truly unique because the teams were not only multidisciplinary, but also multicultural, composed of three or four universities from different countries. Beginners could benefit from the experience of advanced students, and the result was a magnificent experience of know-how transfer and development of multicultural skills.

The impact was also seen in the houses which were the result of a cultural diversity, like the winning team *Interhouse*, which brought together the National School of Architecture of Marrakech, the Caddi Ayyad University of Marrakech, and the Colorado School of Mines from the United States. That dwelling combined natural mud brick and cutting-edge solar technologies. Solar Decathlon Africa was accompanied by several parallel events, such as the Solar Race Challenge, the e-Mobility forum, the conference on sustainable construction, and many other scientific venues, but also diversified music concerts. All this made it possible to combine science with environmental commitment, as well as the arts and culture.

The other success was to adapt the competition in order to have houses that could last and be used beyond the period of the competition for science and recreational activities. The objective was to have this Solar Decathlon Village become the heart of the most important African R&D platform for green buildings. The legacy of Solar Decathlon Africa will always live on in our country.

SDA 2019 took place adjacent to the R&D Center called the "Green Energy Park" in Benguerir, Morocco.

Melissa and Richard King with Badr Ikken and Samir Idrissi Kaitouni at the SDA 2019 awards ceremony; photo credit: SDA team.

Assembling a team to plan, implement, and manage SDA was crucial, so that recruitment of university Decathletes could begin. Samir Idrissi Kaitouni, research engineer at the Green & Smart Building Park, was hired to serve as Competition Project Manager, with support from Oualid Mghazli, his colleague at the Smart Building Park, and other staff members there. In December 2017, a request for proposals was issued.

Richard King with Competition Manager Samir Idrissi Kaitouni at SD Africa Solar Village.

Africa is a huge continent with many countries. Samir Idrissi talks about the intensive effort to announce SDA and encourage broad participation in the event. In his words, "In the process of undergoing a huge communication and outreach strategy to international academic institutions, the organizing team developed a contact list of deans, presidents of universities, and professors with interest in sustainability, climate change, and green buildings. Then personalized emails were sent to each person on our contact list to highlight the perks and interests for participants in the international competition. During the process of the call for teams, in-person presentations were made to more than 30 educational institutions nationally, plus virtual presentations for more than 100 potential teams overseas."

SDA included 18 multidisciplinary academic teams from 54 universities that represented 20 countries: Algeria, Burkina Faso, Cameroon, Democratic Republic of Congo, France, Germany, India, Italy, Malaysia, Mali, Morocco, Nigeria, Lesotho, Senegal, Tunisia, South Africa, Tanzania, Senegal, Turkey, and the United States of America. The following teams were selected to participate:

- *A' Free Home* (Moroccan–Malaysian team): National School of Architecture Fez (Morocco), Universiti Sains Islam Malaysia (Malaysia), EMDD of Ecole supérieure de Technologie de Salé (Morocco), DESTEC of Faculté des Lettres et Sciences Humaines (Morocco)
- *InterHouse* (American–Moroccan–Italian team): Colorado School of Mines (United States), Caddi Ayyad University (Morocco), EMINES of Benguerir (Morocco), Politecnico di Torino (Italy)
- *Jua Jamii* (African Union team): Pan African University of Water and Energy Sciences (Algeria)
- *TADD-ART* (Moroccan team): Mundiapolis University (Morocco), Academy of Traditional Art (Morocco)
- *OCULUS* (American–African team): Worcester Polytechnic Institute (United States), ENSIAS RABAT (Morocco), ENSAM MEKNES (Morocco), AUST ABUJA (Nigeria)
- *PLUG and LIVE* (Moroccan–French team): Université privée de Fes (Morocco), EPF Ecole D'INGENIEUR-E-S (France)
- *SUNIMPLANT* (Moroccan–German team): National School of Architecture Tétouan (Morocco), Ecole Nationale des Sciences

Appliquées Tétouan (Morocco), ADRAR NOUH cooperative (Morocco), Fraunhofer Center for Silicon Photovoltaics (Germany)
- *Bayti-Akhdar* (Moroccan–Senegalese team): University Soultan Moulay Slimane (Morocco), University Cheikh Anta Diop Dakar (Senegal)
- *Africa Golden Ryad* (Moroccan–French team): Ecole Supérieure de Technologie de Fes (Morocco), Université de Pau et des Pays de l'Adour (France)
- *TDART* (Moroccan team): Abdelmalek Essaadi University (Morocco)
- *Mahali* (South African team): Stellenbosch University (South Africa), University of Cape Town (South Arica)
- *SOLARTIGMI* (Moroccan team): Ecole Marocaine des Sciences de l'Ingénieur (Morocco)
- *Afrikataterre* (African–German team): Université Internationale de Rabat (Morocco), Fachhochschule Lubeck (Germany), Academic Institutions of DAKAR (Senegal)
- *BOSPHOROUS* (Turkish team): Yildiz Technical University (Turkey), Istanbul Technical University (Turkey), Istanbul Kültür University (Turkey)
- *DarnaSol* (African team): Al Akhawayn University (Morocco), Helwan Universit (Egypt), International University of Grand-Bassam (Ivory Coast), Faculté des Sciences Appliqués de Kasdi Merbah, Ouargala (Algeria)
- *Solar-ution* (Moroccan team): Moulay Ismail University, Meknes (Morocco)
- *Neopetra* (Moroccan team): Ecole Hassania des Travaux Publics (Morocco), Institut Superieur du Commerce et L'Administration des Entreprises (Morocco), EMINES-School of Industrial Management (Morocco)

SDA 2019 included these contests:

Juried

- Architecture
- Engineering and Construction
- Market Appeal
- Sustainability
- Communication and Social Awareness
- Innovation

Measured

- Comfort Conditions
- Appliances
- Home life and Entertainment
- Electrical Energy and Balance

Jury members and Richard King (back row) inside *A'Free Home*, a team of female Decathletes from Morocco and Malaysian universities.

An aerial view of the solar village in the Green & Smart Building Park, Benguerir, Morocco.

The 2019 competition took place from September 13–27 at the Green & Smart Building Park. Samir Idrissi talks about the pros and cons of the site selected for SDA: "Benguerir, a town of about 100,000 people one hour north of Marrakesh, Morocco, was home to the first Solar Decathlon Africa. On a two-hectare site, over 1000 students, faculty members, industrial partners, volunteers, and organizers in bright yellow vests worked tirelessly in harsh weather conditions to bring this competition to life. Participating Decathletes had to adapt to the limited services and accommodations that a small town can provide. It took three weeks for a solar village of 21st-century green buildings inspired by their African heritage and powered by the sun's energy to rise up in Morocco. Once the construction was finished, the visiting public and stakeholders witnessed firsthand what we are capable of doing when we work together to shift towards carbon-free cities."

The inaugural SDA event was a spirited, friendly competition with dwellings that honored African culture and lifestyle, while integrating energy efficiency and advanced technology with time-tested traditional solutions. This unique exhibit of prototypes showed the public that solar houses can be comfortable, attractive, and affordable. When the final scores for SDA 2019 were compiled, the top three teams were *Solar-ution* in third place, *Bayti-Akhdar* in second place, and *InterHouse* in first place.

Africa Golden Ryad house from Ecole Superieure de Technologie de Fes (Morocco) and Universite de Pau at des Pays de l'Adour (France) represented a typical style of this region with lovely inner courtyard surrounded by functional spaces for daily life.

Melissa and Richard King with Decathlete at the entrance to *Africa Golden Ryad*.

This inside view of *Africa Golden Ryad* shows the influence of Moroccan interior design and motifs.

Beautiful mosaic tiles on the walls of *Africa Golden Ryad*.

Tadd-Art from Mundiapolis University and the Academy of Traditional Art in Morocco made an intentional effort to combine architecture with the fine arts. Deep-set windows in the exterior walls constructed with handmade masonry blocks of local materials helped control interior temperatures in this hot climate.

Tadd-Art Decathlete in front of an interior green wall describing important features of the house.

Some teams at SDA 2019 incorporated art that reflected their heritage. This collage of Africa was created from a collection of recycled materials with colorful paint applied.

Oculus, a team representing Morocco, Nigeria, and the United States, was a circular home constructed with strong, woven material. Due to its shape, solar panels had to be ground mounted next to the dwelling.

Decathletes made the exterior skin for *Oculus* by hand at the competition site.

Beneath the woven exterior of *Oculus* was a strong, watertight membrane to protect the dwelling from the elements.

Mahali from the South African team had an open courtyard with a pool to collect rainwater. The roof was a canvas-covered fabric that contained thin-film photovoltaic panels to produce power for the house.

Mahali embraced the concept of a "house in a box" that was affordable, quick to build, and easy to modify and transport. The house highlighted "Ashanti designs," which honor traditional crafts and resourcefulness with a nod to recycling, upcycling, and "creating something from nothing."

International Expansion | 535

SDA teams were especially adept at utilizing locally sourced materials, such as those pictured here.

The distinctive wavy roof with solar panels on top made *Solar-ution* from Moulay Ismail University in Morocco easily recognizable to all. This team won third place overall in the competition.

The south side of *Solar-ution* featured an open patio with comfortable seating and unique yard art.

Solar-ution's wavy ceiling allowed daylight and fresh air to enter the interior space.

International Expansion | 537

Aerial view of *Bayti-Akhdar*, built by Decathletes from Morocco and Senegal. This dwelling won second place in the competition. It had thick walls and minimal window openings, which made it very energy efficient.

The beautiful interior and furnishings of *Bayti-Akhdar* were a stunning display of African design from top to bottom.

Comfortable seating and traditional elements in the open gathering area of *Bayti-Akhdar*, typical of this region, created a welcoming, friendly atmosphere.

InterHouse, designed and built by Decathletes from Morocco, Italy, and the United States, took first place overall in the competition.

The front entrance to *InterHouse* featured a shaded front porch with stone floor.

Badri Ikken (front row, third from left) and other VIPs celebrating with *InterHouse* Decathletes, first-place winners at SDA 2019.

Reflections from Samir Idrissi Kaitouni, research engineer in renewables, energy efficiency, and zero-carbon buildings, and a doctoral student at Abdelmalek Essaadi University in Tangier, Morocco. Samir specializes in energy simulation and environmental performance of buildings at urban scale, conducts research studies on integration of renewable energy in the urban built environment, and assesses strategies to reach carbon neutrality in Morocco by 2050. He served as Competition Manager for SDA in 2019.

What is your professional background and current position?

I'm from Morocco and work as a research engineer in the field of renewables, energy efficiency, and zero-carbon buildings. I have more than seven years of professional experience in research, mentoring, and consultation in urban building energy performance, energy simulation, and optimization. I specialize in energy and environmental performance of buildings at urban scale and conduct research studies on integration of renewable energy in the urban built environment and assessment of strategies to reach carbon neutrality in Morocco by 2050.

How did you get involved with Solar Decathlon Africa and what was your role?

My three-years working as a member of the organizing team for Solar Decathlon Africa was an invaluable learning experience. Since my first day at IRESEN in 2017 as a research engineer with responsibility for Solar Decathlon Africa, I had to manage all aspects of the program, spanning from the preparation of the call for teams and SDA-related documents, to supervising and supporting the progress of 18 sustainable and solar-integrated building concepts developed by university students in collaboration with industrials, to the preparation of the on-site event where construction of solar-powered green buildings and then the competition took place.

What were the greatest outcomes of SDA?

In the one-and-half-year journey of Solar Decathlon Africa following the selection of participating teams, the competition helped promote student innovation and workforce development opportunities in the building industry and energy transition. Today, the Solar Village acts as the experimental living lab leading the way for all Moroccan and African cities to be carbon neutral by 2050. Solar Decathlon Africa alumni are contributing to the evolution of zero energy buildings, responding to the urgent imperatives for the future of construction around the world.

SDA teams were diverse and represented different cultures. What did you observe about team relationships and communications in the Solar Village? Why is this significant?

In an increasingly globalized society, one of the most valuable lessons we can learn from Solar Decathlon is how to understand and accept perspectives other than our own. In that logic, the Solar Village evolved into a more diverse and global-friendly place where Decathletes from different countries get the opportunity to discover and celebrate other cultures, customs, and traditions through various SDA-related activities, which allows them to cultivate lasting relationships among peers. Such initiatives should also be viewed in a global approach, considering that cooperation in education and training has gradually become an important instrument for the implementation of global warming-related actions, based on universal values and shared knowledge.

IRESEN is a center for solar energy research. How has IRESEN used data from the competition? How has that information been helpful?

After the competition, a digital platform was established for the purpose of securing the collection of continuously monitored data of temperature, humidity, lighting intensity, solar production, and energy consumption. This serves as a major experimental instrument for scientific analysis of the energy and environmental performance of different buildings. This allows us to experimentally evaluate the impact of different passive and active design strategies on the overall energy performance of buildings being studied. Through the living lab of Solar Decathlon Africa, we are implementing a great environment for research and innovation and nurturing an ecosystem that helps to pool peers in academia through research collaboration to consolidate our own practices in renewables and energy efficiency.

How did SDA impact your professional career and influence you as a leader?

As project leader of the first African edition of Solar Decathlon, this program was a turning point in my professional career. It allowed me to meet and exchange with people to push for change and learn from others about the relevance, challenges, and potential of climate protection in transforming the urban buildings and construction sector in the African context. Most importantly, this hands-on experience has allowed me to increase my stock of knowledge with regard to the development of future cities towards climate neutrality. As a result, this experience has enabled me to grow fast and with confidence, acquire the necessary background, tools, and knowledge to alter change in a positive way within my country. Undoubtedly, it's a great legacy for an emerging leader who aims to be in the driver's seat.

Is there anything else you would like to share about SDA?

What an adventure Solar Decathlon Africa has been! We brought it all together in a professional and beautiful way. Although it was so far the toughest job I've ever had, it was very rewarding! Memories of this journey will always be treasured and cherished in my heart.

Chapter 9

The Multiplier Effect

Reflections from Solar Decathlon Founder Richard King: "What started as an unconventional idea in the late 1990s has now come full circle. My initial idea took hold, flourished, and soared to unexpected heights. The dream of a Solar Decathlon design-and-build competition became reality. Across time, each successful edition of this innovative program led to further development and expansion. I am humbled and honored with the amazing outcomes. In my mind, there are several reasons why Solar Decathlon has withstood the test of time and remains viable and valuable in various settings."

"First, this collegiate competition invites open-ended exploration and hands-on experimentation to address real-world challenges. Competitors are encouraged to apply their knowledge and understanding in creative ways. Solar Decathlon is a public testing ground where net-zero-energy houses speak for themselves."

"Second, Solar Decathlon is testimony to the power of human interaction that involves healthy exchange of ideas in a shared setting. Teamwork and collective synergy are meaningful and impactful. Together, we can make a difference."

"Third, our best hope for the future is education. Inspiring the next generation to apply their ingenuity in a tough and challenging competition energizes them to rise above. Decathletes are highly motivated to excel as they showcase their designs for the public.

Solar Decathlon: Building a Renewable Future
Melissa DiGennaro King and Richard James King
Copyright © 2024 Jenny Stanford Publishing Pte. Ltd.
ISBN 978-981-5129-47-2 (Paperback), 978-981-5129-13-7 (Hardcover), 978-1-003-47759-4 (eBook)
www.jennystanford.com

Young people are tomorrow's leaders who shine a light on unforeseen possibilities."

"Fourth, Solar Decathlon has remained true to its mission. However, the competition is adaptable enough to establish new roots in a variety of cultural and geophysical landscapes. As of this writing, the program has been successful for more than 20 years on five different continents."

"Of greatest significance above all else, I have discovered along the way that human relationships are what matter most. I'm immensely grateful for untold blessings from thousands of people who have been part of the Solar Decathlon story, be they mentors, colleagues, organizers, volunteers, university faculty and alumni, students of all ages, Decathletes and their families, sponsors, government leaders, industry representatives, journalists, photographers, media specialists, teachers, hospitality teams, friends, family, and the visiting public. The world is a better place because of those who contributed time and talent to this unique program."

"That said, it seems best to hear directly from a few of those special people who seized the opportunity to immerse themselves in Solar Decathlon experiences. Their voices resonate with personal and professional transformations that go beyond what I might be able to express."

Marianna Angelini is a Master of Architecture student at Delft University of Technology. She studied ancient literature, philosophy, and the history of art in Italy, but then moved to the Netherlands to pursue architecture and immerse herself in another culture. By participating in Solar Decathlon Europe 21/22 as Public Relations Manager of Team VIRTUe, Marianna gained valuable experience in communications, management, and sustainable building design.

Igniting a Movement

Former Decathletes I talked to during my journey to Solar Decathlon 2021/22 told me that this competition was the best experience they'd had during their studies. Their enthusiasm intrigued, inspired, and motivated me. As our SD team worked together towards the competition, my understanding of those words became clearer. Looking back at my own experience, I join that group of former Decathletes to say that the Solar Decathlon is without a doubt the best learning opportunity one may have alongside university studies.

For three years, I was the Public Relations Manager of Team VIRTUe from the Netherlands. My experience was rich in communication with a wide spectrum of people: from professionals, sponsors, and academics, to those who had no background in sustainability or engineering. As I gained confidence with story-making and presenting, I realized that communication has the power to make cutting-edge innovation the norm. Experiencing first-hand the impact of our story on visitors who now see our green technologies as solutions for their lives is a priceless reward.

As the Public Relations Manager of a team that grew from 15 to 70+ members, I gained management and planning skills and had the chance to test and improve strategies for optimal teamwork. Through highs and lows, I learned to be a driving, relatable leader, while giving space to everyone's input, as well as delegating tasks while working toward a shared vision. I am convinced that the Solar Decathlon experience prompted my personal and professional development in the most edifying way I could ever imagine. Taking a leading role within the team and being the face of the project to the outside world meant making decisions through uncertainty, keeping up the team spirit when motivation was lacking, and always putting on a smile to tell our story. This required that I had to believe in what we were doing, but it also prepared me for the world of uncertainty that a professional career entails.

I am a native-born Italian, and as a foreigner living in the Netherlands, I can already sense the benefits of the SD for networking, recognition, and connection with the industry. I am currently proceeding with my architecture studies for a Master's degree, but I am also considering how my communications and leadership skills can expand my career opportunities.

Likewise, thousands of students from all over the world have put their hearts and souls into this competition. If the influence on individual careers is notable, then the impact of our solutions for building and living on the stakeholders we reached is extraordinary. While creating a stimulating environment that prepares students to shape our future, the Solar Decathlon engaged thousands of people and raised awareness about integrated sustainable design and construction.

Participating in Solar Decathlon has given me and my fellow students the agency to contribute to solutions for climate change. Our generation inherited the reality of environmental disasters, and with it, the struggle of eco-anxiety. While many of us feel the strong need to take action in some way, we feel powerless and overwhelmed by the complexity of the situation. To quote an inspiring talk in 2022 by climate activist Clover Hogan, "We need to turn eco-anxiety into agency, and leaders' apathy into action." As we prepare to become designers, engineers, communicators, and parents, I invite my generation to take control of the reality of the climate crisis and reframe it into an opportunity to create a better, more livable world.

This competition brings together students from all over the world driven to work toward the same goal: to normalize sustainability in our dwellings and the building sector. The strong sense of solidarity that dominated Solar Decathlon Europe 21/22 in Wuppertal made an impression on everyone who took part in it, igniting a greater movement for a renewable future.

Bobby Vance leverages his experience in both the profession and academia to bring a unique perspective to the University of Virginia School of Architecture. Bobby was the previous Director of Design at ModularDesign+, a member of the NEXTCouncil, CannonDesign's internal think tank for innovation and actionable solutions, and Program Manager for the Center for Design Research at Virginia Tech's College of Architecture and Urban Studies. During his time at Virginia Tech, he co-led many design-and-build projects, including leading a team to victory in the 2018 Solar Decathlon Middle East competition in Dubai with FutureHAUS, a prefabricated smart home exhibited in Dubai, Times Square, and Washington, DC. He reflects on Solar Decathlon.

A New Way to Build—A New Way to Live

I believe everyone has a few periods of exponential growth in their lives. Some are presented to us. Some are stumbled upon. And some are a product of when preparation meets opportunity.

When I received the call from Joe Wheeler at Virginia Tech in 2016 to turn our research into reality by participating in the inaugural Solar Decathlon Middle East competition in Dubai, I jumped at the opportunity. The goal was simple—to create a House of the Future that would challenge our current idea of both building and living. I had worked on the *FutureHAUS* concept while getting my undergraduate degree at Virginia Tech, but while I was in school, the concept remained just that. After graduation, I began working at Perkins&Will in Dallas and strove to apply the prefabricated and sustainable concepts on real-world commercial projects. I was drawn to the studio as there was a tremendous opportunity to define how Dallas would grow over the next decade, and I wanted to have an influence. And while I had an amazing experience at Perkins&Will, I knew after three years that I needed to apply this research I was so passionate about in a more rigorous capacity. I returned to Virginia Tech to receive my Master of Science in Architecture degree and to design and build the future.

Solar Decathlon was the perfect stage to showcase our research and put it to the test in a completely foreign environment. It was also the perfect setting for me to test myself and learn about the challenges and opportunities of this competition. With 10 different contests came 10 different foci and the necessity to partner with those at the University and with industry leaders to win the competition. It was the greatest feeling to meet with Design Leaders and CEOs and ask "what if" and "what is the future of 'X'"? We used the Solar Decathlon opportunity to create a test vehicle for multiple research projects and a real-time test environment for social interaction through integrated technologies.

As Program Manager for the Center for Design Research and Project Manager for *FutureHAUS*, my focus was on creating successful relationships. I will honestly say that our team became my family over those two years. Everyone involved helped to define our message and then bought into it. We communicated the message to our Hokie alumni base, to our parents and friends, to our research partners, and eventually to the world. The early success of the project idea spurred multiple research initiatives that were realized as a part of the *FutureHAUS* story. When tragedy struck and fire destroyed our initial prototypes and our build facility, there was no doubt in anyone's mind that the project would continue. As much as I tried to inspire those on the team during the early stages, I was equally motivated by our entire group and community to resurrect the project from the ashes and make it even better than the first iteration. Placing in the top three in every contest and winning the Solar Decathlon Middle East competition overall was the happiest moment of my life.

Participating in this competition launched my career in the direction I was hoping for. The project allowed me to touch on many aspects of the competition and ultimately prepared me to start my design consulting business, the Vance Design Company. I also realized that I thrive in the iterative and creative environment that is academia. I have built my career to find the perfect balance between the two. After the competition, I participated in workshops and ultimately served as an Innovation Jury member at the Solar Decathlon Africa competition in Morocco. I loved working with those teams to develop their strategies and gave insights to improve the competition.

When I reflect on how this competition has changed the design and construction industry over the past two decades, I think it comes back to engaging with the public. Somehow the house has become the final frontier for innovation, and holding these competitions around the world allows the public to engage with and be inspired by what these students and universities have created. Universities use their participation as a way to attract talent and provide a real-world project for students to innovate on. As the next generation who is entering the workforce, we need this kind of design-build education to energize future leaders in this industry. Team members go on to influence not only their companies, but also their communities, and they take these insights right into their own homes. The holistic nature of the competition and the passion behind these Solar Decathlon projects will continue to influence how we build and how we live for generations to come.

Dr. Edwin Rodriguez-Ubinas, Principal Researcher for the Energy Efficiency Area at the Dubai Electricity and Water Authority R&D Center in the UAE, was Competition Manager for Solar Decathlon Middle East in Dubai (2018) and the Dubai Expo (2021). Edwin has also been a professor and researcher at the Technical University of Madrid. He is one of the founders of Solar Decathlon Europe (SDE) and served as SDE Competition Manager for several events: 2010 and 2012 (Spain), 2014 (France), and 2022 (Germany).

Closing Thoughts about Solar Decathlon Involvement

I understood the relevance and potential of the Solar Decathlon from the first time I heard about it more than twenty years ago. It has become a significant part of my life, and after dedicating myself through it for seventeen years, I am excited to continue contributing to this program. There is always a lot to learn, give, and receive. There are always new challenges, new opportunities, and new countries. I thank God for permitting me to be a competitor in the U.S. DOE SD, to be one of the founders of SD Europe and SD Middle East, to be the Competition Manager in four countries (Spain, France, the UAE, and Germany), and to be a supporter of the development of SD Latin America and the Caribbean.

My most appreciated rewards from Solar Decathlon are the joy of witnessing the students' progress, seeing its impact on different societal sectors, and making wonderful friends around the world with whom I have stayed in touch over the years. Since my first experience with SD, the times and priorities have changed. Consequently, the Solar Decathlon competition has been evolving; and it will continue adjusting to respond to future local and global needs. However, the relevance of education and knowledge sharing continues to be fundamental for the sustainable development of our societies. Similarly, Solar Decathlon's hands-on approach (learning by doing) continues to be effective.

Finally, I want to acknowledge my mentors in the Solar Decathlon, Richard King, Sergio Vega, and Pascal Rollet. They have guided me along the way in the Solar Decathlon. Thanks also to all my colleagues in the American, European, and Middle Eastern organizations and the professors and students that participated in the Solar Decathlon editions in which I was involved. I am proud of them all.

U.S. DOE Richard King Award

In 2017, the organizers at the U.S. Department of Energy created a special award to recognize the Founder of the Solar Decathlon. In conjunction with the SD competition in Denver, Colorado, Richard King was honored to receive the very first award bearing his own name. After that initial presentation, the U.S. DOE now invites biennial nominations for the Richard King Award. One outstanding faculty advisor and one outstanding student alumnus who meet designated criteria and demonstrate impressive commitment, achievement, and leadership in the design and construction of energy-efficient buildings and the transition to a clean energy economy are selected for recognition.

Richard King sharing a happy moment with Joe Wheeler, recipient of the Richard King Outstanding Faculty Advisor Award in April 2022 in Denver, Colorado.

Richard King congratulating Alex MacDonald as he receives the Richard King Award at the 20th anniversary celebration of Solar Decathlon.

At the 20-year anniversary of Solar Decathlon in 2022, Richard King was pleased to present the outstanding faculty award to his long-time colleague and friend Joe Wheeler, professor of architecture at Virginia Tech. Joe has been an advisor for several SD teams in the United States, Europe, and Dubai and believes in the tremendous power of hands-on learning. The outstanding student alumus award went to Alex MacDonald, former Decathlete and leader for Team Orange County at SD 2015. Alex went on to become a senior program manager at Tesla, where he focused on infrastructure projects that electrify the built environment. The Richard King Award is a worthwhile mechanism to inspire leadership and perpetuate the values and mission of the Solar Decathlon competition.

For Richard King, Solar Decathlon became his life's work. Along the way, many more dreamers and believers who share a common passion became part of that incredible journey. The relentless enthusiasm and dedication of those willing partners made enduring contributions to Solar Decathlon, which set a multiplier effect in motion. Tackling major global problems, such as climate change, calls for inventive thinking. Building a renewable future that rests on sustainable practices requires people working together for that common purpose.

How will *you* pitch in? Your ideas, your effort, your conviction, and your commitment just might spark a movement!

The future is in our hands. Photo credit: Pixabay.

In conclusion, Richard King and Melissa King extend their sincere gratitude to every individual who has been part of this compelling adventure. You have amazed us with your dedication and never-ending hard work. You have astonished us with your unimaginable accomplishments. You have supported the program—and us—with unselfish generosity. You have touched our hearts. Thank you!

Melissa and Richard King

Appendices

A: Invitation to the reception for the 2002 event at the Smithsonian Castle
(inaugural Solar Decathlon)

Imagine a world where energy is abundant and available whenever and wherever you need it. Energy so simple you hardly know it's there. Energy that is clean, safe and secure. That world is solar and it's here today.

Join us as we step into this new world of energy and congratulate our Solar Decathlon participants from 14 universities and colleges for their hard work and enthusiasm in developing effective solar solutions for homes and home businesses.

Auburn University
Carnegie Mellon University
Crowder College
Texas A&M University
Tuskegee University
University of Colorado at Boulder
University of Delaware
University of Maryland
University of Missouri-Rolla & The Rolla Tecnical Institute
University of North Carolina at Charlotte
University of Puerto Rico
University of Texas at Austin
University of Virginia
Virginia Polytechnic Institute and State University

Lawrence M. Small
Secretary of the Smithsonian Institution
and
Spencer Abraham
Secretary of Energy

Invite you to join us in celebrating
SOLAR DECATHLON 2002
THE NATIONAL MALL, WASHINGTON, D.C.

Solar Village Preview Tours
Wednesday, September 25, 2002
5:00 - 6:15 pm

Opening Reception
Wednesday, September 25, 2002
6:30 - 8:30 pm

Smithsonian Castle
900 Jefferson Drive

This reception made possible by the generous support of
BP America

R.s.v.p no later than September 16
410-981-0259
hinklelb@bpsolar.com

Admittance at Jefferson Drive
Business Attire

B: Letter from Heloisa Sabin, wife of visionary medical researcher, Dr. Albert Sabin, who developed the oral polio vaccine, which has saved many lives around the world

Heloisa Sabin
3101 New Mexico Avenue, N.W., #1001
Washington, D.C. 20016

Fax: 202-364-4507

September 28, 2002

Dear Richard J. King

I could not help but remember my late husband dedication to the development of solar energy, while reading Benjamin Forgey's article concerning your brainchild "Solar Decathlon" this morning, in the Post.

Albert Sabin, developer of the oral polio vaccine, that save so many lives, wrote and I quote "the development of an industrially feasible technology for replacing the exhaustible dirty fossil fuels by inexhaustible clean solar energy is the most important contribution - more than any vaccine – to the future welfare of the world."

I do hope my age, almost 85 & great grandmother of 4, even more my everlasting admiration, respect and love for Albert will explain and excuse my impetus to send - for your eyes only – this note applauding your initiative to stimulate more research and development.

Sincerely,

Heloisa Sabin

C: Letter from Sir John Browne of BP Solar, major sponsor of the first Solar Decathlon

bp

Sir John Browne FREng
Group Chief Executive

BP Amoco p.l.c.
Britannic House
1 Finsbury Circus
London
EC2M 7BA
United Kingdom

Switchboard: 020 7496 4000
Central Fax: 020 7496 4630

30 August 2000

Mr Richard J. King
Office of Solar Energy Technologies
Department of Energy
Washington, DC 20585
USA

Direct Tel. 020 7496 4488
Direct Fax: 020 7496 4483
www.bp.com

Dear Mr King

Thank you for your letter on 10th July regarding *The Solar Decathlon*. I have asked Graham Baxter in our Baltimore Solar Headquarters to contact you directly to discuss details of this sponsorship.

Yours sincerely

Sir John Browne

D: Letter from Alice Billmire, elementary student from Arlington County who visited the 2002 Solar Decathlon

Alice Billmire
1806 N. Calvert St.
Arlington, VA
October 7, 2002

Dear Mr. King,

I absolutley, positively, LOVED the Solar Decathlon! It was the best field trip ever! It was really cool to know that college students not much older than us designed and built the houses. The Crowder house was really awesome — and an old lady helped build it! My grandma is going back to college too.... it really shows that you are never too old to learn something!

I think its so cool that I can tell my friends from other schools that my teachers husband invented the solar decatholon! I had so much fun......I can't thank you enough! It really inspiered me to do new things — and think about how solar is our future.

THANK YOU!

Sincerely,
Alice Billmire

E: Letter from Jeff Lyng, team leader for the Colorado Boulder team in the 2005 Solar Decathlon

Jeffrey R. Lyng
P.O. Box 485, Rollinsville, CO
Phone: 303.818.2302, Email: Lyng@colorado.edu, Fax: 303.258.7137

March 17th, 2006

Mr. Richard J. King
2523 Rocky Branch Road
Vienna, VA 22181

Richard:

Over the past few months, I have never felt that I thanked you properly. The Solar Decathlon has shaped my life immeasurably recently and will no doubt serve as a launching point for the rest of my career. I cannot tell you in a letter how appreciative I am for that.

The competition has ignited excitement among current and future students of renewable energy that I witness every time I give a tour of the Bio-S$^{(m)}$IP or presentation. Your dream to create a demonstration of alternative energy problem solving before the policy makers who shape our nation's fate every day has been more than realized. The message has been heard around the world. Moreover, your idea has created a forum for young energy professionals to prove what can be accomplished with enthusiasm, optimism and even a little naivety. In the coming years, I suspect you will be hearing from past competitors who will praise participation in the Decathlon as their "ah-ha" moments and basis for inspiration.

Although I may be perceived by some as one of the sharpest critics of the Decathlon, I remain committed to the competition's potential to foster ingenuity, inspiration and change. I believe that constant feedback from the public, competitors, organizers and judges is crucial to ensuring the competition leads the building industry towards the most sustainable energy future possible.

Competing in the Solar Decathlon, testifying before the House Subcommittee on Energy, meeting with Secretary Bodman, and attending the State of the Union address are experiences for which I am deeply grateful. I hope that you will not hesitate to call on me if I can in any way be of assistance to you or your staff.

Thanks for everything.

Respectfully,

Jeffrey R. Lyng
2005 CU Solar Decathlon Team

F: Note from Dr. Ernest Moniz, U.S. Secretary of Energy, after the 2013 Solar Decathlon

THE SECRETARY OF ENERGY
WASHINGTON, DC 20585

OFFICIAL BUSINESS

Mr. Richard J. King
Office of Energy Efficiency and
 Renewable Energy
950 L'Enfant Plaza, 6th Floor
Office 6059

To Richard King,

Oct. 23, 2013

THE SECRETARY OF ENERGY

Dear Richard, I understand that you were basically a "one-man band" putting on the Solar Decathalon during the shutdown. It seems to have been a great success, and you clearly deserve much of the credit — with of course the great student teams that take part. Thank you. I regret very much that I had to miss it but look forward to participating in the future.

With best regards,

Ernie Moniz

G: Message from Joe Simon, SD Competition Manager, in honor of Richard King's retirement from US DOE in 2016

Feb 2016

Richard,

I just wanted to write and express my thoughts on your upcoming retirement.

Congratulations, it is very deserved. Your impact at the Department of Energy is truly astonishing. The number of futures you have changed is awesome. I still have the solar matchbox car on my desk at home and think of how far everything has come. Students have gone on to do great things, companies have invented and pursued new technologies, and the public has made smarter every decisions because of the solar decathlon.

Personally, I cannot imagine my life without the opportunity to participate in solar decathlon. In school, it taught me to be hands-on, to lead, and to be responsible. It taught me about energy efficiency and smart design. I earned friends-for-life. While I ended up sticking with Solar Decathlon, I know that all of my fellow team members found greater jobs than they would have without Solar Decathlon and all have gone on to do great things.

You may know many of these anecdotes, but hopefully some of them are

New Zealand teammates started (and maintain) architecture firm
Past Illinois team leader as Passive House consultant (including future other teas)
Smart grid consultant for Accenture
Founder of Agrilyst - smarter greenhouses
Founder of Chai Energy - residential energy monitoring & savings
RAD Furniture - modern steel furniture
BDagitz Furniture - wood furniture in Colorado
Founder, net-zero accessory dwelling unit housing
Founder, scientific drone startup company
Founder, architecture firm(s)
Founder, strategic communications consulting (more than one)
Sustainable design manager at chipoltle
Multiple weddings/engagements
Many career changes
Dozen or so people working for national labs
Decathlete>Faculty>juror
visitor>volunteer>decathlete
Decathlete>volunteer>faculty>hired advisor

7,338 people on LinkedIn who list "Solar Decathlon" in their profiles – proud of their involvement with the competition.

Of those, Linkedin considers them to be:

- 2866 for Entry leve
- 2458 for Senior Level
- 671 managers
- 398 Directors
- 394 Owners
- 220 C-suite executives
- 128 Partners
- 90 Vice presidents

And so many more that cannot be easily quantified.

Thanks for it all! Here's hoping the success can continue. Enjoy your retirement.

Joe

--
Joe Simon
Senior Engineer | Solar Decathlon Competition Manager

National Renewable Energy Laboratory (NREL)
15013 Denver West Parkway | Golden, CO 80401
303-275-3910 | M: 303-886-8213
www.nrel.gov | www.solardecathlon.gov

H: Message from Bryan King at Richard King's retirement celebration in 2016

Bryan King is a television editor, producer, and musician living in Los Angeles, CA with his wife and three cats.

My dad is a throwback Renaissance man. He's the only person I know who can dream up something world-changing, fix almost anything, build pretty much everything, talk to just about anyone, but could only cook you scrambled eggs and canned soup. But as his son I'm perfectly fine with that, because besides the fact that I like scrambled eggs, that makes his example slightly less daunting to live up to.

I love you dad, and congrats on an inspiring career! *Bryan*

I: Message from Nathan King at Richard King's retirement celebration in 2016

Nathan King is a corporate attorney who lives in Charlottesville, VA with his wife and four children.

Good evening. Thank you all for being here tonight. For those of you who don't know me, I'm Nathan, Richard's son.

Dad, I have been thinking about what to say up here for weeks. And I knew I couldn't stand up here and ramble all night, especially considering your early bedtime. So, I hope to keep this short.

You've had a successful career, and we are celebrating your retirement tonight. But I hope you know that I appreciate all that you have done outside of the office, too.

One of the most famous businessmen in modern history, Steve Jobs, had a child that he left behind at a young age. What if Mr. Jobs had decided to stay and be there for his young daughter? Maybe he never becomes the savvy CEO of Apple. Maybe the personal computer is never developed. God forbid, maybe the iPhone is never invented. Who knows. But, who cares.

Has the iPhone or personal computer really brought that much more value and meaning to people's lives? Did it end world hunger? poverty? thirst? anything? And what about his daughter who had to grow up without a dad?

For hundreds of years, mankind lived happily without computers, smartphones and tablets. What was most important then is still what's most important today.

My point is not to say that earning a living isn't necessary, that I'm against making a lot of money, or that I don't enjoy my iphone or new technology, but rather that success is not measured by our professional lives. Who we are as human beings is not defined by how high you climb the corporate ladder.

Instead, it's how we carry ourselves in our day-to-day lives. It's how we treat strangers, neighbors, coworkers, and friends. It's how treat our spouses and children. It's what we do for others. It's what we do for our communities, our families, and those we love.

As a father now myself, I understand the enormous sacrifices that it takes to be a real dad, and a real man. Thank you for showing me what that looks like.

In spite of everything, you still managed to become an award-winning scientist and inspirational leader for thousands of people around the world. I have never been more proud, and thankful, to be your son.

Congratulations on your retirement. I love you. *Nathan*

Index

accessory dwelling unit (ADU) 387
ADA *see* Americans with Disabilities Act
ADU *see* accessory dwelling unit
aerogel 97, 155, 157
Affordability Contest 220, 225–226, 248, 384
affordability crisis 365
African team 525–526
AIA *see* American Institute of Architects
AIR House 388–389
Alf House 406
Alliance for Sustainable Energy 191, 229
American Institute of Architects (AIA) 46, 59, 102, 109, 191, 252–253
American Society of Heating, Refrigerating, and Air Conditioning Engineers (ASHRAE) 191, 229
American Solar Challenge 10, 12, 19, 21–22, 75
American Solar Cup 19
American Solar Energy Society (ASES) 69
Americans with Disabilities Act (ADA) 58, 506
architect 61, 63, 66–67, 216, 218, 255, 257, 302–305, 310–311, 315–316, 334, 337, 354–355, 434–435
Architects Pollute 103
architectural beauty 29, 87, 105, 146, 249, 282
architecture 41–42, 52, 61–65, 117–118, 124–125, 169–170, 174–178, 252–255, 257–258, 302–305, 309–313, 319–320, 332, 348–350, 364–365, 367–368, 377–379, 438–439, 456–457, 546
  American 179
  ancient nomadic 315
  award-winning 302
  ecological 303
  energy-efficient 176
  energy-positive 175
  green 52, 56
  parametric 312
  sustainable 179, 218, 305, 307, 311
  vertical collective 308
Architecture Contest 64, 117, 124, 157, 166, 225, 306
Architecture Jury 119, 124, 311, 482
artistic pergola 385
ASES *see* American Solar Energy Society
ASHRAE *see* American Society of Heating, Refrigerating, and Air Conditioning Engineers
assembling 49, 114, 145, 225, 423, 524
audience 81, 86, 218–220, 254, 259, 309, 324, 331–333, 375, 469
awareness 30, 47, 51, 163, 195, 286–287, 482, 545

Baitykool 504–505
bamboo 92, 97, 167, 323, 470, 474, 477, 485, 489
battery 13, 42, 86, 100, 190, 307, 351

Index

battery storage 13, 45, 101
Bayti-Akhdar 526, 529, 537, 538
BeauSoleil 196–198
BIPV *see* building integrated photovoltaics
BP Solar 45–46, 109, 134, 191, 229
BP Solar Trek 8–9
branding 38–39, 122, 413, 415
brick 126, 472–473, 486, 523
budget 21, 25, 71, 133, 136, 269, 284, 286, 288, 522
builder 18, 90–91, 149, 216, 271, 303–304, 310
building 24, 27, 61–63, 102, 179–180, 209, 261–263, 306–307, 312–314, 330, 334–335, 337, 344, 347–348, 436–437, 540–541
  climate-neutral 365
  energy-efficient 176, 337, 364, 515, 549
  equilibrium 348
  green 111, 149, 515, 522–523, 525, 528, 540
  industrial 325
  multi-family 231–232
  multipurpose 340
  net-zero-energy 348
  public 307
  sustainable 306–307, 359, 513
  urban 541
  zero-carbon 540
building code 58, 172, 184–185, 262
building integrated photovoltaics (BIPV) 27, 160, 162, 166, 174, 314, 502, 504
building science 16, 54, 151, 180, 186, 255, 312
building sector 102–103, 546

Canopea 298, 300–301, 303, 306–309, 311–312
Capitol Hill 26, 135–136, 214, 223–224
carbon footprint 393, 424
carbon neutrality 512, 540
Casa Solar 163–164, 263–264, 284–285
Casa UnSolar 474
Casa Uruguaya 468, 478–479
Casa Yarumo 470
CEA *see* Controlled Environment Agriculture
Controlled Environment Agriculture (CEA) 208, 307
COP22 516, 518, 521
COP26 220
CRATerre 302
creativity 27, 92, 103, 146, 184, 231, 261, 286, 292, 336, 386
culture 96, 179, 196, 257, 309, 312–313, 319, 517, 523, 541, 544
  African 517, 528
  American 179
  building 303, 368, 434
curator 61, 218–219
curriculum 39, 53, 177, 259, 312, 376

data analytics 103, 109
daylighting 88, 94, 97, 103, 174, 202, 233, 282, 321, 437, 445
decarbonization 26, 436
decathlete 81–82, 86–88, 91–92, 96, 121–124, 134–135, 137–141, 143–147, 149–151, 186–187, 194–196, 208–210, 220–221, 223–224, 229–231, 331–334, 367–369, 484–485, 527–529, 543–545
DesertSol 390–391
design 21, 27, 29–30, 55, 57–59, 61–67, 72–73, 94, 96, 102–103, 176–178, 218–222, 255–258, 368–369, 376–377, 436–437, 456–457, 465–466, 481–483, 546–549

architectural 96, 139, 173, 178, 183, 290, 349, 368, 376, 419, 519
building 70, 74, 253, 424, 544
climate-responsive 456
communication 354
contemporary open-pavilion 236
corporate 355
creative 101, 111, 146
dog-trot 77, 95
energy-efficient 43, 415, 496
graphic 183
ingenious prototype 502
innovative 91, 182
interdisciplinary 62
interior 131, 166, 178, 257, 390, 502, 530
leading-edge 422
modular 131, 281
net-zero 404
renewable energy 39
schematic 436
silo 204
sliced cylinder 297
tube 208
web 354
designer 18, 29, 62, 91, 124, 196, 334, 355, 369, 546
graphic 40, 413
interior 305
urban 132
design process 16, 22, 172, 177, 331
DEWA *see* Dubai Electricity and Water Authority
diversity 18, 183–184, 191, 257, 368, 376, 419
climatic 520
cultural 257, 523
documentation 52, 58, 60, 101, 117, 145, 259, 320
DSCE *see* Dubai Supreme Council of Energy

Dubai Electricity and Water Authority (DEWA) 254, 416, 495–496, 498, 502, 510, 512–513
Dubai Supreme Council of Energy (DSCE) 496, 512
dwelling 62, 64–65, 87–88, 91–92, 117–118, 201–202, 231, 236–238, 240, 245, 249, 274–275, 319, 502–504, 532–533
compact 173
contemporary 94
creative 81
energy-efficient 92, 507
free-standing 339
futuristic 126
innovative 83, 273
light-filled 246
multi-unit 489
net-zero-energy 207, 231, 344
prototype 243, 350, 421, 468, 485
retrofitted 483
single-family 333
super-strong 411
ultra-efficient 248
unique 55, 81, 202
Dwelling Contest 64, 119, 124–125

EC *see* European Commission
EEF *see* Energy Endeavour Foundation
emissions 25, 102, 387, 437
energy 15, 23–24, 55–56, 85–88, 117–118, 121–123, 176–177, 180, 186–187, 189–190, 192–194, 304, 314, 347–348, 376–377, 424–425, 464–465, 478–479, 481, 516–518
clean 26, 28, 39, 48, 134, 194, 375, 495, 512, 521
electrical 399
green 4, 515

net-zero 65, 86, 118, 177, 180, 437
photovoltaic 269, 464
sun's 259, 528
thermal 173
energy efficiency 39, 109, 111, 117, 122, 124, 136, 138, 213–214, 262–263, 286–287, 413–414, 515–516, 522, 540–541
Energy Endeavour Foundation (EEF) 289, 332–336, 367, 416
energy management 87, 123, 126, 211, 289, 501
energy performance 55, 178, 350, 365, 386, 540–541
energy production 43, 180, 368
energy solutions 28, 136, 143, 189, 229, 375
engineer 10–12, 14–15, 30, 32, 61–62, 66–67, 71, 73, 255, 257, 302–305, 309–310, 328–329, 334, 337
entrepreneur 8, 21, 208, 422–424
environment 96, 102, 303, 305, 314, 319, 336, 434–435, 540–541, 545, 547, 551
   creative 548
   cross-cultural 332
   desert 497
   harmonious symbiotic 504
European Commission (EC) 268, 289, 333–334
evaluation 29, 57–58, 66, 117, 188, 193, 255
evolution 255, 334–335, 520, 541
exhibition 26, 81, 121–122, 218–220, 263, 307, 348, 396, 413, 437, 513
expertise 70, 76, 109, 143, 147, 265, 267, 348, 350, 464, 466, 516, 520

fabrication 122, 255, 449
façade 214, 281, 294, 356, 358, 360, 429, 431, 456

Fibonacci Sequence 130–131
final awards ceremony 106, 279, 324, 327, 331, 413, 416, 480, 489, 495, 502
flooring 97, 103, 109, 129, 173, 203, 275
fossil fuel 1, 4, 12, 23–24, 29, 134, 189, 223, 417
French technostructure 310
Fresnel lenses 296
funding 26, 32, 109–110, 131, 135, 143, 178, 224, 263, 265–266, 423
fundraising 53, 72–73, 133, 136
FutureHAUS 63, 253–254, 508, 513–514, 546–547

Gable House 213–214
garden 117, 236, 239, 315, 351, 360, 381, 393, 437
Golden Ratio 130
gold medal 37, 102, 136
Goldy Gophers 204
Grammy Award 74
Grands Ateliers 303, 309, 312
Great Park 371–373, 375, 395
Great Porch 233–234
Great Recession 189
greenhouse 211, 299, 325, 351
green wall 163, 171, 238–239, 296, 454, 476, 485, 490, 504, 531

heat pump 185–186, 206
heritage 233, 282, 354, 532
   architectural 418
   cultural 231, 333, 419
holistic-systemic approach 304
home 23–24, 27–28, 43–44, 71, 102–103, 122–124, 149, 211, 231–232, 324–325, 386, 404, 418, 446–447, 480–482
   circular 532
   energy-efficient 54, 117
   green 223, 445

innovative 58, 87, 226, 394
island 96
net-zero-energy 30, 41, 65, 69, 81, 88, 225, 229, 290, 387
next-generation 120
single-family 387
smart 157
solar 28, 74, 84, 89, 94, 136, 190, 224, 231, 386
solar-powered 27, 30, 47, 53, 55, 78, 84, 86, 90, 223, 225
state-of-the-art 87
sustainable 326, 386, 481
vacation 406
zero-energy 66
Home+ 280
Home Entertainment 137, 192–193, 204, 213–214, 226, 236, 238, 243, 249, 287, 379
Home Life 398, 413, 439, 457
hot water 56–57, 102, 108, 118, 193, 213–214, 246, 249, 256–257, 263, 277
house 26–29, 42–45, 60–65, 81–82, 92–99, 116–118, 120–124, 126–128, 150–151, 153–158, 167–169, 172–174, 192–193, 201, 209–211, 223–226, 252–254, 378–386, 407, 409–415, 433–434, 440
 disaster-resilient 408
 energy-efficient 95, 167
 extraordinary-looking 63
 flexible 173
 innovative 124, 172, 328, 337
 multi-family 337
 net-zero-energy 144, 269, 377, 543
 off-the-grid 61
 passive 175, 177
 prototype 77, 231, 343, 345–346, 348–349, 355, 430, 442, 455, 485, 512
 smart 485

solar 30, 36, 47, 49, 52, 71, 76, 245–246, 258, 263, 396, 400, 403
solar-powered 39, 108–109, 135, 139, 141, 147, 219, 223, 253–254, 481
state-of-the-art 182
sun-powered 395
sustainable 279
terraced 337
traditional 445
unique 43, 298
winning 85, 301, 306, 511, 513

imagination 84, 88, 109
inaugural event 40, 54, 63, 76, 84–85, 92, 109–110, 124, 270
industry 25, 32, 73, 208, 210, 213, 254, 258, 422–423, 545, 548
 building 39, 151, 461, 465, 517, 540
 electronics 20
 finance 516
 fossil fuel 134
 homebuilding 91, 386
 indoor agriculture 213
infrastructure 46–47, 151, 213, 224–225, 231, 358, 510
ingenuity 14, 84, 86, 103, 121, 124, 143, 213, 261, 269, 279
INhome 247–249
Inhouse 409
innovation 10–11, 65–66, 91, 211, 262–263, 284, 286–287, 319–320, 336, 338, 484, 501–502, 540–541, 546, 548
 cutting-edge 545
 energy-efficient 196
 leading-edge 170
 technical 53, 376
Innovation Contest 482, 484, 488, 492
insulation 23, 43, 155, 160, 202, 278, 295, 314, 322, 412

acoustic 522
closed-cell spray foam 203
cork 238
sheep's wool 247
wood fiber 388
Interhouse 523, 525, 529, 538–539
international competition 179, 302, 416, 525
inverter 28, 167, 191
grid-tied solar 20
non-islanding 28
Sandia-developed 28

juror 58, 60, 68, 101, 118, 148, 150, 193, 310, 379, 398
jury member 58, 60, 96, 101, 117–118, 121, 126, 193, 306, 310–311, 413–414, 482, 484
jury panel 147, 150, 194, 208, 310, 320, 398, 466, 484

LAN *see* local area network
LEAFHouse 169–172
local area network (LAN) 47
LumenHAUS 63, 136, 207, 253, 256, 279, 282–283

Magic Box 114, 135, 262, 283–284
Market Viability 148, 150, 160, 162, 170, 192–193, 196, 201, 208, 214, 287
material 47, 49, 91–92, 131, 153, 155–156, 240–241, 303, 314–315, 349–350, 369, 469, 481, 485, 488–489
building 109, 151, 184, 238
collateral 46, 148
composite 14
construction 360
desiccant 204
ecological 312
energy-efficient 122
industrial 88
phase change 185

photocatalytic 322
recycled 138, 247, 292, 347, 361, 532
roofing 27
semiconductor 419
siding 160
sustainable 149, 350, 353, 377, 481
water-resistant 323
media 84, 89, 91, 110, 262, 354–355, 423
media attention 25, 36, 51, 83, 120, 134, 141, 169, 261
media coverage 137, 265, 267, 466, 502
Memorandum of Understanding (MOU) 264, 285, 289, 418, 420, 479, 496, 518–519, 521
MOU *see* Memorandum of Understanding
mud bricks 522
multidisciplinary approach 172, 434
multidisciplinary team 71–73, 150, 303, 305, 309–310
museum 121–122, 153, 219, 316, 413, 415, 421

NAHB *see* National Association of Home Builders
nanotower 300, 306–308
National Association of Home Builders (NAHB) 149, 224, 229, 285
National Building Museum (NBM) 59, 149, 215–219
National Center for Photovoltaics (NCPV) 31, 109
National Education Association (NEA) 192, 229
National Mall 46–52, 71–80, 89–91, 109, 111–114, 121–123, 134–138, 141, 145–146, 153–155, 157–158,

167–168, 209–210, 218–219, 221–222, 224–225, 254–256, 263–265, 284–285
National Park Service (NPS) 48–49, 74, 153, 221–223, 414
National Renewable Energy Laboratory (NREL) 24–25, 30, 33, 35–38, 40–42, 44, 49, 59, 69, 84–85, 108–109, 143, 423
natural resources 2, 224, 302–303, 488
NBM *see* National Building Museum
NCPV *see* National Center for Photovoltaics
NEA *see* National Education Association
Net Metering 190, 192–193, 201, 213–214, 287
network 84, 136, 138, 170, 211, 289, 365, 419, 423
  communications 365
  distribution 289
  electric 307
  local area 47
New York Institute of Technology (NYIT) 115, 127–129, 134, 145
NPS *see* National Park Service
NREL *see* National Renewable Energy Laboratory
NREL Staff 31, 35, 40, 84–85, 89, 108, 195, 376
NREL team 45, 103, 110
NYIT *see* New York Institute of Technology

observation tower 430, 441
observer 60, 71, 108, 192, 367
OCGP *see* Orange County Great Park
Olympics 26, 32, 136, 156, 344, 367, 423, 454, 513
Olympics of Solar Energy 468

onsite event 45–46, 60, 74, 101, 109, 141, 145, 180, 194, 353, 421
opening ceremony 79, 155, 440
Orange County Great Park (OCGP) 371, 373, 378, 386, 394–396, 399
organizers 38, 60, 63, 136, 141, 143, 149–150, 286–288, 367, 413, 415, 420–421, 454–455
organizing team 270, 289, 291, 420, 435, 510, 525, 540
outsulation 243–244

panel
  folding glass 299
  polycarbonate 124, 157
  PV 13–14, 46–47, 165, 167, 173, 180, 199–200, 276–277, 307–309, 351, 358
  sliding louvered 196
  solar 24, 27, 80, 198, 206, 406, 409, 433, 436–437, 532, 535
  steel 294
  traditional straw 351
  translucent roof-top 204
  translucent skywall 97
  vacuum insulation 281
  wooden 209
Paul Rappaport Award 85
People's Choice Award 106–107, 351, 393, 502
People's Choice Contest 196
performance 7, 57, 61, 105, 110, 172, 180, 187, 388, 395, 540–541
perFORM[D]ance House 236–237
photovoltaics (PV) 4, 13–15, 23–28, 32, 45, 102–103, 111, 173, 178, 361, 368
prototype 57, 74, 130, 132, 160, 164, 168, 339, 341, 351–353, 355–359, 361–362, 448, 456, 459

coLLab 352
contemporary 163
cutting-edge 313
innovative 128, 204, 513
lightweight 363
public awareness 37, 52, 147, 192, 219, 465
public education 4, 26, 51, 79, 83, 222, 269, 271, 457
publicity 73, 90, 109, 229
PV *see* photovoltaics

radio 134, 253, 262
rainwater 117, 160, 170, 249, 356, 407, 534
RASEI *see* Renewable and Sustainable Energy Institute
R&D 12, 24, 32, 52, 56, 91, 109, 348, 464, 510, 520
Reflect Home 404
Refocus 274
Refract House 208–212
regulations 36, 49, 52, 63, 65, 137, 220, 262, 286
Renewable and Sustainable Energy Institute (RASEI) 31, 35, 84
renewable energy 12, 16, 29–30, 39, 42, 88, 90, 263, 265, 413, 424, 464–465, 540
renovation 121, 325, 337, 341, 347–348, 413
Request for Proposals (RFP) 52, 110, 141, 143, 376, 524
RFP *see* Request for Proposals
Rhode Island School of Design (RISD) 65, 115, 126–127, 134, 319, 322
RhOME 326
ribbon-cutting ceremony 79, 120, 168
Richard King Award 549–551
RISD *see* Rhode Island School of Design
RoofKIT 355, 363, 369

rooftop 80, 93, 107, 124, 130, 206, 248, 297, 308, 363, 405

SBEs *see* smart built environments
scoring 41, 56–57, 74, 103, 109, 180, 378
SDA *see* Solar Decathlon Africa
SD China 184–185, 420–421, 423, 426, 435–436, 438, 454, 457, 461–462
SDE *see* Solar Decathlon Europe
SD LAC *see* Solar Decathlon Latin America and Caribbean
SDME *see* Solar Decathlon Middle East
SD Organizers 54, 56–57, 74, 86, 89, 110, 141, 144, 182, 192, 223, 225, 227, 229, 264–265
SD prototype 495, 505
SERI *see* Solar Energy Research Institute
SIP *see* structurally insulated panel
smart built environments (SBEs) 508
Solar Decathlon Africa (SDA) 515–519, 521–525, 528, 540–542
Solar Decathlon Europe (SDE) 262–266, 268–270, 273, 283, 286–289, 309–310, 313, 316, 332–337, 344–346, 348–350, 353–354, 363–367, 369, 416–418
Solar Decathlon Latin America and Caribbean (SD LAC) 462–464, 466, 469, 479–482, 493
Solar Decathlon Middle East (SDME) 495–497, 500–501, 509–513
Solar Energy Research Institute (SERI) 24
Solar Hai 499–500, 502–503
solar power 8–9, 12–13, 24–25, 29–30, 56, 90, 120, 173, 219, 224, 259

Solar Village 46–47, 49, 51, 61–62, 78–79, 81–83, 85–87, 89–91, 102–104, 108–109, 111–113, 115–116, 119–124, 135–137, 152–154, 189–192, 209–211, 222–230, 368–369, 395–396
sponsor 43, 45–47, 49, 108–110, 136–138, 190–192, 209, 211, 222, 225, 229–230, 380, 544–545
sponsorship 44, 109, 136, 146, 178, 191, 286, 328, 375, 466
STILE 405–406
strategy 38, 43–44, 122–123, 138–139, 173, 177, 180, 185, 256–257, 259, 540, 545, 548
   energy 25, 84, 91, 335, 347, 420
   transportation 256, 258–259
structurally insulated panel (SIP) 94, 131, 196, 203, 206, 240, 383
sunlight 8, 10, 13–15, 23, 123–124, 171, 173, 224, 226, 294, 296, 299, 471–472
Sunrayce 19, 25, 495
sunshine 23, 86–87, 196, 229, 324, 381, 387, 399, 515
sustainability 65, 67, 69, 116–117, 121, 168–169, 211, 213–214, 286–287, 289, 318, 385–387, 393–394, 485, 545–546
system 17, 19, 128, 133–134, 170, 173, 180, 182, 184–185, 255, 257, 515, 517
   air-conditioning 399
   air purification 247
   aquaponic 504
   automation 249
   closed-loop smart shower recycling 508
   eco-conscious 204
   energy storage 16, 133
   environmentally-responsible 216
   four-layer shading 282
   geothermal heat pump 256
   greywater recycling 407, 481
   hybrid 107
   innovative PC-controlled 157
   islanding photovoltaic 412
   liquid desiccant wall 169
   low-pressure 324
   mechanical 127, 158, 178, 201
   multi-scenario 286
   passive 173
   PV 13, 23–25, 27, 86–87, 114, 118, 121, 124, 127–128, 240, 248
   rainwater collection 125, 138
   solar 28, 80, 102, 121, 126, 244, 256, 276, 398, 500
   smart building 70
   technical 173
   thermal 59, 129–130, 256
   thermal collection 167
   transportation 271, 308, 319
   ventilation 238
   whole-house control 92

team effort 103, 179, 183, 220, 254
team leader 26, 30, 59, 72, 135, 223
technology 15, 20, 32, 39, 43, 45, 84, 88–89, 304, 307, 525–526, 529
   automotive 14
   battery 12, 133
   building 136, 194, 434
   electric 307
   energy-saving 84
   green 519, 545
   innovation and transformative 415
   innovative 131, 166, 243, 289, 376
   renewable energy 32, 39, 55, 70
   resource-saving 386

solar 12–13, 24, 31, 37, 45, 51, 54, 86, 92, 520, 523
Tidewater House 233

Unplugged 231–232
Unsolar 468, 473–474
U.S. Department of Energy (U.S. DOE) 12–13, 24–27, 31–33, 88–90, 108, 110, 121–122, 135–137, 143, 218–219, 253–254, 289, 371, 376, 394–395, 407–408, 411–413, 418–420, 464–466, 495–496
U.S. DOE *see* U.S. Department of Energy

vehicle 8, 12, 14, 18, 20, 22–23, 75, 224
  aerodynamic 14
  electric 16, 55, 86, 103, 169, 180, 398, 415

solar-powered 9, 11–12
ventilation 55, 87, 103, 299, 314, 341, 356, 386
Video Monitoring Service (VMS) 83
Villa Solar 266, 268–269, 271, 273, 298, 469
VMS *see* Video Monitoring Service
volunteers 74, 79, 90, 96, 108, 136, 138, 141, 229, 231, 329–330

WaterShed 249–252
West Potomac Park 225, 227–229, 240, 250, 253, 371, 413
WHAO House 449, 452
World Expo 344, 496, 511–513
World Solar Challenge (WSC) 9, 11, 14–15, 19, 495
WSC *see* World Solar Challenge

ZEROW 201

For Product Safety Concerns and Information please contact our EU
representative GPSR@taylorandfrancis.com
Taylor & Francis Verlag GmbH, Kaufingerstraße 24, 80331 München, Germany

www.ingramcontent.com/pod-product-compliance
Lightning Source LLC
Chambersburg PA
CBHW071231300426
44116CB00008B/986